INTRODUCTION TO XAFS

A Practical Guide to X-ray Absorption Fine Structure Spectroscopy

X-ray absorption fine structure (XAFS) is a powerful and versatile technique for studying structures of materials in chemistry, physics, biology, and other fields. This textbook is a comprehensive, practical guide to carrying out and interpreting XAFS experiments.

Assuming only undergraduate-level physics and mathematics, the textbook is ideally suited for graduate students in physics and chemistry starting XAFS-based research. It contains concise executable example programs in Mathematica 7. Supplementary material available at www.cambridge.org/9780521767750 includes Mathematica code from the book, related Mathematica programs, and worked data analysis examples.

The textbook addresses experiment, theory, and data analysis, but is not tied to specific data analysis programs or philosophies. This makes it accessible to a broad audience in the sciences, and a useful guide for researchers entering the subject.

GRANT BUNKER is Professor of Physics at the Illinois Institute of Technology. He has over 30 years experience in all aspects of XAFS spectroscopy, from technique development, instrumentation and computation, to applications in biology, chemistry, and physics.

INTRODUCTION TO XAFS

A Practical Guide to X-ray Absorption Fine Structure Spectroscopy

GRANT BUNKER
Illinois Institute of Technology

CAMBRIDGE
UNIVERSITY PRESS

CAMBRIDGE
UNIVERSITY PRESS

University Printing House, Cambridge CB2 8BS, United Kingdom

Cambridge University Press is part of the University of Cambridge.

It furthers the University's mission by disseminating knowledge in the pursuit of education, learning and research at the highest international levels of excellence.

www.cambridge.org
Information on this title: www.cambridge.org/9780521767750

First published 2010

A catalogue record for this publication is available from the British Library

Library of Congress Cataloguing in Publication data
Bunker, Grant, 1954–
Introduction to XAFS : a practical guide to X-ray absorption fine structure spectroscopy / Grant Bunker.
p. cm.
Includes bibliographical references and index.
ISBN 978-0-521-76775-0 (hardback)
1. Extended X-ray absorption fine structure. 2. X-ray absorption near edge structure. I. Title.
QC482.S6B86 2010
537.5′352 – dc22 2009037742

ISBN 978-0-521-76775-0 Hardback

Additional resources for this publication at www.cambridge.org/9780521767750

Contents

Preface

This book is an attempt to fill a gap that has existed since the dawn of XAFS: to provide a broad and sufficiently deep introduction for graduate students and other researchers, to enable them to quickly learn the disparate things that are needed to do XAFS research in a reliable way. The broad applicability of the XAFS technique draws researchers with a variety of backgrounds who are not specialists in the technique. There needs to be an accessible resource by which they can learn the essentials in a reasonably efficient and comprehensive way.

I have been doing XAFS research for more than 30 years, starting out in the laboratory of Edward Stern at the University of Washington. My work has involved applications and methods development; experiment, theory, computation; synchrotron instrumentation, and construction and operation of synchrotron facilities. I have learned a great deal from Ed Stern, John Rehr, Dale Sayers, Steve Heald, Farrel Lytle, Bruce A. Bunker, Gerd Rosenbaum, and many other colleagues and students too numerous to list, to whom I express my gratitude. I also would like to express my gratitude to my family for their patience and support while I have been occupied nights, weekends, and holidays in writing this book; my brother, Bruce, for originally getting me involved in XAFS; and to my parents, now both deceased, for their unwavering support of the study, appreciation, and preservation of nature.

This book covers basic material, occasionally dipping a toe into deeper waters. There are a number of other topics I would have liked to include, but they require greater mathematical and theoretical physics depth than is presupposed here; the material also would take about as much space as this book. A future companion book addressing advanced topics is quite possible if sufficient interest is expressed.

Dale Sayers was one of the founders of the XAFS technique, and a good friend. His untimely death several years ago was a great loss to our community. Dale was always concerned about the problem of education in XAFS, and he strongly supported my efforts along these lines, in particular the writing of this book. I dedicate it to him.

1
Introduction

1.1 What is XAFS?

X-ray Absorption Fine Structure (XAFS) spectroscopy is a unique tool for studying, at the atomic and molecular scale, the local structure around selected elements that are contained within a material. XAFS can be applied not only to crystals, but also to materials that possess little or no long-range translational order: amorphous systems, glasses, quasicrystals, disordered films, membranes, solutions, liquids, metalloproteins – even molecular gases. This versatility allows it to be used in a wide variety of disciplines: physics, chemistry, biology, biophysics, medicine, engineering, environmental science, materials science, and geology.

The basic physical quantity that is measured in XAFS is the X-ray absorption coefficient $\mu(E)$, which describes how strongly X-rays are absorbed as a function of X-ray energy E. Generally $\mu(E)$ smoothly decreases as the energy increases (approximately as $1/E^3$), i.e. the X-rays become more penetrating. However, at specific energies that are characteristic of the atoms in the material, there are sudden increases called X-ray absorption edges. These occur when the X-ray photon has sufficient energy to liberate electrons from the low-energy bound states in the atoms. The cross section, a quantity that is proportional to $\mu(E)$, is shown in Figure 1.1 for the element platinum. Experimental data for MnO and $KMnO_4$ are shown in Figures 1.2 and 1.3.

Absorption edges were first measured in 1913 by Maurice De Broglie, the older brother of quantum mechanics pioneer Louis De Broglie. In 1920, using M. Siegbahn's vacuum spectrograph, Hugo Fricke first observed the "fine structure" – energy-dependent variations in the $\mu(E)$ – in the vicinity of the X-ray absorption edges of a number of elements. Despite some early successes and intermittent progress, for another fifty years the correct

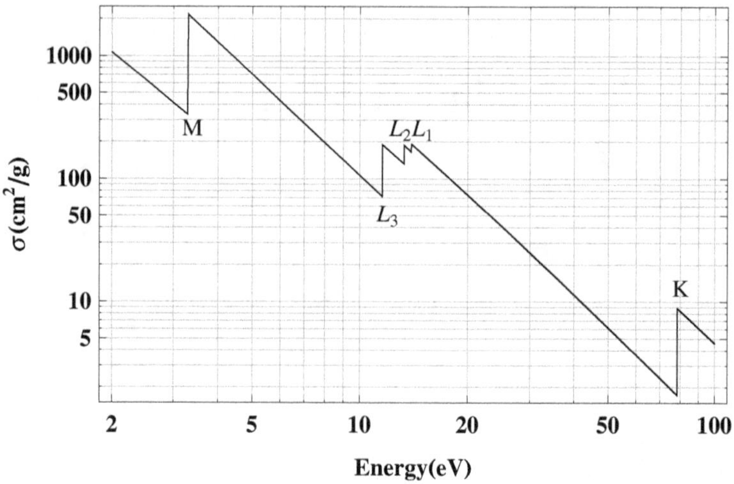

Fig. 1.1 Log-log plot of the (semiempirical) X-ray absorption cross section of platinum (Z = 78) vs. X-ray energy. The K, L_1, L_2, L_3, and M edges are shown; fine structure is *not* shown.

theoretical explanation of XAFS remained obscure. In particular, a controversy existed as to whether a model based on the long-range order (LRO) or short-range order (SRO) in the sample was more appropriate. This confusion was cleared up around 1970 when Stern, Sayers, and Lytle [1, 2, 3, 4] synthesized the essential aspects of a viable theory of XAFS, and further demonstrated that XAFS could be a practical tool for structure determination. A detailed historical account is given in the article by Stumm von Bordwehr [5].

The term "XAFS" is a broad one that comprises several different techniques: EXAFS (Extended X-ray Absorption Fine Structure); XANES (X-ray Absorption Near Edge Structure); NEXAFS (Near Edge XAFS); and SEXAFS (Surface EXAFS). Many papers are published simply as XAS (X-ray Absorption Spectroscopy). Although the basic physics of these techniques is fundamentally the same, different approximations, techniques, terminology, and theoretical approaches may be employed in different situations, particularly in the low-energy (soft X-ray) and high-energy (hard X-ray) regimes. These aspects will be discussed in the experiment and theory chapters of this book.

1.2 Physics of XAFS

XAFS is an intrinsically quantum mechanical phenomenon that is based on the X-ray photoelectric effect, in which an X-ray photon incident on an atom within a sample is absorbed and liberates an electron from an inner

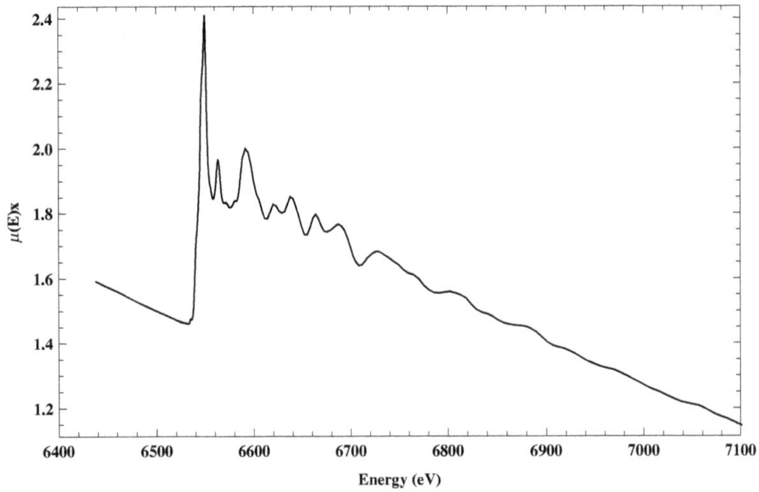

Fig. 1.2 Experimental K-edge XAFS spectrum $\mu(E)x$ of MnO at $T = 80$K.

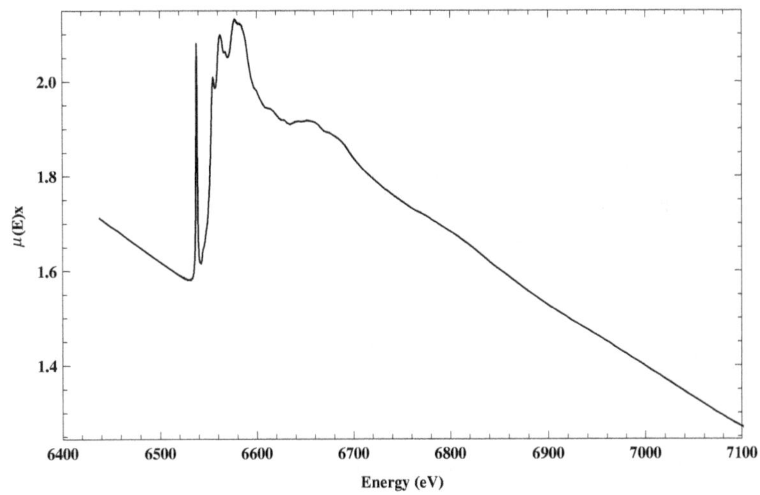

Fig. 1.3 Experimental K-edge XAFS spectrum $\mu(E)x$ of $KMnO_4$ at $T = 80$K.

atomic orbital (e.g. 1s). The "photoelectron" wave scatters from the atoms around the X-ray absorbing atom, creating interferences between the outgoing and scattered parts of the photoelectron wavefunction. These quantum interference effects cause an energy-dependent variation in the X-ray absorption probability, which is proportional to the X-ray absorption coefficient, a measurable quantity. When properly decoded these modulations provide information about the structure, atomic number, structural disorder, and thermal motions of neighboring atoms.

In addition, the XAFS technique is an *electronic* spectroscopy that measures transitions between bound initial states and bound and continuum final states. This has much in common with familiar techniques such as UV-Vis spectroscopy. As such, it also can be used to probe the nature of bound states near the Fermi level, or in molecular orbital language, the partially occupied HOMO[1], and LUMO orbitals. Use of polarized X-ray beams (which are easy to come by) permits dichroism experiments, much like those in optical spectroscopy.

1.3 Synchrotron radiation

XAFS spectroscopy has developed hand in hand with the growth of synchrotron radiation research. The first useful synchrotron X-ray facilities were developed around 1970, about the time of Stern, Sayers, and Lytle's modern synthesis. XAFS requires an X-ray beam of finely tunable energy; although it is possible to do limited experiments with a laboratory X-ray source, most experiments benefit enormously from the availability of synchrotron radiation. As a consequence, nearly all modern XAFS experiments are performed at Synchrotron Radiation Sources (SRSs).

SRSs are shared regional research facilities that, with few exceptions, are operated under government support. At present there are more than 60 rings of various sizes more or less equally distributed between the Americas, Europe, and Asia; and Australia has its own. A listing of these sources can be found at www.lightsources.org/. Often complementary research facilities such as nanotechnology centers are adjacent to the facilities.

SRSs are based on technology originally developed for high-energy physics experiments, but subsequently they have been adapted to reliably produce high-energy electromagnetic radiation such as X-rays with desirable spectral characteristics. Electrons moving close to the speed of light within an evacuated pipe are guided around a closed path of 100–1000 meter circumference by vertical magnetic fields. Wherever the trajectory bends, the electrons accelerate (change velocity vector). Accelerating charged particles emit electromagnetic radiation, and the fact that the electrons are moving at nearly the speed of light implies that relativistic effects are important. In this case they profoundly affect the properties of the emitted radiation: the average energy of the X-rays and the total radiated power are greatly increased, and the radiation pattern becomes more directional, making it much easier to employ X-ray optics such as monochromators. Often

[1] Highest Occupied and Lowest Unoccupied Molecular Orbitals.

"insertion devices" such as "wigglers" and "undulators" also are used to further enhance the characteristics of the emitted radiation.

Beamlines are large (usually tens of meters in length) and complex instruments that include a variety of components such as X-ray monochromators; X-ray mirrors; slits; ultrahigh vacuum pumps, valves and gauges; photon shutters; shielding; safety interlocks; X-ray detectors; UHV chambers; cryostats, furnaces, reaction cells, magnets; motorized sample stages; goniometers; and data acquisition systems. All of these are under remote computer control from outside the experimental radiation enclosure. Remote control is necessary when the beam is on because ambient radiation levels are too high for experimenters to be present inside the experimental radiation enclosure.

Since the facilities are shared resources, policies are established that regulate access for experimenters. Individual beamlines may be operated by the institution that operates the ring. Alternatively they may be operated through an arrangement with one or more separate institutions that contribute resources. Generally there is some mechanism by which an outside experimenter (general user) can submit a beamtime proposal and, if it is approved, obtain days or weeks of beam time per year for their project. In most cases, if the scientific results are to be published in the open literature (i.e. the work is not proprietary), only minimal (or no) user fees will be applied for beamtime. This mechanism makes access to these outstanding research instruments available to a wide community of scientists. Usually the facilities run 24 hours per day (three 8 hour shifts) when they are operating so experimenters must have a team of several workers in order to use the beamtime effectively.

The usual mode of experimentation is as follows: samples are prepared ahead of time at the experimenter's home institution, they are transported to the beamline, spectra are measured over a period of minutes, hours, or days, and the spectra are analyzed in detail during and after the experiment. In simple cases data analysis can be done immediately, even while acquiring data, but for complex samples, analysis may take considerably more time and effort. For some systems, it may be necessary to prepare the samples on-site, or even prepare them *in situ* – i.e. right in the beam. Usually preparing and characterizing the samples, and analyzing the data are steps that take much longer than acquiring the data at the beamline. The care taken on sample preparation and measurement can have a great impact on how long it takes to interpret the data, however.

The multidisciplinary nature of XAFS and synchrotron radiation techniques, in combination with long hours at the beamline often has an unexpected benefit: experimenters working in widely different areas find themselves colliding over coffee and talking with each other. This has generated many productive collaborations.

1.4 Purpose of this book

This book has several aims. First, it is to provide an accessible, practical introduction to the basics of X-ray Absorption Fine Structure (XAFS) spectroscopy for beginning graduate students and researchers with diverse backgrounds who have a substantial undergraduate preparation in chemistry or physics. To keep the book accessible, the use of advanced mathematics has been kept to a minimum. Second, it is to provide clear explanations of certain topics that I think are important to grasp if one is to do XAFS reliably, and to remind a newer generation of researchers of useful lessons that otherwise may be forgotten. Third, it is to provide useful reference information, resources, and links for further exploration. The applications and techniques of XAFS are diverse, and it is not possible (or probably, desirable) to comprehensively cover all variants without being both tedious and instantly outdated. Programs written in the *Mathematica 7* language[2] are included to provide illustrative (and I hope, useful) examples of computations. Additional code and supplementary information will be posted from time to time, on my web site `http://gbxafs.iit.edu/` or `www.cambridge.org/9780521767750/`.

Following the Chapter 1 introduction, Chapter 2 describes the basic physics of the interaction of X-rays with matter. The same physical processes apply to the workings of the X-ray optics that are used to prepare the X-ray beams, the interaction of the beam with the sample, and the detectors that are used to measure the signals. Chapter 3 provides a description of the experimental apparatus – sources, optics, detectors – and experimental techniques. Those wishing for more extensive coverage of these topics (or lusting for more equations) may wish to consult the recent books by Als Nielsen and McMorrow [7] and Duke [8]; the venerable and still useful book edited by Koningsberger and Prins [10]; the Handbook on Synchrotron Radiation series [6]; and the X-ray data booklet from Lawrence Berkeley Laboratory [9].

Chapter 4 provides a minimal introduction to the basic theory of XAFS, delves more deeply into the computational methods that are used to calculate theoretical XAFS spectra, using as an example FEFF8. Chapter

[2] *Mathematica 7* is used because it is powerful, self-contained but extensible, interactive, cross-platform, widely available, has numerous data sources built-in, and powerful programs can be expressed compactly. It is unfortunately not free, but much of the code can be converted to other languages by the motivated reader without great effort.

5 describes the architecture of XAFS spectra, with the goal of developing some intuition for the spectra. It then presents the principles and rationale behind XAFS data analysis methods, and some examples with code to implement it. Detailed description of the use of various analysis packages is not attempted, since that information has a limited shelf life, there are too many from which to choose, and they are generally well described in the user guides to the programs. Chapter 6 briefly describes a variety of related techniques such as Diffraction Anomalous Fine Structure (DAFS) and Inelastic X-ray Scattering (IXS), and a conclusion. The book concludes with Appendices on Fourier methods, cumulants, X-ray filters, reference spectra, and tables of X-ray properties.

2

Basic physics of X-ray absorption and scattering

2.1 X-rays

This chapter provides a brief description of the basic X-ray physics needed to design XAFS experiments. We start with the basics.

X-rays are short-wavelength electromagnetic (EM) radiation; except for their wavelength, they are essentially the same as radio waves, microwaves, infrared, visible, ultraviolet, and gamma radiation. The frequency f is related to the wavelength λ by $f\lambda = c$, where c is the speed of light, $\approx 3 \times 10^8$ m/s.

In free space, EM waves are transverse: the electric and magnetic field vectors of the wave are perpendicular to each other, and also to the direction of propagation. The electric and magnetic field vectors oscillate in phase, and their magnitudes are proportional to each other. The direction of the electric field is described by the "electric polarization vector" $\hat{\epsilon}$, which is a unit vector in the direction of the wave's electric field vector. The direction of wave propagation is given by the wave vector $\vec{k}$, where $k = |\vec{k}| = 2\pi/\lambda$.

From a quantum perspective, the electromagnetic waves of classical physics consist of swarms of photons, which carry energy, linear momentum, and angular momentum. Such a wave is illustrated in Figure 2.1. The wavelength λ of all particles, including photons and electrons, is related to their momentum p through the De Broglie relation $\lambda = h/p$, where h is Planck's constant. Similarly, the particle frequency f is related to the energy E by $f = E/h$. Defining the angular frequency $\omega = 2\pi f$ and $\hbar = h/2\pi$, for any nonrelativistic quantum particles we have the relations $E = \hbar\omega$, $\vec{p} = \hbar\vec{k}$. The energy and momenta of photons are related by $E = pc$, which gives the relation $E = hc/\lambda \approx 12398.4\,\mathrm{eV\,Å}/\lambda \approx 1240\,\mathrm{eV\,nm}/\lambda$ for photons, a relation worth committing to memory. One eV (electron-volt) is the energy acquired by an electron as it moves between potential difference of 1 volt.

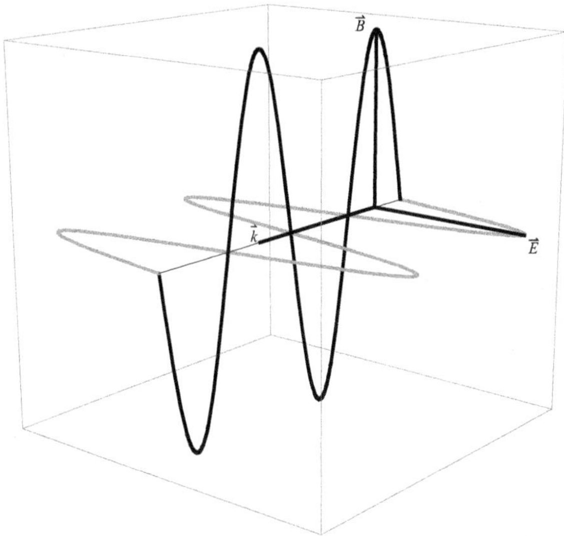

Fig. 2.1 Electric and magnetic fields in a plane EM wave in free space. The direction of propagation is along $\vec{k}$.

The wavelengths of X-rays are on the order of Ångstroms (1Å = 0.1 nm), while the wavelengths of visible light are nearly 10^4 times longer: nominally ≈ 0.4–0.7μm. The relation $E = hc/\lambda$ indicates that the energies of visible photons are a few eV (red is about 2 eV, green about 2.5 eV, blue about 3 eV), while X-rays typically have energies on the order of 10 KeV. The energies of interest for "hard" X-ray XAFS span the range from a few KeV up to about 100 KeV. "Soft" X-rays cover the range from hundreds of eV to a few KeV.

2.2 The structure of atoms

We assume the reader has a basic understanding of the architecture of atoms. Here we briefly review it from the point of view of (nonrelativistic) quantum mechanics – it is essential for understanding the interactions of X-rays and electrons with atoms.

The structure of many-electron atoms is similar to the structure of the hydrogen atom, with a few important differences. Solving Schrödinger's equation for the hydrogen atom (or one electron "hydrogenic" atoms of nuclear charge Z) can be done exactly, something that is worked out in quantum mechanics books [17], and we will not repeat it here.

The various bound state orbitals in a spherically symmetric potential can be written as the product of radial functions $R_{nl}(r)$, which depend on the potential, and universal angular functions that have the same form for all central (spherically symmetric) potentials. The detailed shapes and scales of the functions $R_{nl}(r)$ differ as a function of atomic number, but they have certain general properties and shapes. The spherical harmonics $Y_{lm}(\theta, \phi)$ are complex functions of the angular spherical coordinates θ and ϕ, and they are natural forms in which to express the angular part of the wavefunction. They are simultaneously eigenfunctions of the operators for the squared angular momentum $\hat{L}^2$ (with eigenvalue $l(l+1)\hbar^2$) and the z component of the angular momentum $\hat{L}_z$ (with eigenvalue $m\hbar$). In this case the orbital is written $\Psi_{nlm} \propto R_{nl}(r)Y_{lm}(\theta, \phi)$. They can also be expressed in other forms, such as the real-valued s, p, d orbitals familiar from basic chemistry; the converse is evidently true also.

The quantum numbers n, l, m are integers with $n \geq 1$, $l < n$, and $-l \leq m \leq l$. The principal quantum number n is one greater than the number of "nodes" (zeros) of the radial wavefunction. For example, in the $n = 1$ shell, there are no radial nodes; the $n = 2$ shell has one radial node. The reason this is significant is that the more nodes there are, the more rapidly in r the wavefunction must oscillate. The average kinetic energy is related to the mean square slope of the wavefunction, so more nodes translates to a higher kinetic energy. Similar trends apply to the shape of the angular functions, so that larger l implies a greater number of nodes in the angular part of the wavefunction, and usually a larger energy.

Solving for the hydrogenic wavefunctions is one thing – exactly solving the complete many-electron wavefunction of a heavy atom is quite another. Fortunately, mean-field methods, in particular "self-consistent field" (SCF) methods, are sufficiently accurate for our present purposes.[1] These allow us to approximate the potential energy experienced by an electron in an atom as due to the attraction of the nucleus of charge Z, the repulsion of the other $\approx Z-1$ electrons, and quantum statistical effects owing to the Pauli Exclusion Principle and collective motions of the electrons which introduce a contribution (called the "exchange correlation potential" (static case) or "self-energy" (dynamic case)) to the effective one-electron potential. This allows us to approximate the many-electron problem as a one-electron (or at least one quasiparticle) problem.

Since we may approximate the many-electron atom problem as that of a single electron in an effective central potential, the orbitals have the same product form familiar from the hydrogen atom, $\Psi_{nlm} \propto R_{nl}(r)Y_{lm}(\theta, \phi)$.

[1] It must be mentioned however that multielectron excitations do play an important role in the quantitative theory of XAFS.

The kinetic energy consists of a term corresponding to radial motion, plus terms corresponding to tangential motion. Conservation of angular momentum (which follows from the spherical symmetry) implies that the tangential terms in the kinetic energy are proportional to $1/r^2$ where r is the radial coordinate; these effectively act as a repulsive "centrifugal potential" contribution that becomes more important as angular momentum increases.

Hydrogenic atoms have the unusual property that all the orbitals with the same principal quantum number n, but different l, have the same energy; they are "degenerate",[2] with energy levels

$$E_n = -\frac{1}{2}\alpha^2 mc^2 \frac{Z^2}{n^2} = -\frac{me^4}{2\hbar^2}\frac{Z^2}{n^2},$$

where (in SI units) m and q are the mass and charge of the electron, c is the speed of light, the fine structure constant $\alpha = e^2/\hbar c \approx 1/137$, and $e^2 = q^2/4\pi\epsilon_0$. The potentials of other atoms do not have this simple $1/r$ form, and the orbitals with the same n but different l are nondegenerate, generally increasing in energy as l increases. Levels with the same n, l but different m are degenerate, as a consequence of the spherical symmetry, unless something such as an applied external electric or magnetic field breaks the spherical symmetry.

In heavy atoms, the kinetic energies of the bound inner electrons can be large enough that relativistic effects must be considered, adding terms to the Hamiltonian that split different orbitals corresponding to different total angular momentum $\hat{J} = \hat{L} + \hat{S}$. For this reason, the $2p$ levels are split into $2p_{1/2}$ and $2p_{3/2}$ levels, corresponding to the L_2 and L_3 absorption edges. The binding energies for $n = 1$ (K-edge) and $n = 2$ (L-edge) levels are plotted vs. atomic number in Figures 2.2 and 2.3.

2.3 Atomic units

"Atomic Units" make up a system of units that is very useful in practical calculations in atomic and molecular physics, because quantities expressed in this way are of a magnitude on the order of one, and it is unnecessary to look up physical constants to get a numerical answer. At a minimum it is necessary to know about atomic units so that one knows what an "a.u." is, when reading the literature.

In this unit system, the electron charge e,[3] mass m, and $\hbar$ are all equal to 1, which implies that $c \approx 137$. The unit of length is the "Bohr" $a_0 =$

[2] This degeneracy is a consequence of a symmetry connected with the $1/r$ potential that results in conservation of the Laplace–Runge–Lenz vector in the classical Kepler problem.

[3] $e^2 = q^2/4\pi\epsilon_0$, where q is the electron charge in SI units, and ϵ_0 is the permittivity of free space.

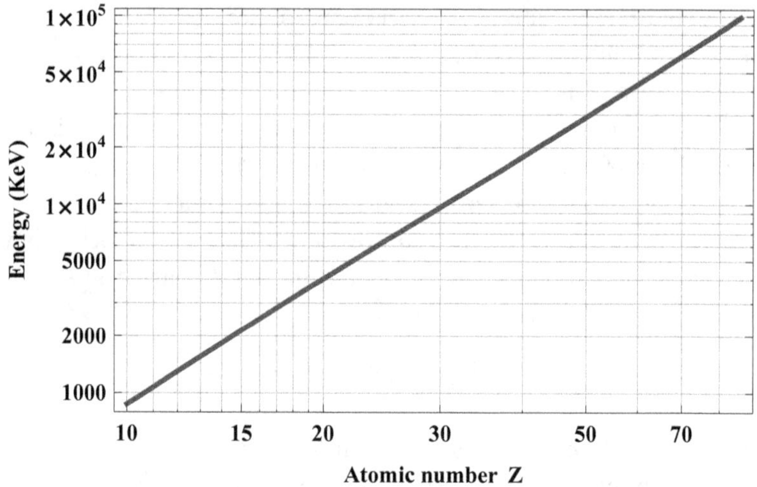

Fig. 2.2 Log-log plot of K-edge energies vs. atomic number Z. The K-edge energies scale approximately as $Z^{2.16}$. Data are from Deslattes *et al.* [12].

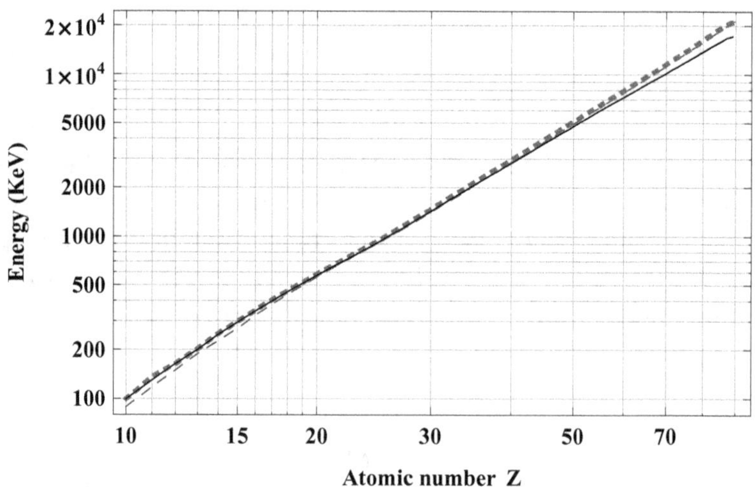

Fig. 2.3 Log-log plot of L_1 (dashed line), L_2 (thick dotted line) and L_3 (solid line) edge energies vs. atomic number Z.

$\hbar^2/me^2 = 1$ au ≈ 0.529Å and the unit of energy is the Hartree $= me^4/\hbar^2 = 1$ au ≈ 27.2 eV $= 2$ Ryd. With this substitution the Schrödinger equation for a hydrogenic atom of nuclear charge Z, $-\frac{\hbar^2}{2m}\nabla^2\psi - \frac{Ze^2}{r}\psi = E\psi$ becomes simply $\nabla^2\psi + 2(E + \frac{Z}{r})\psi = 0$. As a practical example, for a free electron of kinetic energy E we have $E = \hbar^2k^2/2m$ or $k^2 = 2E$ in atomic units. Converting from atomic units of (Bohr, Hartree) to (Å eV), we find $k^2 =$

$E(2/(0.529^2 \times 27.2)) = 0.263E$, which is a much-used equation that is basic to XAFS.

2.4 Types of interactions

We are concerned with several kinds of interactions of X-rays with matter, specifically X-ray absorption, elastic scattering, and inelastic scattering. At very high energies, greater than 1.022 MeV, production of electron–positron pairs out of the kinetic energy of colliding particles (e.g. X-ray/electron) becomes possible by virtue of $E = mc^2$; this is how high-energy physicists create new particles by colliding existing ones. Pair production is negligible in the energy range of interest for XAFS, however.

Absorption events occur when a photon incident on a sample interacts with electrons and gives up all its energy, disappearing in the process. It does so by causing transitions of electrons from lower atomic energy levels to higher levels. This leaves vacancies in the lower levels, which are filled a short time later by transitions of electrons from higher levels down to lower ones. The stored energy may be released radiatively, by emitting fluorescence radiation at a lower energy than the incident, or non-radiatively, by kicking off an electron in an "Auger process," named after the French physicist Pierre Auger. For example, in a "KLL" Auger transition, a K-shell excitation might be followed by an L→K electronic transition while simultaneously ejecting another L shell electron. By measuring the kinetic energy of escaping electrons, the energy levels can be probed through Auger Spectroscopy. The cascade of de-excitation following an absorption event can be complicated, because a sequence of many single- and multi-electron processes are possible. Photons generally are emitted at X-ray, UV-visible, and infrared wavelengths, electrons may be ejected from the atoms, and heat pulses are produced following the primary absorption event. These decay modes afford a number of possible detection methods, the most common of which are fluorescence and electron yield detection (Chapter 3).

Elastic scattering events occur when a photon of specific ω, $\vec{k}$ impinges on a sample, interacts with the electrons, and a photon of the same frequency but different direction $\vec{k}'$ is emitted. This scattering geometry is illustrated in Figure 2.4. From a classical point of view, an incident electromagnetic wave causes the electrons bound to an atom to oscillate back and forth, and these accelerating electrons generate their own secondary wave field that has the same frequency as the incident wave.

Inelastic scattering events differ from elastic scattering events in that the frequency of the emitted photon is not the same as the incident photon;

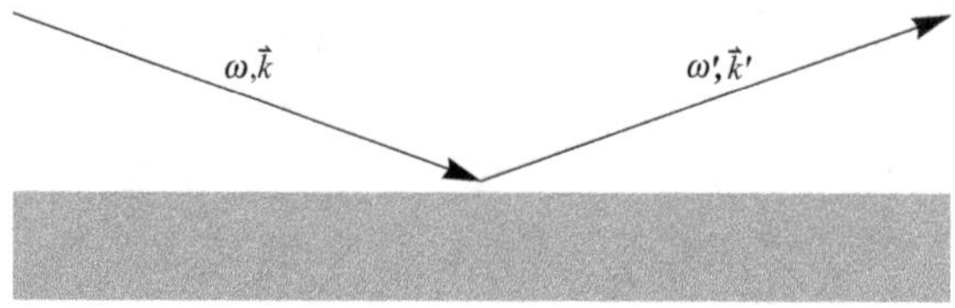

Fig. 2.4 Scattering geometry: If $\omega' = \omega$ the scattering is elastic (the energy of the scattered photon is the same as the incoming one); if $\omega' \neq \omega$, it's inelastic.

it is generally of lower frequency, i.e. lower energy. The energy difference between the scattered and incident photons is transferred to the sample by inducing electronic transitions in the same manner as absorption events.

It will come as no surprise that Elastic Scattering, Inelastic Scattering, and Absorption events are closely related, and in fact can be viewed as different aspects of the same phenomena. Experimentally we tend to focus on one process or another, but they are intimately connected. In X-ray absorption spectroscopy we focus on the energy dependence of the total absorption, without measuring the scattering intensity versus angle, as is done in scattering experiments. However XAFS-equivalent information and more can be obtained by Inelastic X-ray Scattering. This is described in Chapter 6.

2.5 Cross sections

The strength of an interaction such as absorption or scattering can be quantified in terms of the "cross section" σ, which has units of area, and generally is a function of energy. A special unit of area, the "Barn" ($10^{-28}\,\mathrm{m}^2$), is sometimes used because barns are of a convenient size to describe nuclear scattering, nuclear dimensions being a few femtometers. A related quantity, the differential scattering cross section $d\sigma/d\Omega$, is the cross section per unit solid angle[4] Ω into which the particles are scattered.

Operationally the cross section can be defined in terms of a uniform beam of particles interacting with an object, such as a collection of atoms. The number of interactions (e.g. absorption or scattering events) per unit time R (the rate) is equal to the product of the incident beam intensity (particles/time/area) i_0 and the total cross section of the target particles: $R = i_0\sigma_{\mathrm{tot}}$. If the volume density of target particles is ρ, the probability of a particle interacting in a thin layer δx is then $\rho\sigma\delta x = \mu\delta x$, where μ is

[4] The solid angle subtended by a small (planar) patch of area $d\vec{a}$ at position $\vec{r}$ is $\vec{r}\cdot d\vec{a}/r^3$, where the direction of $d\vec{a}$ is normal to the area. The solid angle subtended by a large surface is the integral of this expression over the surface. The solid angle Ω is expressed in steradians, which, like radians, is dimensionless.

the linear attenuation coefficient. This quantity has dimensions of inverse length; it is of central importance in XAFS, and it is important to know how to calculate it. It comprises several contributions: the photoelectric absorption, coherent (elastic) scattering, and incoherent (inelastic) scattering, but the photoelectric cross section is dominant in the energy range of interest to XAFS. The photoelectric cross sections depend strongly on energy, roughly as $1/E^3$ between absorption edges. The quantity μ^{-1} is called the "absorption length" – it denotes the characteristic length over which the beam intensity is attenuated to $1/e \approx 0.37$.

Consider the case of X-ray absorption: a photon beam of flux (photons/sec) is incident normally onto a homogeneous sample of thickness x. Consider the sample to be composed of a large number n of identical thin layers of thickness $\delta x = x/n$. The probability of absorption in one layer is $\mu\delta x$, so the probability of transmission through one layer is $1 - \mu\delta x$, and the probability of transmission through all n layers is then $(1-\mu\delta x)^n = (1-\mu x/n)^n \rightarrow \exp(-\mu x)$ in the limit of large n. Alternatively, one realizes the decrease in intensity dI between depth x and $x + dx$ is proportional to the intensity I at that depth times the absorption probability μdx: $dI = -I\mu dx$; integrating, we find $I/I_0 = \exp(-\mu x)$. This is the equivalent of Beer's law in UV-visible spectroscopy. It applies only for uniform homogeneous samples.

Cross sections can be specified on a per-atom basis (e.g. barns/atom), or a per-mass basis (e.g. cm^2/g). In either case, they represent an effective area per amount of the element. If the cross sections are expressed on a per-mass basis, they are often called "linear mass attenuation coefficients" and represented as $[\frac{\mu}{\rho}]$. For many purposes the cross section for a mixture of different species i, each of cross section σ_i and partial density ρ_i (mass of the ith species in the sample volume) can be approximated as $\mu = \sum_i \rho_i\sigma_i = \rho_{\mathrm{tot}} \sum_i \phi_i\sigma_i$, where ρ_{tot} is the density of the sample, ϕ_i is the number fraction N_i/N of species i if the cross sections are on a per atom basis; ϕ_i is the mass fraction M_i/M of species i if the cross sections are on a per mass basis, i.e. linear mass attenuation coefficients. If there is any confusion, check the units. This expression is very useful for calculating macroscopic absorption, but it is only approximate, because it neglects interactions between different atoms, the very phenomena that are of interest to XAFS.

A wealth of X-ray data is available online, pdf, and convenient booklet form in Lawrence Berkeley Labs X-ray data booklet (http://xdb.lbl.gov/). The URL www.csrri.iit.edu/periodic-table.html is another convenient (although dated) online source of X-ray absorption and scattering cross sections. After the user specifies the energy in and the element name, the

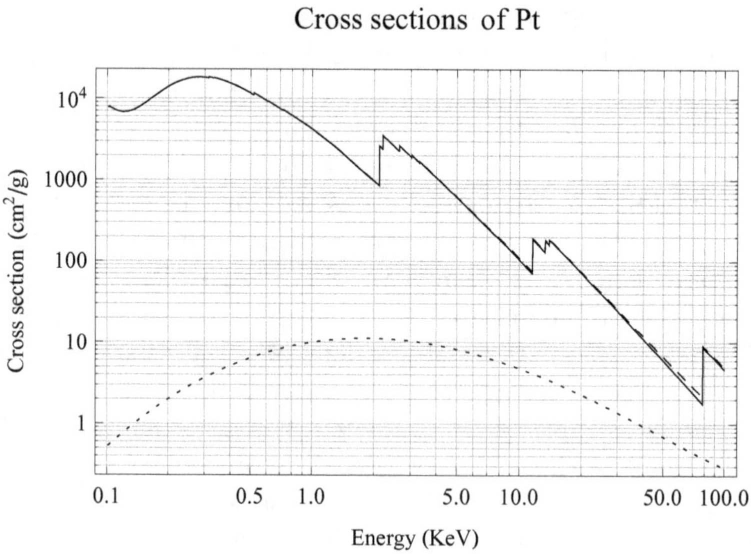

Fig. 2.5 Photoelectric absorption (solid), elastic + inelastic scattering (dotted), and total (dashed) cross sections, from Chantler *et al.* [26].

program generates an html form containing cross sections, edge energies, and other useful information, based on the fits to experimental data tabulated by McMaster *et al.* [25]. Another is the FFAST and XCOM tabulations at the National Institute of Standards and Technology which is appropriate for the energy ranges of interest in XAFS [26, 28]. The *Mathematica* 7 program shown in the listing can be used to generate figures such as Figs. 2.5–2.11; (note that a backslash at the end of a line indicates that the line and the following one should be combined, and the backslash deleted).

Mathematica 7 program to download and plot FFAST cross sections

```
GetData[name_,emin_,emax_]:=
(template="http://physics.nist.gov/cgi-bin/ffast/ffast.pl?&Formula=<name>\
&gtype=4&range=S&lower=<emin>&upper=<emax>";lbl=name;qstring=\
StringReplace[template,{"<name>"->name,"<emin>"->ToString[emin],\
"<emax>"->ToString[emax]}];dat=Import[qstring,"Table"];
If[!StringFreeQ[dat[[1,1]],"Error"],Return[ "Error getting data from Server"]];
nskip=Position[dat,"Photoelectric"][[1,1]]+1;{e,f1,f2,photo,allscatter,\
total,kphoto,lambda}=Transpose[Take[dat,{nskip+1,-2}]];)

PlotIt:=ListLogLogPlot[{Transpose[{e,photo}],Transpose[{e,allscatter}],\
Transpose[{e,total}]},Joined->True,Frame->True,GridLines->True,\
FrameLabel->{"Energy (KeV)","Cross Section (cm^2/g)"},PlotStyle->\
{Black,{Black,Dotted},{Black,Dashed}},PlotLabel->"Cross Sections of "<>lbl];
(*the following code executes these functions*)
GetData["Pt",1,100]; SystemOpen[qstring]; PlotIt
```

Cross sections of Te

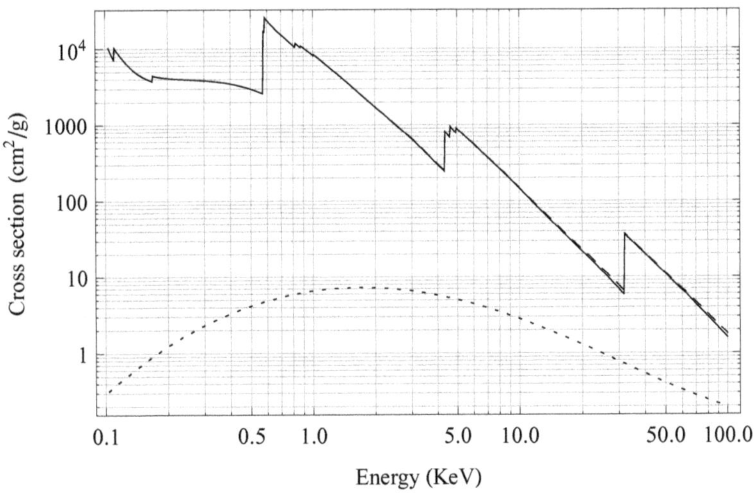

Fig. 2.6 Photoelectric absorption (solid), elastic + inelastic scattering (dotted), and total (dashed) cross sections, from Chantler *et al.* [26].

Cross sections of Mo

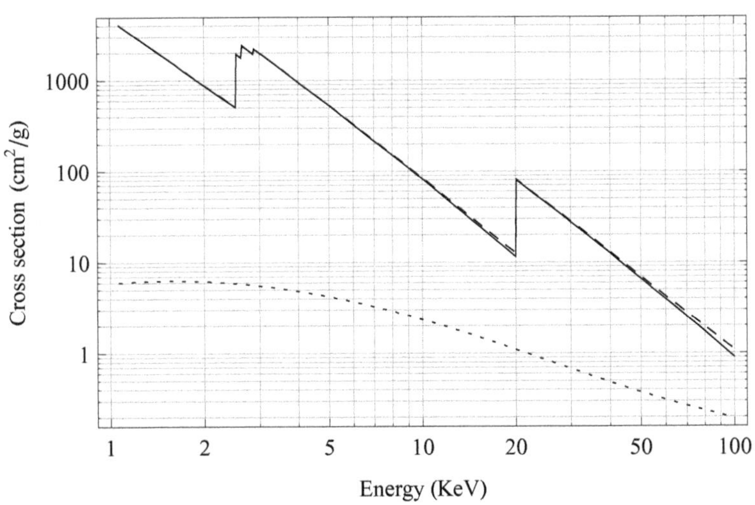

Fig. 2.7 Photoelectric absorption (solid), elastic + inelastic scattering (dotted), and total (dashed) cross sections, from Chantler *et al.* [26].

2.5.1 Interpolating power laws

Often one needs values of absorption coefficients or other quantities when only tables at specific values are available. Many of the quantities with which we are concerned approximately follow a power law, so linear interpolation

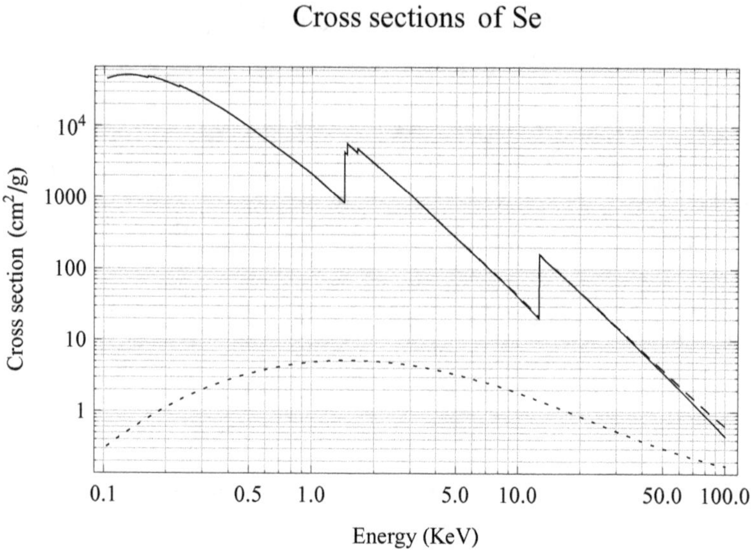

Fig. 2.8 Photoelectric absorption (solid), elastic + inelastic scattering (dotted), and total (dashed) cross sections, from Chantler *et al.* [26].

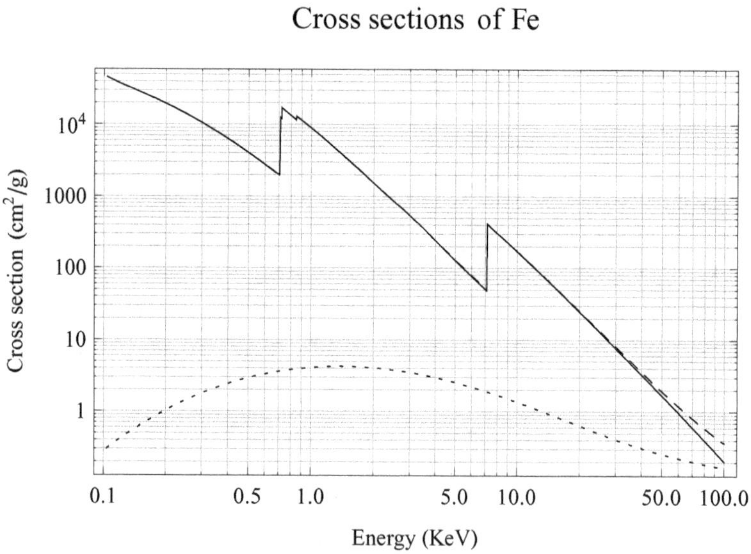

Fig. 2.9 Photoelectric absorption (solid), elastic + inelastic scattering (dotted), and total (dashed) cross sections, from Chantler *et al.* [26].

is not accurate. A better approach is power-law interpolation, which is only slightly more complex than linear interpolation. Suppose that one knows the values of a function $f(E)$ at energies E_1 and E_2; further suppose that the function $f \propto E^b$ over the range of interest (so make sure there

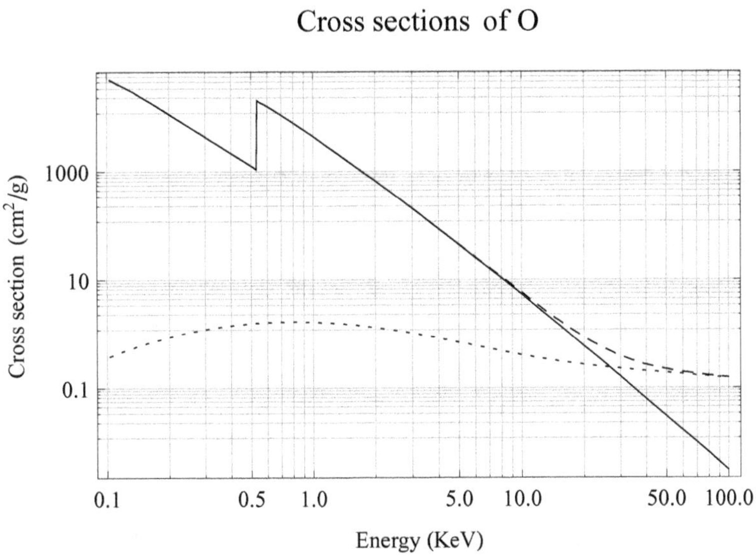

Fig. 2.10 Photoelectric absorption (solid), elastic + inelastic scattering (dotted), and total (dashed) cross sections, from Chantler *et al.* [26].

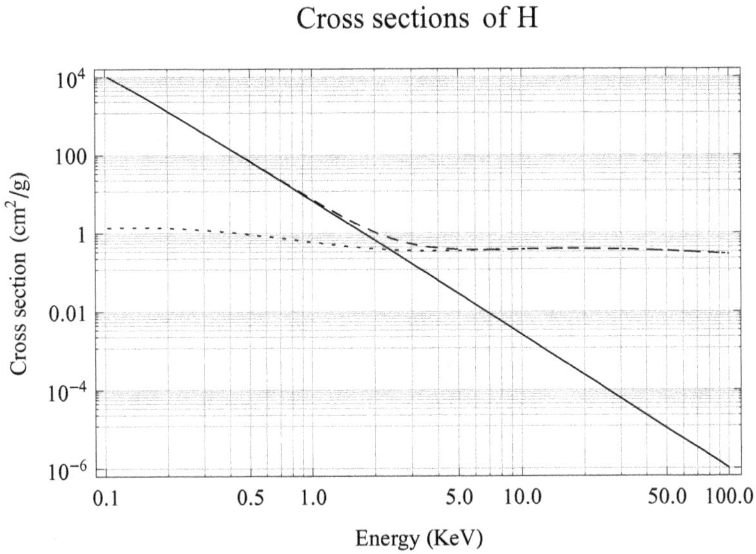

Fig. 2.11 Photoelectric absorption (solid), elastic + inelastic scattering (dotted), and total (dashed) cross sections, from Chantler *et al.* [26].

are no absorption edges in the interval). Then over that interval $f(E) \approx f(E_1)(E/E_1)^b$, where the exponent $b = \ln(f(E_2)/f(E_1))/\ln(E_2/E_1)$.

For example, consider iron, which has a photoelectric absorption coefficient of approximately 300 cm^2/g at 8 KeV and 221 at 9 KeV (and no absorption edges between 8 and 9 KeV); then $b = \ln(221/300)/\ln(9/8) \approx -2.6$, and the absorption at 8.5 KeV is $300(8.5/8)^{-2.6} \approx 256$ cm^2/g. At 10 KeV the absorption coefficient extrapolates to $300(10/8)^{-2.6} \approx 168$ cm^2/g, which agrees very well with the tabulated value. Similarly, power law interpolation of absorption edge energies and fluorescence energies vs. atomic number Z is very useful. For example K-edge energies vary approximately as $Z^{2.156}$ between Ca ($Z = 20$) and Mo ($Z = 42$). Using this approach Fe ($Z = 26$) is predicted to have its K-edge at $4038 * (26/20)^{2.156} \approx 7109$ eV, very close to its nominal value of 7112. This simple scaling rule allows one to remember only three numbers (the exponent 2.156, and one atomic number/edge energy pair) to estimate all the others to reasonable accuracy; it is often helpful when planning and executing experiments.

2.6 Elastic scattering from free electrons

For simplicity we treat the incident electromagnetic wave as a classical field; the results are substantially the same as in a quantized field picture. A classical electromagnetic wave that is incident on a (classical) free electron will exert a Lorentz force $\vec{F} = q(\vec{E} + \vec{v} \times \vec{B})$ on the electron where $\vec{E}$ and $\vec{B}$ are the electric and magnetic field vectors of the EM field, $\vec{v}$ is the electron's velocity, and q is the electron charge. The oscillatory electric field causes the electron to accelerate in the direction of $\vec{E}$, and the velocity it acquires produces a force perpendicular to both $\vec{v}$ and the oscillating $\vec{B}$, which is small if $v << c$. The interplay between the two terms in the Lorentz force delivers a small time average momentum to the electron which is described as "radiation pressure."

If $v << c$, for the purposes of scattering, we can neglect the magnetic force and write $\vec{F} = q\vec{E}$. The radiated power per solid angle $d\Omega$ ($= \sin\theta d\theta d\phi$ in spherical coordinates) is given by the Larmor formula [19]:

$$\frac{dP}{d\Omega} = \frac{e^2 a^2 \sin^2\theta}{4\pi c^3}$$

where a is the magnitude of the electron acceleration, c is the speed of light, and $e^2 = q^2/4\pi\epsilon_0$. Integrating the solid angle over all 4π steradian solid angle gives the total power

$$P = \frac{2}{3}\frac{e^2 a^2}{c^3}.$$

A relativistic generalization of this will be given in the discussion on synchrotron radiation.

Because of its acceleration, the oscillating electron radiates a spherical wave that is of the same frequency as the incident wave. The radiation pattern varies with angle as $\sin^2 \theta$, where θ is the angle between the acceleration $\vec{a}$ (which is essentially parallel to $\vec{E}$) and the vector from the electron's position to the observation point. The radiation pattern (in the nonrelativistic case) is roughly toroidal, as shown in Figure 2.12. Notice in particular that there is no radiation along the direction of the electron's acceleration. As a consequence, initially unpolarized radiation acquires a polarization upon scattering from an object. We note in passing that many animals (e.g. insects, arthropods, amphibians, cephalopods) use this effect in the visible wavelength region to help them navigate using sunlight and moonlight. It is also normal to exploit this effect in fluorescence mode XAFS experiments. Since most X-ray beams produced by synchrotron radiation are plane-polarized, undesired scattered background is minimized by centering the fluorescence detector at the minimum of the radiation pattern (the hole in the doughnut), which lies along the electric polarization vector of the X-ray beam.

2.7 Elastic scattering from electron distributions

Elastic scattering occurs when X-rays impinge on a material and interact with its electrons. The waves radiated by electrons in different locations within the scattering volume interfere with each other and produce a diffraction pattern that depends on the electron distribution.

Elastic scatter makes a significant contribution to the undesired background in fluorescence XAFS experiments. On the other hand, it is very useful, because elastic scattering in the form of Bragg diffraction from monochromator crystals is generally used to pick out specific energy photons from the polychromatic X-ray beams that are provided by the source. In this section we derive some simple results that have very broad utility.

In the case of weak X-ray scattering, only single scattering events are important, and simple expressions for the scattered radiation pattern can be derived. The idea is that an X-ray photon enters the material or volume of interest, scatters once, and leaves the volume without scattering again. This approximation ("kinematical" diffraction) is often a good one for atoms, multilayers, polymers, fibers, and most crystals.

Conversely, when many multiple scattering interactions occur in the volume of interest, "dynamical" scattering/diffraction must be considered. In this case, when treated classically, Maxwell's equations are solved for the complete X-ray wave-field within the crystal. Many new and interesting effects such as anomalous transmission (Bohrmann effect) occur in this limit.

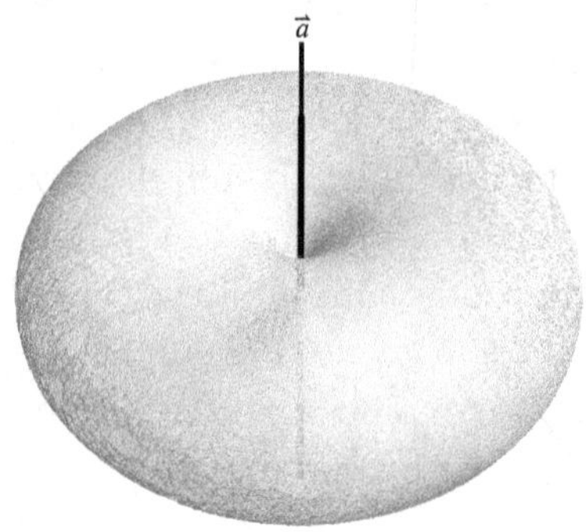

Fig. 2.12 Angular pattern of radiated (scattered) wave. The acceleration vector of the electron, $\vec{a}$ is along the direction of the electric polarization vector.

Dynamical diffraction is beyond the scope of this book; it is well treated elsewhere [21, 22, 23].

Here we confine our attention to the kinematic case which is simple and broadly applicable to many types of waves. For simplicity, but without losing any essential physics, we further approximate the form of the waves by treating them as scalar waves instead of the exact vector solutions.

The scattering geometry is shown in Figure 2.13. An incident wave encounters a point scatterer of strength η at position $\vec{r}_0 + \delta\vec{r}$ and then emits a spherical wave that is detected at $\vec{r}$. It is convenient to represent waves as complex exponentials: a plane wave has the form $e^{i\vec{k}\cdot\vec{r}}$, and a spherical wave centered at $r = 0$ has the form $e^{i|\vec{k}||\vec{r}|}/r = e^{ikr}/r$. If a phase change occurs upon scattering, η will be a complex number.

We will calculate the form of the scattered wave ψ_{scat}, neglecting the incident wave, which only contributes to the scattering pattern at zero angle. The wave intensity then can be calculated as $I \propto |\psi_{\text{scat}}|^2$.

The scattered wave then has the form

$$\psi_{\text{scat}} \propto \eta e^{i\vec{k}\cdot(\vec{r}_0+\vec{\delta}r)} \frac{e^{ik|\vec{r}-\vec{r}_0-\delta\vec{r}|}}{|\vec{r}-\vec{r}_0-\delta\vec{r}|} = \eta e^{i\vec{k}\cdot(\vec{r}_0+\vec{\delta}r)} \frac{e^{ik|\vec{s}-\delta\vec{r}|}}{|\vec{s}-\delta\vec{r}|}.$$

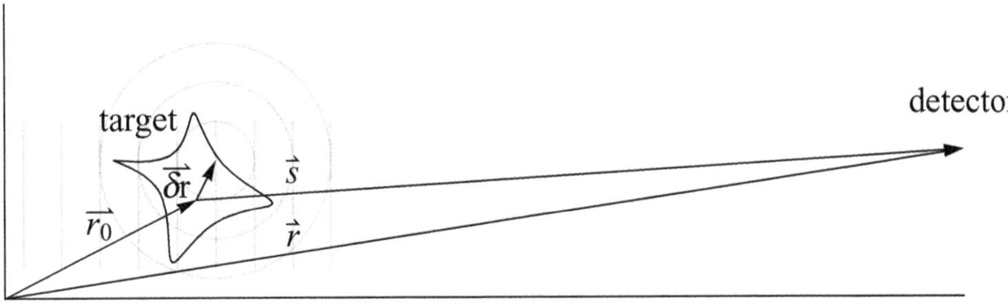

Fig. 2.13 Scattering geometry: $\vec{r}_0$ is a reference point within the target, which is represented by the oddly shaped object; $\delta\vec{r}$ denotes the position of a point-scatterer within the target; the scattered X-rays are detected at position $\vec{r}$. We make the approximation $|\delta\vec{r}| << |\vec{s}|$. The incident plane wave and scattered spherical waves are shown.

For $\delta r << s$, $|\vec{s} - \delta\vec{r}| = ((\vec{s} - \delta\vec{r}) \cdot (\vec{s} - \delta\vec{r}))^{1/2} = s(1 - 2\vec{s} \cdot \delta\vec{r}/s + (\delta r/s)^2)^{1/2}$. Neglecting the quadratic term and using the approximation $(1+\epsilon)^n \approx (1 + n\epsilon)$ for $\epsilon << 1$, which follows from a Taylor expansion, we find $|\vec{s} - \delta\vec{r}| \approx s - \delta\vec{r} \cdot \vec{s}/s$. We then substitute this into the exponential.

Next, defining the wave vector in the scattered direction $\vec{k}' \equiv k\vec{s}/s$; the scattering vector $\vec{q} \equiv \vec{k}' - \vec{k}$,[5] where $\hbar\vec{q}$ is the momentum transfer in the scattering process; and neglecting terms of order $|\delta\vec{r}|/s$ in the denominator, the scattered field at $\vec{r}$ then reduces to

$$\psi_{\text{scat}} \propto e^{i(\vec{k}\cdot\vec{r}_0 + ks)} \frac{\eta}{s} e^{-i(\vec{k}' - \vec{k})\cdot\delta\vec{r}} \propto \frac{\eta}{s} e^{-i\vec{q}\cdot\delta\vec{r}} .$$

The imaginary exponential $e^{i(\vec{k}\cdot\vec{r}_0 + ks)}$ is independent of $\delta\vec{r}$ and merely contributes a phase factor that does not affect the scattered intensity, so we will suppress it for simplicity. The s in the denominator simply gives the familiar $1/s^2$ falloff of intensity which is also of little interest, so we take this as an approximately constant scale factor.

So far we have calculated the scattered wave from a single point scatterer in the volume of interest. If there are multiple point scatterers located at $\delta\vec{r}_j$, and if multiple scattering can be neglected, the total scattered field will simply be the sum of contributions from the various scatterers j:

$$\psi_{\text{scat}}(\vec{q}) \propto \sum_j \eta_j e^{-i\vec{q}\cdot\delta\vec{r}_j} .$$

Note that the dependence on the observation point enters entirely through $\vec{q}$, which depends only implicitly on $\vec{k}$, $\vec{r}$, $\vec{r}_0$, $\delta\vec{r}$. For simplicity we therefore will write the scattering pattern only in terms of $\vec{q}$.

[5] This is the conventional notation; q should not be confused with the electron charge.

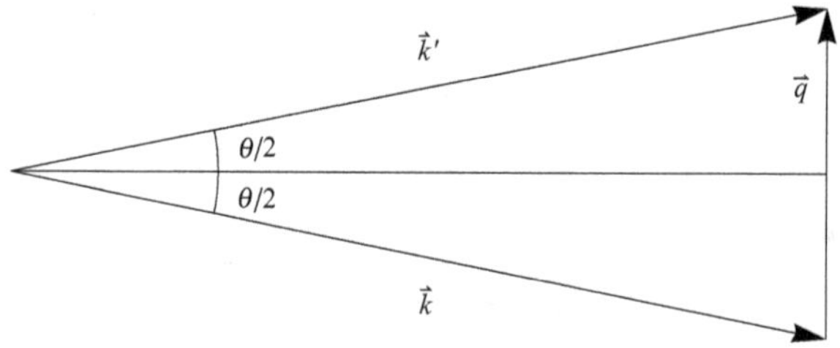

Fig. 2.14 Relation between $\vec{k}$, $\vec{k}'$, and scattering vector $\vec{q}$. The magnitude of the scattering vector $\vec{q} = \vec{k}' - \vec{k}$ evidently is $q = 2k\sin(\theta/2)$.

A continuous distribution of scattering strength density $\eta(\vec{r})$ within a volume τ gives a scattered wave:

$$\psi_{\text{scat}}(\vec{q}) \propto \int \eta(\vec{r}) e^{-i\vec{q}\cdot\vec{r}} d\tau,$$

where for simplicity of notation we take $\vec{r}$ to be a dummy integration variable. This integral has the form of a Fourier Transform (see Appendix 1) of the function η from $\vec{r}$ space to $\vec{q}$ space, which leads us to an important conclusion: *the scattered wave is proportional to the Fourier Transform of the scatterer density* $\eta(\vec{r})$*, and the intensity (diffraction pattern) is proportional to its squared modulus:*

$$I(\vec{q}) \propto |\psi_{\text{scat}}(\vec{q})|^2 \propto |\int \eta(\vec{r}) e^{-i\vec{q}\cdot\vec{r}} d\tau|^2.$$

Interestingly, a quantum mechanical treatment in the weak scattering limit gives a very similar expression, but with a quantum interpretation: the integral is the matrix element of the scattering potential between plane wave initial ($e^{i\vec{k}\cdot\vec{r}}$) and final ($e^{i\vec{k}'\cdot\vec{r}}$) states, and the scattering probability per unit time is proportional to the squared modulus of the matrix element.

The scattering strength for X-rays is proportional to the electron density ρ_{elec}:[6]

$$I(\vec{q}) \propto |\psi_{\text{scat}}(\vec{q})|^2 \propto |\int \rho_{\text{elec}}(\vec{r}) e^{-i\vec{q}\cdot\vec{r}} d\tau|^2.$$

2.8 Elastic scattering from atoms with no excitation

The expressions derived above have a particularly simple form when applied to a spherical distribution of electron density, which we will consider

[6] The nuclear charge does not contribute significantly to the scattering for the following reason: the mass of a proton is ≈ 1836 times greater than that of an electron, and since the scattered intensity is proportional to the square of the oscillating charge's acceleration, the scattered intensity varies inversely as the square of the mass.

atoms approximately to be. In this section we also ignore excitations of the electrons in the atom by the incident X-ray photon, which requires energy-dependent corrections, that are in fact closely related to XAFS. To treat this case it is best to use the quantum mechanical methods introduced later.

In the spherical atom approximation, the electron density within an atom depends only the radial distance r. Using spherical coordinates r, θ, ϕ, the volume element is $r^2 \sin\theta dr d\theta d\phi$, and the integral becomes:

$$\int_0^\infty dr \int_0^\pi d\theta \int_0^{2\pi} d\phi \; \rho(r)\; r^2 \sin\theta.$$

Integrating over ϕ and choosing the coordinate axes so that the z axis ($\theta = 0$) is oriented along $\vec{q}$, we then have

$$\int_0^\infty dr \int_0^\pi d\theta \rho(r)\; e^{-iqr\cos\theta} r^2 \sin\theta.$$

Substituting $u = \cos\theta$ ($du = -\sin\theta d\theta$) we have

$$\int_0^\infty dr\; 2\pi r^2 \rho(r) \int_{-1}^{1} du\; e^{-iqru} = \int_0^\infty dr\; 2\pi r^2 \rho(r) \frac{e^{iqr} - e^{-iqr}}{iqr}$$

Simplifying, we find the "form factor" $f(q)$:

$$f(q) = \int_0^\infty dr\; 4\pi r^2 \rho(r) \frac{\sin(qr)}{qr}.$$

As is evident from Figure 2.14, $q = 2k\sin(\theta/2)$, and therefore as $\theta \to 0$, $q \to 0$. In this limit $\sin(qr)/qr \to 1$, and $f(0) = \int_0^\infty dr\; \rho 4\pi r^2 = Z$, the atomic number of the atom. This shows that scattering in the forward direction is proportional to Z^2.

For example, consider a hydrogenic atom in its ground state. In this case the wavefunction is of form $\psi \propto e^{-Zr/a_0}$. Evaluating the integral we obtain $f(q) = Z/(1 + (\frac{qa_0}{2Z})^2)^2$, a quantity called the X-ray scattering "form factor," which is shown in Figure 2.15. Heavier atoms have more of their electrons in the core of the atom, which occupy a smaller volume, and therefore scatter more strongly at large q and large θ.

The electron scattering form factor f_{el} can be obtained approximately from the X-ray form factor using the Mott–Bethe formula

$$f_{\mathrm{el}}(q) = \frac{2me^2}{\hbar^2} \frac{Z - f(q)}{q^2}.$$

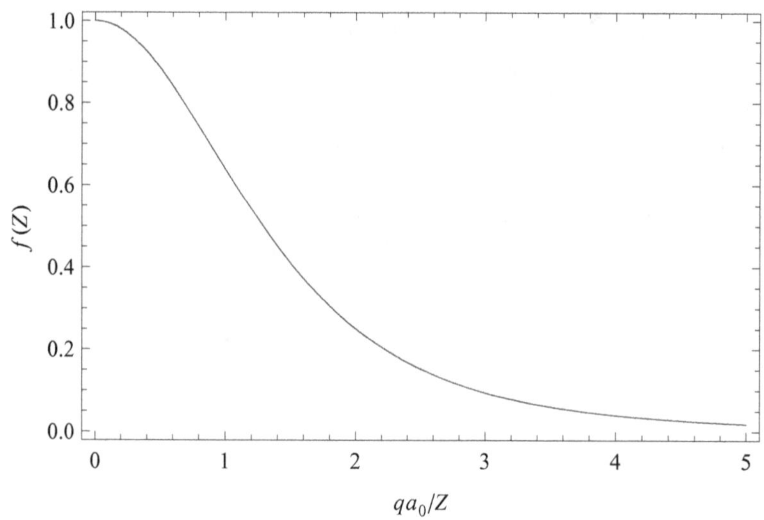

Fig. 2.15 Graph of $f(q)/Z$ vs. qa/Z for a hydrogenic atom, i.e. a single electron atom of nuclear charge Z, in its ground state. As Z increases, the wavefunction contracts by the factor of $1/Z$, and the form factor expands in q-space by the scale factor Z, broadening the diffraction pattern.

2.9 Elastic scattering from crystals

Consider an ideal crystal, which we can imagine constructing in the following way. We define a set of fixed basis vectors $\vec{a}$, $\vec{b}$, $\vec{c}$, which are not necessarily perpendicular, but no one of them can be coplanar with the other two. Then we can generate a 3D "Bravais" lattice of points through the coordinates $\vec{r}_{lmn} = l\vec{a} + m\vec{b} + n\vec{c}$, where l, m, n are integers.

If we plant an atom at each of these points $\vec{r}_{lmn}$, we get a simple ideal crystal. Better, imagine planting a *pair* of atoms, say Na and Cl in a salt crystal, at defined locations with respect to each lattice point. Mathematically this is a convolution. We have now constructed a "lattice with a basis." The "unit cell" consists of the two atoms. If we plant a complicated arrangement of atoms (or any other collection of scattering objects) at each lattice point, we have generated an idealization of a general crystal. Real crystals of course have vibrational motion, defects, dislocations, etc, but we will leave those complications to books on crystallography.

The convolution theorem (see Appendix 1) then tells us that the Fourier transform of the convoluted structure is just the product of the Fourier transforms of the two functions that are convoluted (lattice, and atom/unit cell). Convolution of a function with a lattice is illustrated in Figure 2.16. Since the diffraction pattern is just the squared modulus of the Fourier transform, this implies *the diffraction pattern is simply the product of the diffraction*

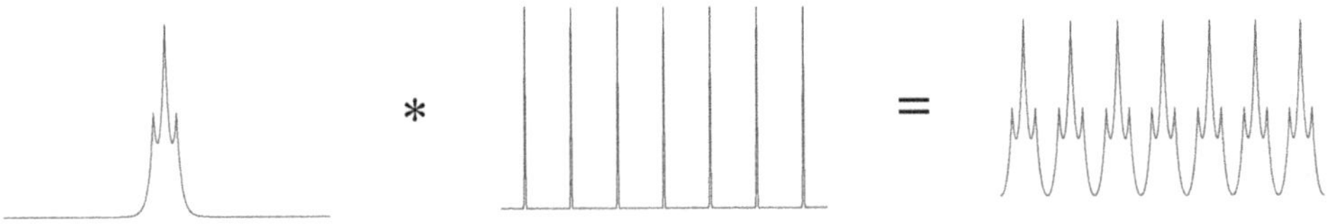

Fig. 2.16 This illustrates mathematical convolution, represented by *. The peculiar multipeaked function on the left (which could represent an atom, or a group of atoms), when convoluted with the lattice of delta functions, plants a replica of the function at each of the lattice points. In this manner a complex crystal structure can be generated by convoluting a pattern of atoms (e.g. a unit cell) with a simple ("Bravais") lattice.

patterns of the lattice and the unit cell. This is a great simplification of the problem – it permits a divide-and-conquer strategy by which we can separately find the Fourier transform of the unit cell and the Bravais lattice, then just multiply them afterward to find the total diffraction pattern. Similarly, if we were to pack together a cluster of identical crystals (with the same orientation) and wanted to calculate the diffraction pattern, we would just then multiply the previously calculated diffraction pattern by the square of the cluster structure factor. Since many real objects consist of hierarchical arrangements of simpler parts, this is an extremely useful approach.

Colloquially the diffraction peaks of different h, k, l are often called "reflections," presumably because the angle of incidence and diffraction are the same, as they are in reflection from a mirror. A key difference however is that the "mirror" only "reflects" when the diffraction conditions are satisfied.

We can represent the arrangement of atoms in the unit cell by their coordinates $\vec{r}_j$ and their scattering (form) factors f_j, which will be different for different elements. In this case the Fourier transform is called the "Structure Factor"

$$S(\vec{q}) = \sum_j f_j e^{-i\vec{q}\cdot\vec{r}_j}.$$

If the Structure Factor is zero, which often occurs because of symmetries within the unit cell, the total diffraction pattern for that value of $\vec{q}$ will likewise be zero, even if the Fourier transform of the Bravais lattice does not vanish there. In such conditions the "missing" diffraction peak is called a "forbidden reflection."

An example of practical importance for XAFS (because of the pervasive use of Si crystal monochromators) is the Diamond structure, which consists of a face-centered cubic (FCC) unit with atoms at coordinates

$$(0, 0, 0); (0, 1/2, 1/2); (1/2, 0, 1/2); (1/2, 1/2, 0),$$

plus another FCC unit displaced from those positions by $(1/4, 1/4, 1/4)$, giving eight atoms in the unit cell. Here the coordinates are expressed as fractions of the unit cell size. The structure factor (which is complex) and its modulus are easily calculated; the following *Mathematica 7* code[7] generates the lowest-order "allowed reflections," i.e. the sets of hkl for which the structure factor is nonzero. The output is presented in Table 2.1. The last column is an approximate measure of the integrated reflectivity of the reflection, using results from dynamical diffraction [21, 7, 24]. In general higher index reflections (ones for which $h^2 + k^2 + l^2$ is larger) have sharper rocking curves, and provide lower intensity, but better energy resolution. Notice that the structure factor is zero for certain harmonics: for example, the 222 second harmonic (at twice the energy) of the 111 reflection is forbidden, whereas the third harmonic is allowed.[8]

Mathematica 7 program to calculate "allowed" reflections for silicon

```
fcc={{0,0,0},{0,1/2,1/2},{1/2,0,1/2},{1/2,1/2,0}};
diamondstructure=Join[fcc,fcc+Table[1/4,{4},{3}]];
expterm[atomcoord_]:=Exp[-2 Pi I {h,k,l}.atomcoord];
SF=Total[Map[expterm,diamondstructure]];
Flatten[Table[{h,k,l,Abs@SF,NumberForm[N@Abs@SF,{5,3}],
NumberForm[N@Abs@SF/(h^2+k^2+l^2),{5,3}]},{h,1,6},{k,0,h},{l,0,k}],2];
TableForm@Select[%,#[[4]]>0&]
```

Silicon (and occasionally diamond) monochromator crystals used for XAFS usually employ the (111), (220), (311), or (400) reflections. The last column in the table is roughly proportional to the integrated reflectivity, indicating that the (111) reflection is the brightest. These topics will be covered in more detail in the next chapter.

2.10 Fourier transform of the lattice

As described above, the points of a Bravais lattice are given by $\vec{r}_{lmn} = l\vec{a} + m\vec{b} + n\vec{c}$, where l, m, n are integers. For simplicity we consider a parallelepiped-shaped crystal. The most important results do not depend on this assumption. We then need to calculate

$$\sum_{lmn} e^{-i\vec{q}\cdot\vec{r}_{lmn}} = \sum_{lmn} e^{-i\vec{q}\cdot(l\vec{a}+m\vec{b}+n\vec{c})} = (\sum_{l=0}^{L-1} e^{-il\vec{q}\cdot\vec{a}})(\sum_{m=0}^{M-1} e^{-im\vec{q}\cdot\vec{b}})(\sum_{n=0}^{N-1} e^{-in\vec{q}\cdot\vec{c}}).$$

[7] *Mathematica* tip: "$f@x$" applies the function f to x. "$f\backslash@x$" maps the function across each of the elements of a list x, equivalent to Map[f,x].

[8] Our convention in designating harmonics is the same as in music. The second harmonic (which is the first overtone) is twice the frequency of the fundamental, the third harmonic is three times the frequency, etc.

Table 2.1 *Allowed reflections for silicon*

h	k	l	$\|SF\|$	$\|SF\|$	$\frac{\|SF\|}{(h^2+k^2+l^2)}$
1	1	1	$4\sqrt{2}$	5.657	1.886
2	2	0	8	8.000	1.000
3	1	1	$4\sqrt{2}$	5.657	0.514
3	3	1	$4\sqrt{2}$	5.657	0.298
3	3	3	$4\sqrt{2}$	5.657	0.210
4	0	0	8	8.000	0.500
4	2	2	8	8.000	0.333
4	4	0	8	8.000	0.250
4	4	4	8	8.000	0.167
5	1	1	$4\sqrt{2}$	5.657	0.210
5	3	1	$4\sqrt{2}$	5.657	0.162
5	3	3	$4\sqrt{2}$	5.657	0.132
5	5	1	$4\sqrt{2}$	5.657	0.111
5	5	3	$4\sqrt{2}$	5.657	0.096
5	5	5	$4\sqrt{2}$	5.657	0.075
6	2	0	8	8.000	0.200
6	4	2	8	8.000	0.143
6	6	0	8	8.000	0.111
6	6	4	8	8.000	0.091

Each of these sums is of the form $s = \sum_{n=0}^{N-1} e^{inx} = \sum_{n=0}^{N-1} z^n$, with $z = e^{ix}$, and x is of the form $-\vec{q} \cdot \vec{a}$, etc. We then have the series $s = 1 + z + z^2 \cdots z^{N-1}$; multiplying by s, $zs = z + z^2 + \cdots z^N$, and subtracting we find

$$z = \frac{1 - z^N}{1 - z} = \left[z^{(N-1)/2}\right] \left(\frac{z^{N/2} - z^{-N/2}}{z^{1/2} - z^{-1/2}}\right) = \left[e^{i(N-1)x/2}\right] \left(\frac{\sin(Nx/2)}{\sin(x/2)}\right).$$

The last factor in square brackets on the right is a phase factor of modulus 1 that disappears when calculating diffraction pattern, so we ignore it henceforth.

The quantity of interest is the interference function $\left(\frac{\sin(Nx/2)}{N\sin(x/2)}\right)^2$, shown in Figure 2.17. The peaks in the interference functions form a one-dimensional lattice vs. x of period 2π, which implies that the peaks occur when $\vec{q} \cdot \vec{a} = 2\pi h$; $\vec{q} \cdot \vec{b} = 2\pi k$; $\vec{q} \cdot \vec{c} = 2\pi l$, where h,k,l are integers.[9] Since the 3D diffraction pattern is proportional to the product of all three interference functions, peaks only occur in the diffraction pattern if all three conditions

[9] The indices are conventionally written as hkl; the integer k should not be confused with the magnitude of the wave vector.

are satisfied. This implies that each diffraction spot corresponds to a specific triplet of integers h, k, l.

The integral over an interval $\pm\pi$ around the center of a peak is $2\pi/N$. The height of the peaks in the one-dimensional diffraction pattern is proportional to N^2 times the interference function. The width of the peaks decreases as $1/N$, while the height increases as N^2, so the peak areas scale as N, the number of scatterers in one dimension.

The full 3D diffraction pattern is a product of the squares of the three interference factors:

$$I(\vec{q}) \propto N_{\mathrm{tot}}^2 \left(\frac{\sin(L\vec{q}\cdot\vec{a}/2)}{L\sin(\vec{q}\cdot\vec{a}/2)}\right)^2 \left(\frac{\sin(M\vec{q}\cdot\vec{b}/2)}{M\sin(\vec{q}\cdot\vec{b}/2)}\right)^2 \left(\frac{\sin(N\vec{q}\cdot\vec{c}/2)}{N\sin(\vec{q}\cdot\vec{c}/2)}\right)^2,$$

where the total number of lattice points is $N_{\mathrm{tot}} = LMN$, which is proportional to the crystal volume.

It is useful to define an alternative set of vectors that are complementary to the original set a, b, c:

$$\vec{\alpha} = \frac{\vec{b}\times\vec{c}}{\vec{a}\cdot\vec{b}\times\vec{c}};\ \vec{\beta} = \frac{\vec{c}\times\vec{a}}{\vec{a}\cdot\vec{b}\times\vec{c}};\ \vec{\gamma} = \frac{\vec{a}\times\vec{b}}{\vec{a}\cdot\vec{b}\times\vec{c}}.$$

These have the useful properties:

$$\vec{\alpha}\cdot\vec{a} = 1 \quad \vec{\alpha}\cdot\vec{b} = 0 \quad \vec{\alpha}\cdot\vec{c} = 0,$$

$$\vec{\beta}\cdot\vec{a} = 0 \quad \vec{\beta}\cdot\vec{b} = 1 \quad \vec{\beta}\cdot\vec{c} = 0,$$

$$\vec{\gamma}\cdot\vec{a} = 0 \quad \vec{\gamma}\cdot\vec{b} = 0 \quad \vec{\gamma}\cdot\vec{c} = 1.$$

These vectors allow us to expand the scattering vector as $\vec{q} = (\vec{q}\cdot\vec{a})\vec{\alpha} + (\vec{q}\cdot\vec{b})\vec{\beta} + (\vec{q}\cdot\vec{c})\vec{\gamma}$, which, by the conditions $\vec{q}\cdot\vec{a} = 2\pi h$, $\vec{q}\cdot\vec{b} = 2\pi k$, $\vec{q}\cdot\vec{c} = 2\pi l$ implies that at a diffraction peak we must have $\vec{q} = 2\pi(h\vec{\alpha} + k\vec{\beta} + l\vec{\gamma})$. Since h,k,l are integers, this set of points itself defines a lattice that is "dual" to the real-space lattice. The "reciprocal lattice" vectors are defined as $\vec{g}_{\mathrm{hkl}} = 2\pi(h\vec{\alpha} + k\vec{\beta} + l\vec{\gamma})$. This shows that to get a strong diffraction peak the scattering vector $\vec{q} = \vec{k}' - \vec{k}$ must lie on some point on the reciprocal lattice.

Consider the diffracting planes of atoms shown in Figure 2.18. The displacement vector between any two lattice points has the form $\Delta\vec{r}_{\mathrm{jmn}} = j\vec{a} + m\vec{b} + n\vec{c}$, where j, m, n are integers. Since $\vec{g}_{\mathrm{hkl}} = 2\pi(h\vec{\alpha} + k\vec{\beta} + l\vec{\gamma})$, we have $\vec{g}_{\mathrm{hkl}}\cdot\Delta\vec{r}_{\mathrm{jmn}} = 2\pi(hj + km + ln) = 2\pi\nu$, where ν is an integer. It is clear from the figure that the scattering vector $\vec{q}$ is normal to the diffracting planes (for a specific reflection hkl) and therefore so is $\vec{g}_{\mathrm{hkl}}$. The perpendicular distance

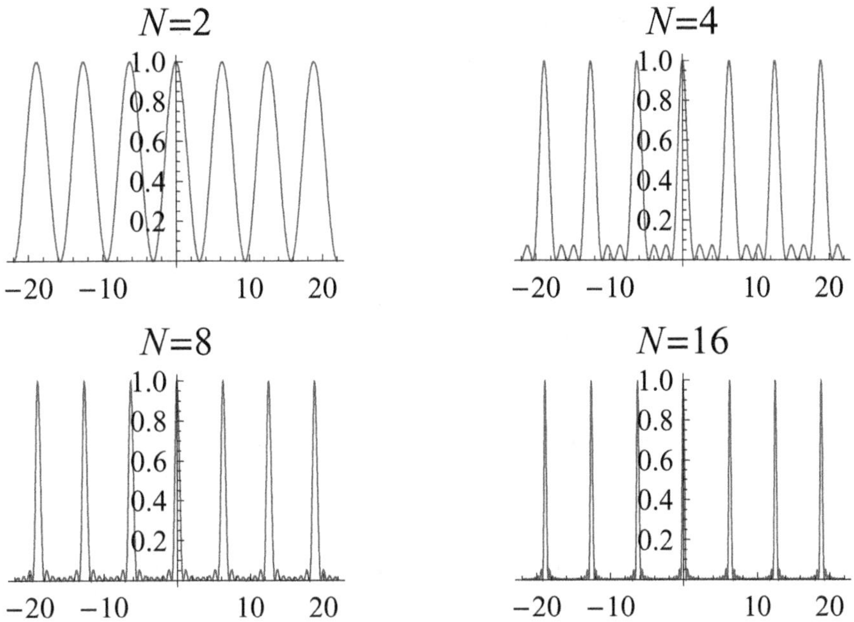

Fig. 2.17 The interference function $\left(\frac{\sin(Nx/2)}{N\sin(x/2)}\right)^2$ for various values of N. It is periodic, exhibiting sharp peaks at $x = 2\pi n$, with n integer. As N increases, the peaks become sharper. For large N, the peaks after scaling by $N/(2\pi)$ approximate a lattice of Dirac delta functions.

between the planes is just $d_{\text{hkl}} = \Delta\vec{r}_{\text{jmn}} \cdot \vec{g}_{\text{hkl}}/g = 2\pi\nu/g = 2\pi\nu/(2k\sin\theta_B)$, yielding

$$n\lambda = 2d\sin(\theta_B),$$

where we have suppressed the indices on d_{hkl} and substituted the integer n for ν to put the equation into canonical form. This is known as "Bragg's law."

Bragg's law has the following interpretation: it simply says that if waves scattering off adjacent planes of atoms are to interfere constructively, the difference in their path lengths ($2d\sin\theta$) must equal an integer number of X-ray wavelengths ($n\lambda$).

Consider the coordinates of three atoms in a diffracting plane, $\vec{r}_1$, $\vec{r}_2$, $\vec{r}_3$. Each of these can be expressed as distinct integer multiples of the basis vectors $\vec{a}$, $\vec{b}$, $\vec{c}$. The normal vector to the plane is in the direction $(\vec{r}_2 - \vec{r}_1) \times (\vec{r}_3 - \vec{r}_1)$. Expanding this expression in terms of $\vec{a}$, $\vec{b}$, $\vec{c}$, we obtain linear combinations of the vectors $\vec{\alpha}$, $\vec{\beta}$, $\vec{\gamma}$, where the coefficients are in integer ratios. This implies the plane normals are proportional to reciprocal lattice vectors g_{hkl}, and that the *diffracting set of planes can be indexed by* hkl, the "Miller indices" of the plane, which are related to the inverses of the intercepts of the planes with the coordinate axes.

Since $E = hc/\lambda$ for X-rays, we have

$$E \sin\theta_B = hc/2d_{\rm hkl} \approx 1977.1 \text{ eV} \sqrt{(h^2+k^2+l^2)/3}$$

for silicon, because in the diamond structure $d_{\rm hkl} = a_0/\sqrt{h^2+k^2+l^2}$ and a_0 is the unit cell dimension ($a_0 = 5.43095$Å for Si at 300K). This expression shows there is a lower limit to the energy range of diffraction for each reflection, an important fact for monochromator design.

2.11 Compton scattering from free electrons

Consider the collision of a photon and a free electron, in the rest frame of the electron. The photon carries momentum $\hbar\vec{k}$, and after scattering through nonzero angle θ, conservation of momentum requires the electron to recoil, so that it acquires kinetic energy. Conservation of energy then implies the scattered photon must have a lower energy (i.e. longer wavelength) than the incident photon, by an amount equal to the recoil energy of the electron. This why the scattering is called "inelastic" – the projectile (a photon) loses energy upon scattering.

Using the relativistic expressions for momentum and energy of the photon and electron, it is not hard to derive the following simple and important equation for the difference in wavelengths of scattered and incident photon: $\Delta\lambda = \lambda_c(1 - \cos\theta)$, where the Compton wavelength is $\lambda_c = h/mc = hc/mc^2 \approx 12.4$ KeV Å/511 KeV ≈ 0.0243Å. A photon of wavelength λ_c has an energy $mc^2 = 0.511$MeV, which is half of the minimum energy needed to create electron-positron pairs. The Compton shift $\Delta\lambda$ is evidently zero in the forward direction $\theta = 0$, and maximum in the backscattered direction $\theta = \pi$. It is easy to show that since $E = hc/\lambda$, the maximum energy loss, which occurs in the backscattered direction, is given by $E\frac{(2E/mc^2)}{1+(2E/mc^2)}$, where E is the initial photon energy. In the energy range of interest for XAFS the Compton shift (Figure 2.19) is of the order of hundreds of eV (about 200eV right above the Fe K-edge at 7.3 KeV) to thousands of eV, and it grows approximately quadratically (exponent ≈ 1.95) with the X-ray energy.

2.12 Inelastic scattering from atoms

When X-rays are scattered from atoms, only the outer electrons can be sensibly approximated as free electrons. The inner electrons are strongly bound

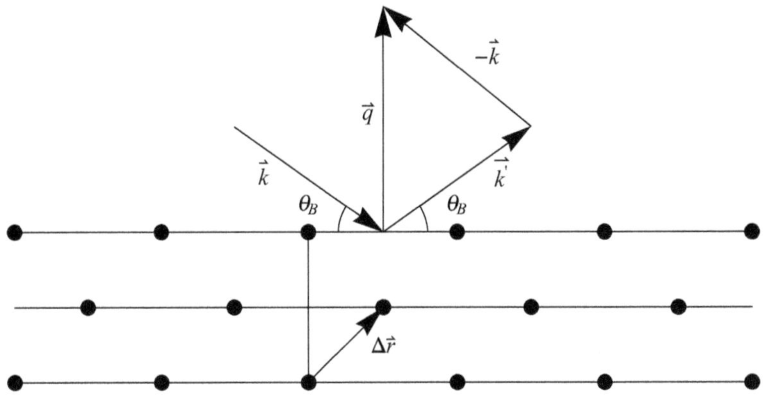

Fig. 2.18 "Bragg diffraction" from planes of atoms. The angle of the X-ray beam with respect to the sample surface is the "Bragg angle" θ_{B}.

to the nucleus, and their recoil is strongly suppressed because they can transfer momentum to the nucleus, and indeed the whole crystal, which, being very massive, recoils to a negligible extent. Those bound electrons scatter, but they scatter elastically, because there is little recoil. Thus we can expect a range of scattered energies from atoms, with a strong elastic peak at the incident energy, to a Compton backscatter peak at hundreds to thousands of eV lower energy, and a continuum of energies in-between, depending on the structure of the atom. That is indeed what is measured. Compton scattering also is important when interpreting energy spectra of particles of sufficiently high energy that they deposit only part of their energy into the detector before escaping.

If the energy of the X-ray is close to that of an absorption edge, resonant effects occur that strongly affect the intensity of the scattering. These can be interpreted to give information about the electron structure of the scatterer. This Resonant Inelastic X-ray Scattering is intimately connected with XAFS. Further discussion of the physics of X-ray absorption will be deferred until Chapter 4, Theory.

2.13 Core-hole lifetime

The XAFS spectra are intrinsically affected by core-hole lifetime broadening, which gives a finite energy width to the energy levels. Energy levels have a precise sharply defined energy only if the states last for an infinitely long period; anything that limits their lifetime introduces a broadening to the spectral features. The "core hole" is the vacancy in the orbital from which the photoelectron is excited. For example in K-edge excitation, a 1s core

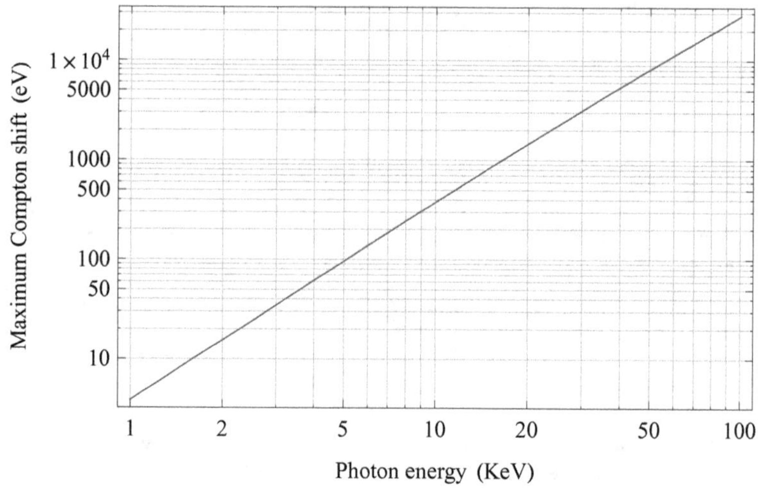

Fig. 2.19 Compton backscattering energy shift vs. energy.

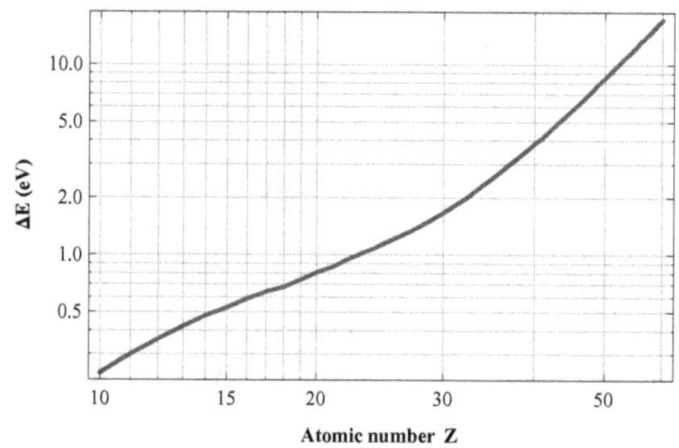

Fig. 2.20 Log-log plot of the K-edge core-hole broadening (FWHM) as a function of atomic number Z of the X-ray absorbing atom. The level width over the interval $Z = 16-60$ is well represented by $1.0(Z/42)+2.1(Z/42)^2-4.6(Z/42)^3+6.0(Z/42)^4$.

hole is produced because a single electron is removed from that orbital. An atom with a core hole is in an excited, unstable state, and it has a finite lifetime because electrons in higher energy states rapidly make transitions to fill the core hole, either emitting fluorescence radiation in the process, or undergoing nonradiative decay. The Heisenberg uncertainty relation implies that the finite lifetime τ produces an uncertainty (broadening) of the energy $\Delta E \approx \hbar/\tau$. The lifetime τ decreases rapidly with atomic number of the central atom, so the core-hole broadening ΔE increases rapidly with Z. These values [11] are plotted in Figures 2.20 and 2.21.

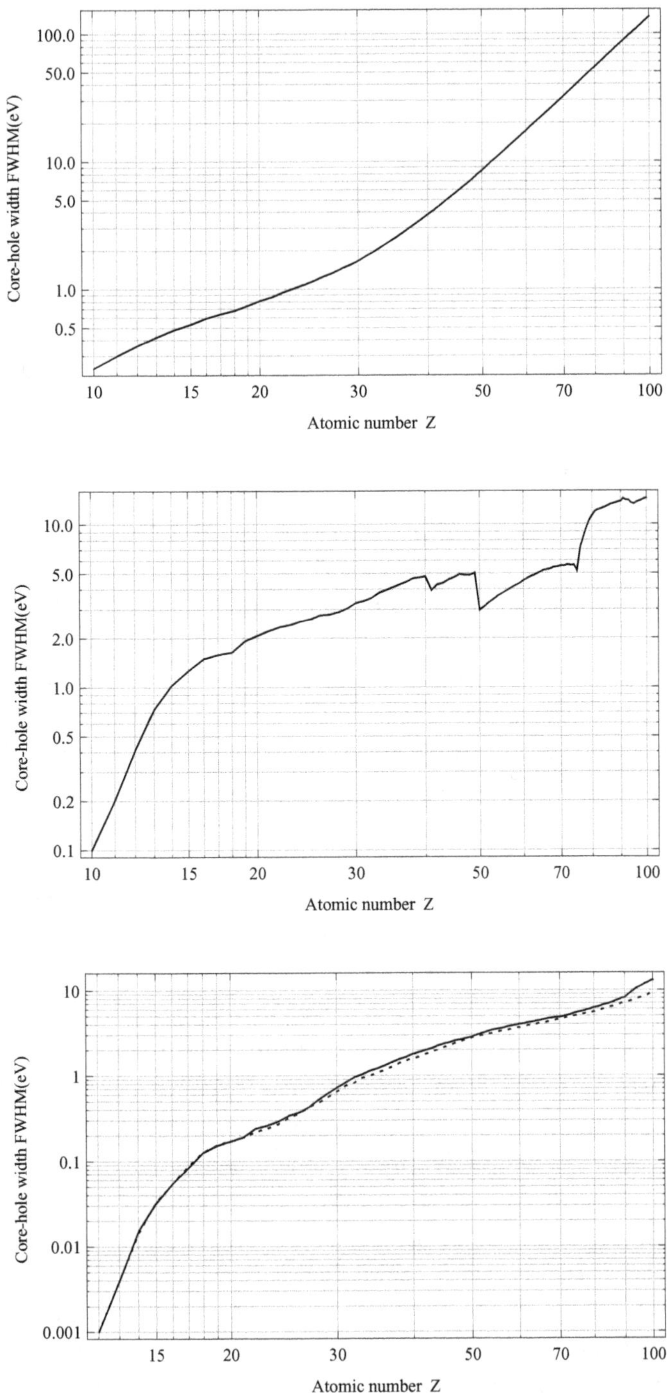

Fig. 2.21 Log-log plot of the (from top) K; L_1; and $L_{2,3}$ core-hole broadening (FWHM) functions as a function of atomic number Z of the X-ray absorbing atom. The L_2 curve is shown as solid line; the L_3 curve is shown as a dotted line.

3
Experimental

3.1 Requirements for XAFS experiments

The features of interest in XAFS spectra consist of small variations in the absorption coefficient $\mu(E)$, which can be determined directly in a transmission experiment, or indirectly by measuring the variation in the intensity of specific fluorescence emission lines as the energy of the incident beam is scanned over an absorption edge. Sometimes useful information, such as edge shifts, can be obtained from XANES spectra that are noisy, but in general very good signal to noise ratio (S/N) is required for EXAFS analysis and detailed XANES analysis. For EXAFS measurements one requires S/N ratios better than 10^3 in order to determine the spectra accurately enough in the region $\approx$ 600–1000 eV above the absorption edge. It is generally necessary to measure spectra at least this far above the edge in order to get adequate spatial resolution. The EXAFS signal may damp out quickly above the edge because of disorder and/or low-Z scatterers, rapidly plunging the signal into the noise, unless the noise level is kept very low.

An intense beam is required to obtain good data in a reasonable time frame (minutes to hours): on the order 10^{10} photons/sec or better within the required energy bandwidth of an eV or so. For this reason, synchrotron radiation sources are by far the preferred sources of X-rays for XAFS experiments. The X-ray beam should have an energy bandwidth of approximately 1 eV or less in order to resolve the XANES features, or at most a few eV, in order to resolve the XAFS right above the edge.[1] It is not difficult to produce such a narrow bandwidth using crystal monochromators, but in addition it is necessary to accurately and stably scan the energy of the incident beam over a range of approximately 1KeV above the edge. Furthermore one must

[1] It is possible to use energy widths larger than this, but at the cost of introducing more noise through the need to numerically deconvolute the spectra.

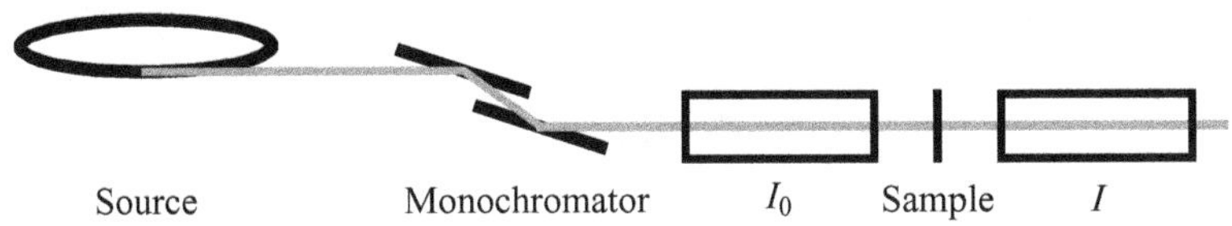

Fig. 3.1 Transmission XAFS experiment.

do so in a time frame of seconds to minutes. Nor should the intensity of the beam vary much over the scan because such variations will affect the data through residual nonlinearities in the detection system. These requirements impose significant demands on the design and quality of the mechanics and control system of any beamline that is to be used for XAFS. Most samples of interest are not physically large, so beams of several centimeters in size down to submillimeter dimensions are typically used; these can be effectively produced on many modern beamlines through the use of focussing optics. Using more specialized optics, beams of much smaller (e.g. submicron) dimensions can be prepared for the most demanding experiments. These methods are discussed below.

3.1.1 Experimental modes

XAFS is typically measured in several modes: transmission, fluorescence, and electron yield. These experimental modes are illustrated in Figures 3.1–3.3. The choice of the most appropriate one to use in an experiment is a crucial one. A number of factors must be considered in doing so. This process will be described more fully after describing experimental apparatus; here we briefly describe the basic experimental modes of measuring XAFS.

Transmission mode is the most straightforward: it simply involves measuring the X-ray flux before and after the beam is passed through a uniform sample. Under these conditions, the transmitted flux I is related to the incident flux I_0 by $I/I_0 = \exp(-\mu(E)x)$, or $\mu(E)x = \ln(I_0/I)$.

In fluorescence mode one measures the incident flux I_0 and the fluorescence X-rays I_f that are emitted following the x-ray absorption event. Usually the fluorescence detector is placed at 90° to the incident beam in the horizontal plane, with the sample at an angle (usually 45°) with respect to the beam. Fluctuations in the number of elastically scattered X-rays are a significant source of noise in fluorescence XAFS, so the position of the detector is selected to minimize the elastically scattered radiation by exploiting the polarization of the X-ray beam.

It is common for experimenters to informally use the terms I_0 and I to

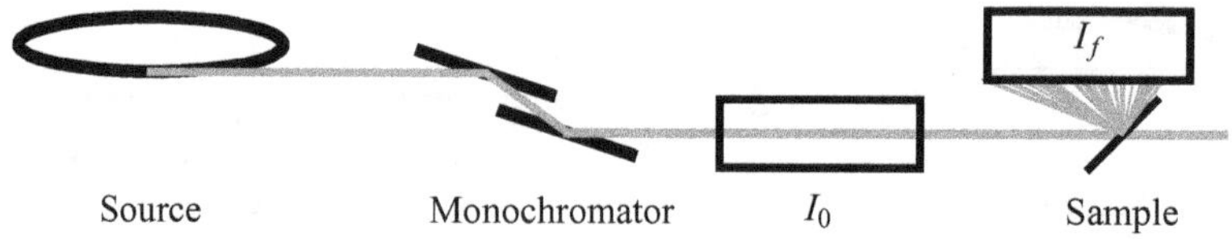

Fig. 3.2 Fluorescence XAFS experiment.

refer to the ionization chamber signals, rather than the beam fluxes. This imprecision is of no consequence for transmission experiments, but it does matter for fluorescence and electron yield. If the detectors are linear, the detector signal will be proportional to the flux, however the output will be multiplied by an energy-dependent detector sensitivity that is a consequence of the declining fill gas absorption as a function of energy. The measured absorption coefficient also will have contributions from the air and windows that are in the beam path between the two detectors. Since a logarithm of the detector signals is taken, the log transforms the multiplicative energy dependent detector sensitivity factor into an additive slowly varying function of energy that can be readily subtracted out. The normal XAFS data reduction methods automatically take care of such "background."

This situation is different in fluorescence and electron yield, however, because no logarithm is taken: the I_f signal (or equivalent electron yield) is simply divided by I_0. Since I_0 decreases with energy, this is equivalent to multiplying the data by a smoothly increasing function of energy. This directly affects EXAFS amplitudes and should be accounted for. In general such slowly varying multiplicative functions will appear as a slight correction to the EXAFS Debye–Waller factors, as explained in the section on data reduction.

Electron yield detection is similar to fluorescence yield, in that it indirectly measures absorption events by detecting the decay products when the core-hole is refilled. In the case of fluorescence one measures photons; in the case of electron yield one measures the electrons that are emitted from the surface of the sample. The relatively short path length (≈ 1000 Å) makes the technique surface-sensitive, which can be beneficial if one is interested in near-surface phenomena. It also can be beneficial for avoiding "self absorption" effects that can occur in fluorescence mode.

A peculiar feature of electron yield methods is that the sample in effect is situated inside the detector. In total electron yield, which is used for soft X-ray work, the sample is under ultra-high vacuum, and the electrons are collected by electrodes. At higher energies, conversion electron detection

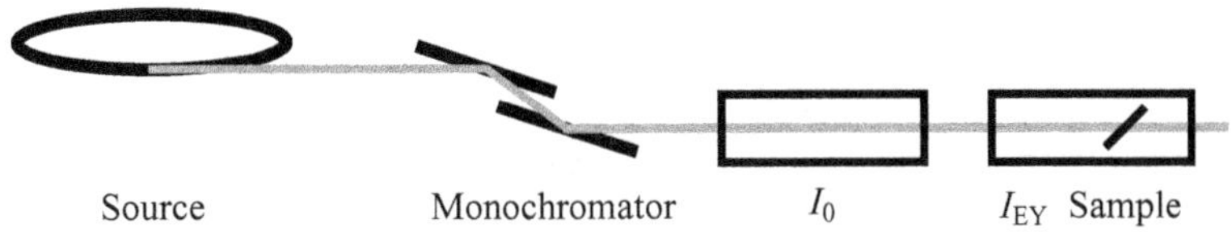

Fig. 3.3 Electron yield/conversion electron yield XAFS experiment.

can be used. In this case, the electrons that are emitted by the sample collide with helium gas within the detector, producing a number of secondary electrons that can be collected as in ionization chambers. In this case, the detector effectively is an ionization chamber with the sample in electrical contact with the cathode. This approach only works if the sample is, or can be made to be, sufficiently electrically conductive, otherwise sample charging effects will distort the spectra. Conversion electron detection may also suffer from nonlinearities if the incident beam flux is too large. Conversion electron detection can be a very effective way to acquire data for some specimens however.

Other methods for detecting the energy released after decay of the core hole are feasible, e.g. detection of visible light, but are seldom used.

3.1.2 Dispersive and time-resolved XAFS

Most XAFS scans are carried out over time scales of seconds to minutes. Rapid scan, or "Quick XAFS" also is feasible if the monochromator and data acquisition electronics are designed for the purpose. This can allow one to perform measurements in the sub-second time range. However if one is interested in time-dependent phenomena on shorter time scales, different approaches are needed. One is dispersive XAFS, in which many simultaneous XAFS measurements at different energies are carried out at the same time, at slightly different angles. Another approach is time-resolved XAFS, which has several sub-varieties.

An important benefit of rapid data acquisition is that it can support modulation experiments, in which one perturbs the sample in some manner at a specific frequency, and then measures synchronously the changes to the data that occur at the modulation frequency. This method allows one to reject noise contributions at other frequencies, and thereby permits very high precision measurements. There are many potential applications of this approach, one of which is X-ray magnetic circular dichroism (XMCD). ESRF's ID24 is a good example of such a facility [58].

Dispersive XAFS is generally done in transmission mode. A polychromatic X-ray beam (often from a bend magnet or wiggler, so that a fan of radiation is available) is made to impinge onto a bent polychromator crystal, so that a range of Bragg angles is obtained, each brought to a focus on the sample. These beams simultaneously pass through the sample at different angles, and the transmitted beams are measured using a position-sensitive detector such as a photodiode array. This approach has no advantage over conventional monochromators as far as total flux is concerned, because each small segment of the crystal only diffracts a single energy. However, it does have the virtue of requiring no mechanical motion to do a scan. Therefore, if the X-ray flux is sufficient, and the detection system is fast enough, rapid time slicing becomes possible, allowing the study of a wide variety of fast reactions and other time-dependent phenomena.

In its usual form, dispersive XAFS is not applicable to fluorescence, because it would not be clear which of the beams of different energy that are simultaneously hitting the sample would have induced the measured fluorescence. H. Oyanagi has proposed that a fast-moving mask could be used to "chirp" the beam, and with a sufficiently fast detection system, perform fluorescence measurements in a dispersive configuration.

Another approach to address rapid data acquisition is time resolved XAFS, of which one mode is time-slicing. A sample is perturbed in some manner, for example by an intense laser pulse, pressure jump, or by rapidly mixing with a reactant, and the time course of the XAFS signal (usually fluorescence) is recorded. This process is repeated at the different energy points to make up XAFS signal vs. time. It is essential that the sample fully "resets" before the next stimulus (or that it be replaced by fresh sample) for this approach to work reliably. Another approach, which is suitable for ultrafast measurements, is pump-probe. This is similar to time-slicing, but the signal is acquired during a precisely defined time interval following the stimulus. Since the synchrotron light itself comes in pulses of a few hundred picosecond duration, one can arrange to first pulse a laser, and then, after a defined time delay, acquire XAFS data as an X-ray pulse is exciting the sample. Since light travels at about 30 cm per nanosecond, the physical length of each X-ray pulse is about 5 cm, and typically are separated by perhaps 45 meters (APS, 24 bunch mode). By varying the delay time between the laser (pump) and X-ray (probe) pulse, one can map out the time dependence of the signal. This process is repeated at different energies. Fast beam choppers, or fast gateable detectors must be used in this case. Diffractive analyzers may be used to good effect in such cases for reducing background. Fast plastic scintillators and photomultipliers, or avalanche photodiodes (APDs) are useful detectors for such high rate applications. A new class of devices called Silicon Photomultipliers may also prove useful for synchrotron radiation applications. The limitations of pulse counting at low bunch repetition rate (see Deadtime Corrections) must be heeded in such experiments, however.

3.2 X-ray sources

The traditional source of X-rays for laboratory experiments are X-ray tubes. These consist of an evacuated enclosure (tube) in which a source of electrons, such as a heated tungsten filament or other material is caused to emit electrons, which are accelerated through a large voltage difference ΔV, on the order of tens of kilovolts. The electron emitter is the cathode, and the electrons are attracted by the anode. Upon falling through this large voltage difference, the electrons acquire $q\Delta V$ (tens of KeV) kinetic energy, where q is the electron charge, which they abruptly deposit through collisions with the atoms in the anode. If the kinetic energy of the electrons is sufficiently high, they can excite electrons out of core levels in the atoms; when these are filled by higher-energy electrons re-filling the core level, they produce fluorescence at energies characteristic of the anode material. Generally the K_α lines are used because the fluorescence yields are maximal. Typical target materials are Cu, Mo, Ag, and W, with fluorescence energies of $\approx 8.0, 17.5, 22.2$, and 59.3 KeV respectively.

When the electrons hit the target, they decelerate rapidly and therefore produce X-rays through another mechanism: bremsstrahlung, or "braking radiation." In contrast to the X-ray fluorescence, which are concentrated at specific energies, the bremsstrahlung spectrum is a continuum from low energies up to a maximum which is the kinetic energy of the electron ($q\Delta V$, where q is the electron charge).

Only a fraction of the total power consumed is turned into useful X-rays; the rest is dissipated as heat, which is removed, usually by water or air cooling. It is no accident that anode materials either are good conductors or are difficult to melt. There are serious limits to the amount of power that can be dissipated in this manner, and for high-power applications X-ray generators with rotating anodes are often used. This mechanically spreads the heat load over a larger area while keeping the X-ray source spot fixed. Unlike fixed anode generators, which are simple and reliable, rotating anode generators are complex instruments that require diligent maintenance to work reliably.

Since XAFS experiments require X-ray beams of tunable energy, fluorescence lines from X-ray tubes are not useful for XAFS. However, the lower intensity broad-spectrum bremsstrahlung can be used if a suitable scanning (and preferably, focussing) monochromator is employed. A number of laboratory-based XAFS spectrometers have been built and successfully used; even commercial units are available. Nevertheless, despite their higher output, even rotating anode generators are usually inadequate for XAFS experiments on dilute species, and synchrotron radiation sources usually are clearly preferred over laboratory sources. It remains to be seen what impact the new "fourth-generation" sources such as free-electron lasers will have on XAFS.

3.3 Synchrotron radiation

Synchrotron radiation is so named because it was first observed at synchrotrons, which are large and complex machines built for high-energy physics experiments. They are used to accelerate charged particles up to relativistic speeds, i.e. the particle velocity v approaches c, the speed of light. Accelerating charged particles generate electromagnetic radiation, and they do so as they are directed around curved paths by magnetic fields. When the charges move at speeds approaching c, the spectrum and angular emission pattern of the radiation are dramatically altered from those at lower speeds, as a consequence of relativistic length contraction and time dilation. One consequence is that large amounts of power are radiated, the spectrum of the electromagnetic radiation is shifted to much higher frequencies (energies), and the radiation pattern becomes highly directional. Particle physicists generally regard the radiation as a nuisance, because it steals energy from their particle beams, but over the last few decades synchrotron radiation has been found to have extremely beneficial properties for X-ray experiments.

Synchrotron radiation research has evolved substantially over the last few decades. In the early 1970s the new field of synchrotron radiation research was entirely "parasitic" on the high-energy physics experiments. Following the early success of synchrotron radiation research, dedicated "second-generation" facilities such as the NSLS were built. The charged particles used in these machines are generally electrons, which radiate much more efficiently (for a given particle energy) than the much more massive protons that are preferred in high-energy physics experiments. These were oriented toward using the radiation produced at bend magnets. After the development of "insertion devices," e.g. wigglers and undulators, "third-generation" sources such as the APS, ALS, ESRF, and SPRING8 were built specifically to accommodate insertion devices. Fourth-generation sources are under development; their utility for XAFS remains to be seen.

3.3.1 Flux and brilliance

We use the term "flux" to represent the total number of photons/sec in a beam, and "intensity" to represent the photons/sec/area incident on a target. There is another quantity, "brilliance," that is particularly useful for describing X-ray sources and optics. Brilliance has dimensions photons/sec/(source area mm^2)/(source angular divergence $mrad^2$). The reason brilliance is useful to describe beams is that it is an invariant in ideal optical systems, and it can only be degraded by imperfect optical systems. This is a consequence of Liouville's theorem in classical mechanics. It expresses the basic fact that, whatever the complexity of an optical system, it can reduce the spatial size at the cost of increasing the angular divergence, or vice versa, but not both at once, unless one is willing to discard photons (e.g. by slitting down the beam). Brilliance is an excellent figure of merit for a source, and the brilliance of synchrotron radiation is many orders of magnitude greater than X-ray tubes.

A brilliant source emits many photons per second, with a small source size, and a small angular divergence. The small angular divergence allows one to collect the photons on X-ray optics of reasonable size, such as mirrors and monochromators. The small source size allows focussing optics to better image the source, because it is more point-like.

3.3.2 Storage rings

Synchrotrons increase the kinetic energy of the beam by passing groups of particles (called "bunches") through conducting cavities ("RF cavities") in which a radio frequency electromagnetic standing wave is maintained. The timing is carefully controlled so the bunches "surf" through the oscillating wave field which gives the particles a push forward. Booster synchrotrons accelerate the electrons up to the desired energy, after which they are transferred to "storage rings." These consist of evacuated pipes under ultra-high (10^{-12} torr) vacuum through which the relativistic electrons travel in straight paths until they encounter a strong magnetic field in the vertical direction which is produced by bend (dipole) magnets. While in the magnetic field, they experience a force that causes them to move in an arc with a radius of curvature on the order of tens of meters, after which they continue down another straight section. This continues until their path closes, so the path is more like a polygon with rounded corners than a circle. The same electrons recirculate for hours or days, all the while generating X-rays.

Wherever the path of the electrons bends, they accelerate, and therefore they radiate. The particle energy is replenished by passing the beam through

RF cavities. Relativistic effects cause the radiation pattern to strongly tilt into the forward direction, i.e. in the same direction as the electron's instantaneous velocity. As the electrons sweep around their path the radiation is emitted in a fan that is of order $1/\gamma$ (radians) in angular width (perpendicular to the orbital plane), where $\gamma = E/mc^2$, where E is the electron energy, and mc^2 is the electron's rest energy (0.511 MeV). For example, a machine of energy 5 GeV produces bend magnet radiation with an angular width in the vertical direction of order 5 GeV/.5 MeV $\approx 10^{-4}$ radian. This matches well the angular acceptance of silicon monochromator crystals. In the horizontal plane the beam usually is on the order of a few milliradians, which, at tens of meters downstream gives an X-ray beam of a few mm to a few centimeters length in the horizontal direction.

There is of course considerably more to a storage ring. Quadrupole and sextupole magnets are used to keep the beam orbit stable in the directions transverse to the motion as well as the longitudinal direction along the beam path. The electron beam varies in size and transverse momentum in a controlled manner[2]; these deviations from the ideal path are described by "beta functions" for that configuration of magnets, the "lattice."

Radiation shielding is a major consideration for high-energy (several GeV) rings, not only because of the copious production of high-energy radiation through synchrotron radiation, but even more the extremely high energy bremsstrahlung that is occasionally produced by collisions of the electrons with residual gas atoms in the ring. The worst case situation of a beam dump at an unplanned location around the ring requires the complete enclosure of the ring by massive shielding, as well as extensive safety interlocks.

Once generated, the radiation produced by bend magnets and insertion devices travels tangentially to the beam path and exits in a controlled manner through the shield wall to the beamlines, which tailor the beam properties for the experiments. At a typical light source there are dozens of beamlines optimized for various purposes, positioned at specific locations around the storage ring.

[2] Liouville's theorem from classical mechanics tells us the phase space volume – the product of the position and momentum spreads – are constant around the path. The conserved volume is called the emittance.

3.3.3 Bend magnet radiation

The properties of the radiation are given by Lienard's relativistic generalization of the Larmor formula [19]:

$$P = \frac{\mu_0 q^2 \gamma^6}{6\pi c}\left(a^2 - \left|\frac{\vec{v}\times\vec{a}}{c}\right|^2\right).$$

In this expression m and q are the mass and charge[3] of the particle (electron), $\vec{v}$ and $\vec{a}$ are its velocity and acceleration, $\gamma = E/mc^2$ where E is the particle energy (rest energy plus kinetic), and μ_0 is the magnetic permeability of free space. Applying this expression to a relativistic particle moving in a circular arc one calculates the power radiated as a function of angle. Note that the total power varies as the sixth power of the particle energy, which has important implications for beamline design at high energy rings.

The frequency dependence (and hence, photon energy spectrum) can be calculated by Fourier transforming the time dependence of the wave field detected at the observation point [20]. The energy spectrum of bend magnet radiation, integrated over the vertical opening angle, is proportional to a universal curve $g_1(x)$, where $x = \epsilon/\epsilon_c$, ϵ is the X-ray energy, and ϵ_c is the "critical energy" parameter, which depends on bend radius and electron beam energy. This function is defined as $g_1(x) = x\int_{t=x}^{\infty} K_{5/3}(t)dt$, where K_n is the modified Bessel Function of the second kind, of order n. A least-squares fit[4] shows that it can be well-approximated by the easily remembered $g_1(x) \approx 1.8\ x^{0.3}e^{-x}$. The function is rather broad, rising to its maximum of 0.918 at $x = 0.2858$, and decreasing exponentially for $x > 1$; the peak ϵ is about 2/7 of ϵ_c. The exact curve and this approximation are shown in Figure 3.4.

The beam is highly collimated in the horizontal plane; its root mean square angular width in the vertical direction can be approximated by $(0.57/\gamma)(\epsilon_c/\epsilon)^{0.43}$, so that the higher-energy photons are more strongly confined to the orbital plane than are the low-energy photons.

The universal curve allows us to calculate the flux that can be expected from a bend magnet. The number of photons/sec/mA/mrad horizontal acceptance in $10^{-3}\Delta\epsilon/\epsilon$ bandwidth is $1.256 \cdot 10^7 \gamma\ g_1(x)$. This expression can also be used for wigglers (but not undulators), after multiplying the result by the number of wiggler poles. Windows in the beamline may reduce

[3] The prefactor in this equation can be written $2e^2\gamma^6/3c^3$ where $e^2 = q^2/r\pi\epsilon_0$. This (gaussian units) electric charge e should not be confused with the number 2.71828...

[4] A slightly more accurate least-squares fit over the range $x = 0.01 - 8$ gives $g_1(x) \approx 1.721e^{-0.9694x}x^{0.2824}$.

the flux substantially at lower energies and attenuation from them must be considered to calculate the spectrum accurately.

The critical energy is $\epsilon_c = \gamma^3(3hc)/(4\pi\rho)$ where $\gamma = E/mc^2$, E is the electron beam energy, ρ is the bend radius of the electron path, m is the electron mass, c is the speed of light, h is Planck's constant ($hc \approx 1.24 \cdot 10^{-9}$ KeV meter). The bend radius ρ (meters) is related to the magnetic field strength (Tesla) and the beam energy (GeV) by $\rho \approx 3.336E/B$.

For example, consider a 7 GeV beam energy, 100 milliamps of beam current, with a 0.6 Tesla bend magnet field, and we collect 2 cm horizontal width of the beam at 20 meters (1 mrad), and all of the beam in the vertical direction. The bend radius is then $3.336 \cdot 7/0.6 \approx 39$ meters; with $\gamma \approx 7000$ MeV/0.511 MeV $\approx 1.37 \cdot 10^4$, we have a critical energy of $(1.37 \cdot 10^4)^3(3 \cdot 1.24 \cdot 10^{-9})/(4\pi \cdot 39) \approx 19.5$ KeV. Then the flux at 10 KeV in a 10^{-3} bandwidth (e.g. 10 eV @ 10 KeV) can be calculated as $1(\text{mrad}) \cdot 100(\text{mA}) \cdot 1.256 \cdot 10^7 \cdot 1.37 \cdot 10^4 \cdot 1.8 \cdot (10/19.5)^{0.3} e^{-10/19.5} \approx 1.5 \cdot 10^{13}$ photons/sec. In a 1 eV bandwidth ($10^{-4}\Delta\epsilon/\epsilon$) it would be $1.5 \cdot 10^{12}$ photons/sec.

The polarization content of the beam depends on observation position relative to the orbital (horizontal) plane. Precisely in the orbital plane, the X-ray polarization of bend magnet radiation is entirely horizontally polarized. Away from the orbital plane the X-ray beam acquires some degree of circular polarization, which can be usefully exploited in X-ray MCD experiments.

3.3.4 Insertion devices

Dipole bend magnets are required to deflect the beam around a closed path; as an additional benefit they produce useful synchrotron radiation. Another type of magnetic structure, an "Insertion Device," can be inserted into the straight sections between the bend magnets. A planar wiggler/undulator is illustrated in Figure 3.5. These have the great advantage that they can be optimized to produce radiation specifically tailored to experimental needs.

As discussed in the previous section, the critical energy ϵ_c at a bend is proportional to the magnetic field used to produce it. Wigglers and undulators consist of arrays of magnetic pairs of alternating orientation, so that electrons passing through the device feel a magnetic field of alternating polarity. The alternating field generates a force perpendicular to both the field and the electron velocity, which causes the path to oscillate back and forth in an approximately sinusoidal manner, producing synchrotron radiation at each bend. Usually the magnetic field is in a vertical direction, so that the

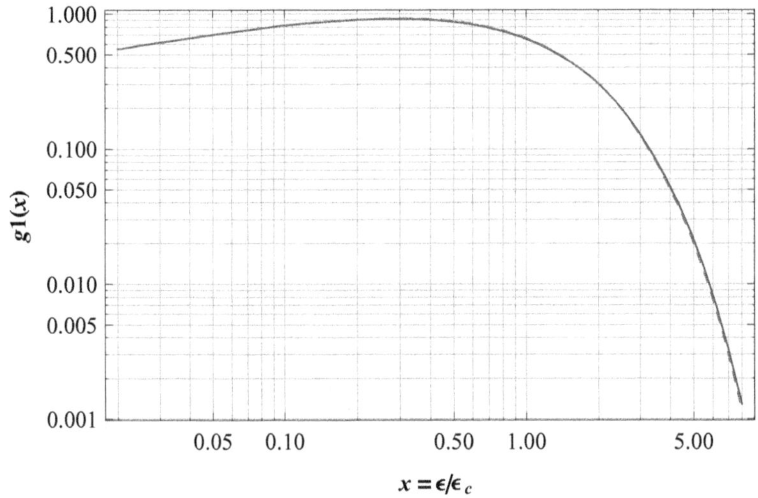

Fig. 3.4 Spectral density function $g(x)$ emitted by bend magnets and wigglers. The solid line is the numerically evaluated Bessel function integral and the dashed line shows the approximation $g_1(x) \approx 1.8\ x^{0.3}e^{-x}$.

beam oscillations are in the horizontal direction, which produces horizontally polarized X-rays. In special devices such as helical wigglers the electron motion is designed to produce other X-ray polarizations, such as vertically plane-polarized X-rays, or left- and right-handed circularly polarized X-rays.

The configuration of the magnets in wigglers and undulators is basically the same, but the devices are operated under different conditions. The radiation pattern from a single electron is strongly collimated around its instantaneous velocity because of relativistic effects, with an angular width of $\approx 1/\gamma$. The difference between wigglers and undulators comes down to the extent of the angular deflection that is produced by the magnetic field, compared to this intrinsic radiation width. Wigglers have a large deflection, and undulators have small deflection; real devices often have properties of both. The "deflection parameter" K is defined as $K = \gamma\delta_w$, where $\delta_w = \lambda_0/(2\pi\rho_0)$, ρ_0 is the bend radius corresponding to the magnetic field (as discussed in the previous section), and λ_0 is the spatial period (i.e. distance over which it repeats) of the magnetic structures in the device.

In a wiggler, the angular excursion of the electron beam is large compared to the intrinsic $1/\gamma$ angular width. The spectrum is essentially the same as bend magnet spectra of different radius, because the radius of the bends in the wiggler is often much shorter than that in bend magnets. Since the

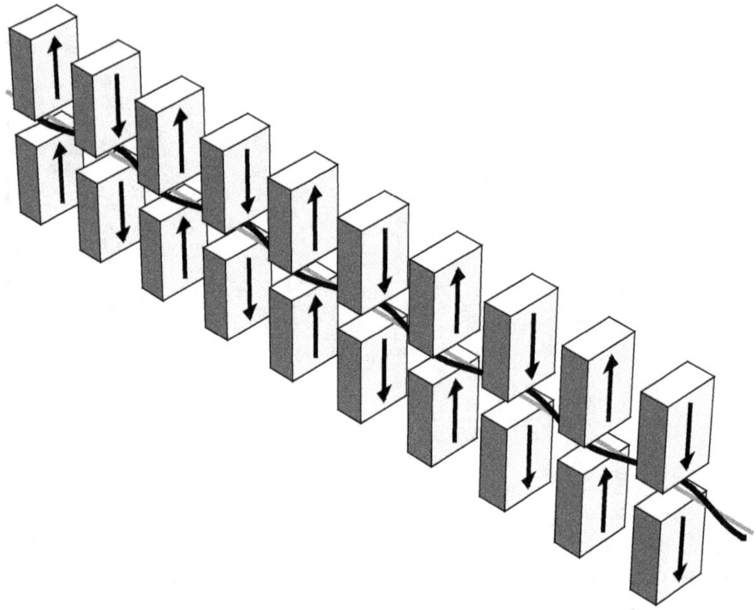

Fig. 3.5 Undulators and wigglers consist of magnetic poles of alternating polarity, between which the electron beam travels.

wiggler field can be independently controlled, the critical energy can be made much higher (or lower, if desired) than typical bend magnets. Also, each pole contributes incoherently to the emitted radiation so that, for N poles, the number of photons emitted is increased by the factor N. Wigglers produce a fan of radiation in the horizontal plane much like bend magnets, with a horizontal angular width of order K/γ, and a vertical angular width of order $1/\gamma$. Planar wigglers do not produce the out-of-plane circular polarization that bend magnets do, which is good for most experiments, but precludes those that rely on circular polarization. Helical wigglers, however, may be ideal for such experiments.

In contrast, in an (ideal) undulator, the angular excursion is much less than the intrinsic radiation width $1/\gamma$. This has important effects: the light emitted at each of the poles of the undulator can then interfere coherently with the light produced at the other poles. This causes the energy spectrum to bunch up into peaks at specific energies, and the angular pattern becomes collimated in the horizontal direction as well as the vertical. Because of the coherent interference between the radiation emitted at each pole, the angular width in both horizontal and vertical directions is of order $1/(\gamma\sqrt{N})$ – the more poles, the narrower the angular width. Similarly the peaks in the energy spectra are sharper when there are more poles, much like the narrowing of diffraction peaks as the number of scatterers increases. The $1/\sqrt{n}$ scaling assumes perfect coherent interference between all poles; coherence lengths are finite for real-world devices, however.

This pencil beam of X-rays of well-defined energy is ideal for many experiments such as diffraction experiments. The complication for XAFS experiments is that this energy must be scanned, which requires a way to tune the spectrum. Fortunately there are tricks for doing this.

First let's understand the undulator spectrum for a horizontal planar undulator. As described above, the alternating magnetic field produces an oscillating force in the horizontal direction as the electron moves through it. In the lab frame, the undulator has spatial period λ_0, but because of the Lorentz contraction, the spatial period of the undulator is contracted to $\lambda_1 = \lambda_0/\gamma$ in the rest frame of the electron's average motion. The electron moves along the axis of the undulator at very nearly c (since the speed of light is the same in all inertial frames). As a consequence it experiences a force that oscillates with the frequency $c/\lambda_1 = \gamma c/\lambda_0$, producing radiation (in its average rest frame) of that frequency. Back in the laboratory frame, near the forward direction, that emitted radiation is relativistically doppler shifted by the factor $\sqrt{(1+v/c)/(1-v/c)} = \sqrt{(1+v/c)^2/(1-v^2/c^2)} = \gamma(1+v/c) \approx 2\gamma$. Combining these observations gives a fundamental frequency of $2\gamma^2 c/\lambda_0$, and a fundamental energy of $2\gamma^2 hc/\lambda_0$.

The electrons move at speeds just barely slower than the speed of light, and they do this along their curved trajectory. Therefore their average speed along the undulator axis is less than c. The larger is the deflection, the slower the electrons travel along the undulator axis, the lower the frequency of oscillation in the rest frame of the electrons' average motion, and the lower is the energy of the undulator peak. The deflection parameter is controlled by varying the magnetic field experienced by the electrons. In an electromagnetic device the field can be controlled by the magnet current. In a permanent magnet undulator, this can be done by controlling the separation ("gap") between the north and south poles. The smaller the gap, the larger the magnetic field. Reducing the gap lowers the harmonic energy, but there are practical limits of tunability. For this reason it is normal to use different undulator harmonics to cover different energy ranges, and adjust the undulator gap to cover each energy range.

Accounting for this effect of deflection parameter on the energy, and geometric effects, the X-ray frequency of the undulator fundamental peaks is given approximately by $2\gamma^2 c/\lambda_0(1 + K^2/2 + \gamma^2\theta_0^2)^{-1}$, K is the deflection parameter, and θ_0 is the angle between the forward direction of the electron beam and the vector from the source to the observer. Since the radiation

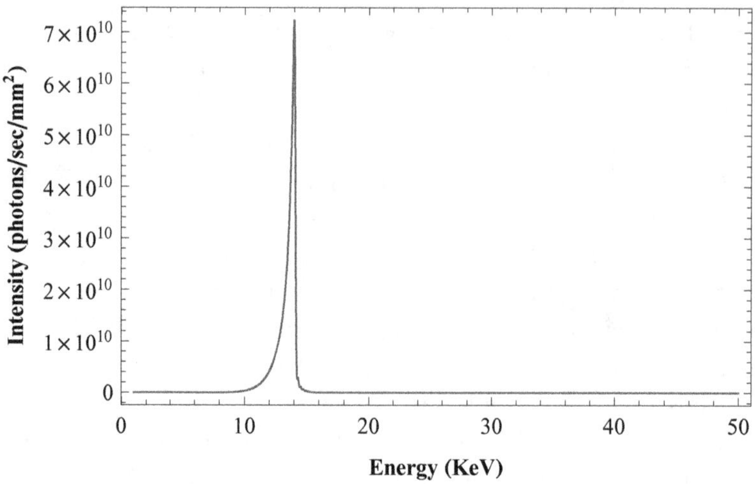

Fig. 3.6 Intensity of undulator with small deflection parameter ($K = 0.01$) for a 7 GeV beam and 3.3 cm period. This gives a single undulator peak at 14.1 KeV.

pattern is highly collimated, only small angles are important, and the expression above makes a small-angle approximation.

For example, consider an APS undulator A, with period $\lambda_0 = 3.3$ cm, and 7 GeV beam energy, or $\gamma = 7\text{GeV}/.511\text{MeV} \approx 1.37 \times 10^4$. The fundamental energy is then $2hc\gamma^2/\lambda_0(1 + K^2/2)^{-1}$ on axis. With $hc \approx 1.24 \cdot 10^{-9}$ KeV meter this gives a fundamental energy of 14.1 KeV for $K = 0.01$, and 2.94 KeV for a deflection parameter of $K = 2.76$. Spectra calculated with the program US are shown in Figures 3.6 and 3.7.

At high deflection parameters the motion becomes more complicated in the average rest frame of the electrons, and consequently so does the spectrum. The spectra from high field undulators consists of a fundamental frequency, and multiples of that frequency called the harmonics. In addition, a wiggler-like background spectrum emerges beneath the undulator peaks. The odd-order harmonics are intense, even on-axis, in the forward direction. In contrast, the even-order harmonics are of low intensity, particularly on-axis, but they can contain substantial intensity off-axis.

To perform XAFS scans with an undulator, there are two basic approaches. For XANES it may be possible to use the undulator spectrum as-is, provided that it is tuned to the desired energy. For EXAFS, however, one needs to perform energy scans over about a KeV. Since the energy width of an undulator is typically on the order of 100 eV, it must either be scanned, or broadened. The peaks can be broadened by "tapering" the undulator, i.e. choosing a different gap at the two ends of the undulator, which produces

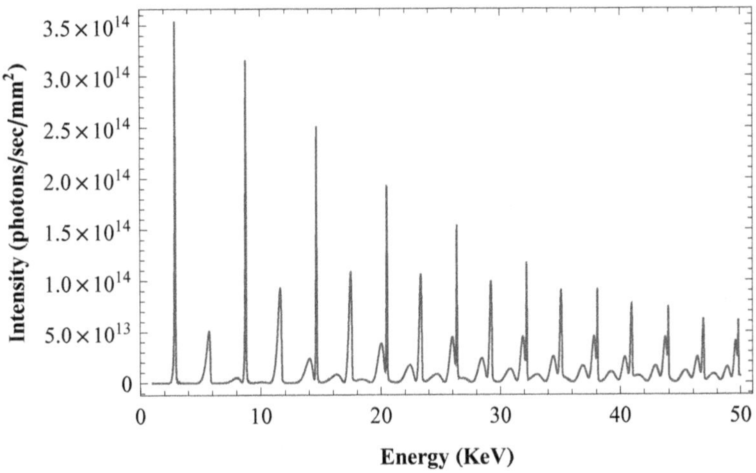

Fig. 3.7 Intensity of undulator with large deflection parameter (K=2.76) for a 7 GeV beam and 3.3 cm period. This gives a series of undulator peaks with a fundamental at 2.94 KeV.

a gradient in deflection parameter as electrons travel through the device. Alternatively the undulator gap can be adjusted during the XAFS scans. With suitable computer control systems, it is often feasible to coordinate scans of the undulator gap with scans of the beamline monochromator.

3.3.5 Time structure

The electrons in the ring are not spaced uniformly around its periphery, but rather are concentrated in "bunches," typically numbering from one to several hundred. The bunch structure is under the control of the accelerator operators and it can have an important impact on experiments, especially if pulse-counting detectors such as photomultipliers or Si/Ge detectors are used. Each bunch generates a short (typically a few hundred picoseconds) pulse of X-rays. The separation in time between the pulses is determined by the bunch structure and the orbital period of the ring. Pulse-counting detectors generally will require dead-time corrections which may depend on the time structure that is in effect during the experiment. Even if it is not used, the time structure cannot be ignored, unless integrating detectors such as ionization chambers are used.

The time structure can offer important benefits for certain kinds of experiments. The beam is like a stroboscope – the time dependence (if there is any) is sampled at times determined by the flashes of the beam. If the sample can be periodically stimulated, for example by a laser, and the sample fully "resets" between flashes, a detailed time dependence can be measured by systematically changing the time difference between the stimulus ("pump") and measurement ("probe").

3.3.6 Fourth-generation sources and free electron lasers

Accelerator technology has produced new methods of extending the brilliance of X-ray beams even further by using free electron laser technology. The characteristics of fourth-generation sources, such as the LCLS (Linac Coherent Light Source) and the European XFEL, although offering substantial promise for new types of experiments, do not appear to be compatible with XAFS experiments. Unlike synchrotrons, which offer a wide energy bandwidth, FELS are narrow bandwidth devices, producing coherent X-ray pulses of extremely short duration (tens of femtoseconds), high intensity, and relatively low repetition rate, making them ideal for one-shot interference and diffraction experiments, even on single molecules. The intensity is high enough that the samples typically are instantly destroyed, but not before the diffraction pattern or hologram is generated.

XAFS, in contrast, requires a beam of wide bandwidth from which a narrow slice can be extracted, or a tunable energy. Beam coherence is not necessary, and indeed is a potential complication when the coherent fraction of the beam cannot be neglected.

3.4 Beamlines and optics

After the X-rays are produced by the source, they travel tangentially to the beam path in a fan or pencil of radiation through heat filters, masks, stops, shutters and the like, emerging through beryllium windows into awaiting beamlines. Beryllium is used because of its very low atomic number and consequent low absorptivity, sufficient strength to resist vacuum forces, and good thermal conductivity.

Beamlines are complex instruments that perform a variety of functions: safely conveying the X-rays to the experimental area by employing appropriate shielding to synchrotron radiation and electron beam bremsstrahlung; providing interlocked access to experiments to prevent users from becoming accidentally irradiated; preparing the beam for the experiment by precollimating, filtering, splitting, monochromating, focussing, and shuttering the beam using X-ray optics; performing precisely timed data acquisition that is synchronized to the motions of beamline optics; controlling ancillary equipment such as cryostats and lasers (perhaps synchronized with the beamline optics); and providing an effective interface for the user to set up and control the experiments, evaluate data, and archive it.

3.4.1 Monochromators

Monochromators are essential for XAFS experiments because an X-ray beam of approximately 1 eV bandwidth FWHM[5] is needed, and synchrotron sources generally produce beams with bandwidths of tens to hundreds of eV (undulators), or thousands of eV (bend magnets and wigglers). Approximately 1 eV bandwidth is needed so that near-edge (XANES) features and the EXAFS modulations right above the edge can be resolved. Ideally the monochromator resolution width will be several times smaller than the intrinsic lifetime broadening (see Chapter 2) of the absorption edge under consideration, so the data quality is not limited by the monochromator.

Double crystal monochromators are most commonly used in XAFS spectroscopy. Four-crystal monochromators have the benefit that they strongly reduce the intensity in the tails of the monochromator resolution function, but they are more complex, difficult to align, and more expensive than two-crystal designs, and therefore are not often used. The precise rotational motions are carried out by a precision goniometer or rotation stage that keeps the diffracting surface of the first crystal fixed. Usually the crystals are contained within a stainless steel vacuum chamber a meter or so in size. Vacuum enclosures are preferred for the following reason: if the polychromatic "white" beam passes through air, ozone is produced, which may damage the optics. Furthermore, air scatters and absorbs the beam, degrading the beam brilliance. Although helium helps cool the optics and does not absorb the beam much at X-ray energies greater than a few KeV, helium does scatter the beam, and therefore vacuum is generally preferred to helium-filled monochromators.

The energy of the beam is selected by setting the angle of incidence of the beam onto a crystal (hard X-rays) or a diffraction grating (soft X-rays). Silicon (occasionally germanium) is commonly used for energies above approximately 2 KeV (Bragg's law cannot be satisfied at energies lower than 1.977 KeV for Si) because it has suitable diffraction and thermal properties, and extremely high quality crystals are readily available from the semiconductor industry. The silicon is cut so that specific diffraction planes (e.g. $hkl = 111, 220, 311, 400$) are parallel to the crystal surface – these are called symmetric "Si(111)" etc. crystals. Asymmetric crystals, which have diffraction planes that are not parallel to the crystal surface, can be used to

[5] FWHM = Full Width Half Maximum (of the resolution function).

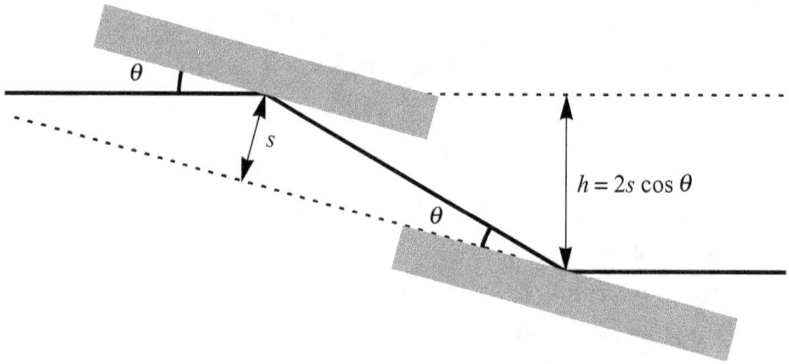

Fig. 3.8 Schematic of a downward reflecting double-crystal monochromator. The beam enters from the left. The perpendicular separation of the crystals is s; h and b are the horizontal and vertical displacements of the beam. Defining $l = s/\sin(\theta)$, $h = l\cos(2\theta) = 2s\cos\theta$ and $b = l\sin(2\theta)$.

compress or expand the beam, but they are not often used in monochromators for XAFS.

The incident beam from the synchrotron has a very wide range of wavelengths in it. Although the lower-energy photons usually are absorbed by the beamline windows, or filters specifically inserted for that purpose, the beam falling on the first monochromator crystal can carry significant (e.g. tens of kilowatts) power that must be handled. The beam at the desired energy is diffracted because it meets the Bragg condition $n\lambda = 2d\sin\theta_B$, or equivalently $E = (hc/2d_{\rm hkl})(1/\sin\theta_B)$. Multiples of that energy ("harmonics") generally will also be diffracted. Photons of other energies are absorbed, which deliver a substantial amount of heat to the crystal. This is removed directly by passing coolant through channels in the crystal, or indirectly by conduction to a cooled heat sink (typically a water-cooled copper block), or other means.

3.4.1.1 Heat load

Thermal drifts can be a serious problem for precision optics such as monochromators, requiring careful design of the cooling within the monochromator. The precision needed for monochromators to work properly is striking. The widths of silicon rocking curves (i.e. curves of reflectivity vs. incidence angle for a monochromatic beam) depend on hkl and energy, but generally they are on the order of about 10 arc-seconds. Since an arc-second is 1/3600 of a degree, or $\approx 5\,\mu$-radians, a rocking curve width of 10 arc-sec $\approx 50\,\mu$-rad or $50\,\mu$meter per meter. A small dust particle under a crystal can destroy the reflectivity; other small effects, such as thermal expansion and drifts, vibrational coupling, gravitational deformation of components, heating of components due to scattering off the first crystal – all must be carefully managed by the monochromator design.

Insertion device sources can cause special problems. Under some operating conditions undulators can produce a power density on the first crystal that is so great that the power deposited in the first silicon crystal cannot conduct away quickly enough, and a temperature gradient in the crystal causes a distortion ("thermal bump") in the first crystal which degrades the brilliance of the beam. To avoid this problem one needs to optimize the ratio of the thermal conductivity to the thermal expansion coefficient. One successful solution is to use a different crystal material of higher thermal conductivity than room temperature silicon, such as synthetic diamonds, which are commonly used at the third-generation sources ESRF and Spring8. Another approach is to operate the silicon crystals at a temperature at which the thermal conductivity is high, and the thermal expansion coefficient is low, so that the ratio is large. Interestingly both of these conditions are simultaneously satisfied for silicon near liquid nitrogen temperature (77K), and for this reason cryogenically cooled monochromators are commonly used at other third-generation sources, such as the APS.

Wigglers present their own challenges. The total power emitted by a high field wiggler can pose extreme challenges for optics. There are several ways to limit the heat load on the monochromator through the judicious use of slits, water-cooled X-ray filters, or cooled high conductivity mirrors upstream of the monochromator. These mirrors essentially act as energy-dependent filters, reflecting the low-energy X-rays and transmitting the desired high-energy X-rays.

3.4.1.2 Detuning

The diffracted beam is deflected from the original direction by an angle $2\theta_B$. To compensate, a second crystal is placed parallel to the first so that the doubly diffracted beam comes out parallel to original beam. For the second crystal to meet the Bragg condition, it must be precisely aligned with the first crystal. This takes some precision alignment of the second crystal to the first, which is normally accomplished using piezoelectric transducers "piezos," which change their dimensions very slightly when a voltage is applied to them. A commonly used approach to keep the crystals aligned is to apply a small oscillating voltage to the piezo, which causes a small variation in the intensity at the modulation frequency. This can be sensitively detected with a lock-in amplifier that only picks up variations at the modulation frequency and used in a feedback circuit.

There is a simple alternate method to align the second crystal to the first: make both diffracting surfaces out of the same crystal by milling a uniform channel through it, while keeping the crystal in one piece. The parallel sides of the channel then constitute the two parallel (first and second) crystals. This configuration is called a "channel-cut" crystal; it is simple, and it does work. It has a drawback though: often you may want the crystals to be slightly non-parallel, for the following reason. As previously mentioned, not only the fundamental energy is diffracted: the (usually undesired) harmonics are also. These harmonics can be a serious problem for XAFS measurements, and it is necessary to get rid of them (or make detectors insensitive to them). One standard way to do this is to intentionally misalign the first and second crystals slightly. The "rocking curves" (reflectivity curves vs. crystal angle) of the harmonics are substantially narrower than that of the fundamental, so slightly misaligning ("detuning") the crystals reduces throughput of the fundamental only slightly, but it reduces the harmonics considerably. A drawback to this approach is that if the misalignment ("detune") varies during a scan, the proportion of harmonics also can vary, potentially generating artifacts in the data. This can be prevented by continuously monitoring the harmonic content of the beam during scans and adjusting the tune accordingly. It also should be noted that changing the tune of the second crystal does cause a displacement of the beam, which may be a concern for some experiments.

3.4.1.3 Fixed exit monochromators

Scanning the monochromator causes a vertical motion of the beam that should be compensated. As shown in Figure 3.8, for a downward reflecting monochromator, the beam is displaced vertically relative to the incident beam by an amount $h = 2s\cos\theta$, where s is the perpendicular distance between the crystals. Because θ varies over a scan, the beam will move unless the motion is compensated in some manner. Failure to do so can produce artifacts in the XAFS data if the sample is inhomogeneous on length scales comparable to the amount of beam travel. A simple way to compensate for beam motion is to place the slits, sample, and detectors on a computer-controlled table that moves to keep the sample fixed relative to the beam. This approach generally works well, but in some cases it may not be practical to vertically move the whole experimental setup downstream of the monochromator. In such cases a fixed exit monochromator can be used which changes s during a scan so as to keep h constant.

Another point to consider is that if the crystals are in fixed positions relative to each other, the point at which the beam lands on the second crystal

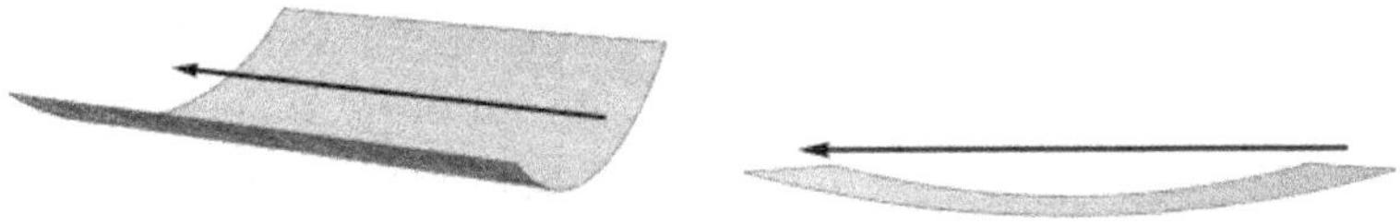

Fig. 3.9 Sagittal and meridional bending of optics. The arrow indicates the x-ray beam direction.

varies over a scan, because of the changing θ. For this reason, the second crystal must either be sufficiently long to accommodate the beam travel over the surface, or else the second crystal must be translated (changing b) during a scan to keep the beam centered. If the second crystal is sagittally bent to accomplish focussing it will be desirable to keep it centered.

3.4.1.4 Focussing conditions

It is possible to focus X-rays by bending crystals or X-ray mirrors to appropriate radii of curvature. The geometry of focussing gives equations for sagittal bending and meridional focussing which are illustrated in Figure 3.9. In one successful design, sagittal bending of the second monochromator crystal provides horizontal focussing and meridional bending of an X-ray mirror provides vertical focussing. Alternatively, in Kirkpatrick–Baez focussing, separate sequential mirrors in meridional focussing conditions are used for both horizontal and vertical focussing. Focussing in both directions also can be carried out using a single doubly bent toroidal mirror, again in meridional focussing conditions.

The equation for the sagittal focussing condition is $2\sin(\theta)/R_s = (1/u + 1/v)$, where θ is the local angle of incidence of X-ray beam on the optic; R_s is the local radius of curvature where the beam hits the optic; and u and v are respectively the source to optic distance, and the optic to focal point distance. The angle θ is determined by the Bragg condition (for crystals) or the condition for total external reflection (for mirrors). The equation for meridional focussing is very similar, except that the $\sin(\theta)$ is in the denominator instead of the numerator: $2/(R_m \sin(\theta)) = (1/u + 1/v)$. The beam size in the direction of focussing is (de)magnified by the ratio v/u.

3.4.2 X-ray mirrors

X-ray mirrors typically serve several functions in XAFS beamlines: harmonic rejection, focussing, and collimation of the beam that is incident on the monochromator. Any material interface can act as an X-ray mirror if the angle of incidence of the X-ray beam onto the material is sufficiently shallow (typically a few milliradians at the energies that concern us). From optics we are familiar with the phenomenon of total internal reflection, in which a ray of light that is propagating in a dense (high index of refraction) material is incident on an interface with a material of lower index of refraction. If the angle of incidence is low enough, the light is totally internally reflected back from the interface, and the light beam remains confined to the high index material. Familiar examples of this effect are fiber optic light guides, and mirages.

The same thing occurs at X-ray energies, but with a twist: the index of refraction of materials at X-ray energies is slightly less than 1.0, the index of refraction of vacuum (and approximately, air), and it is a complex number. In contrast, indices of refraction at optical wavelengths are normally[6] greater than 1.0. For this reason, at X-ray energies, one gets total external reflection rather than total internal reflection at air–material interfaces.

An effective X-ray mirror must be locally flat and extremely smooth, preferably with an RMS roughness of a few Å or less, and RMS slope errors of approximately 1 μrad. If it is not, one obtains not only the desired specular reflection, but also scattered X-rays over a spread in angles, which degrades the effectiveness of the mirror as an optic. Fortunately the technology for producing such mirrors is now well developed. Adaptive "Bimorph" mirrors are also available, which allow the user to fine-tune the shape using a number of integral piezo-electric transducers.

The index of refraction $\tilde{n}$ of materials at X-ray energies can be expressed as $\tilde{n} = 1 - \delta - i\beta$, where $\delta = ne^2\lambda^2/2\pi mc^2$ and $\beta = \mu\lambda/4\pi$. The real and imaginary parts of $\tilde{n} - 1$ respectively describe dispersion and absorption of the X-rays. Here, n is the number density of dispersive electrons in the material, $e^2 \equiv q^2/4\pi\epsilon_0$; q and m are the charge and mass of the electron; λ is the wavelength; and μ is the X-ray absorption coefficient. In pure elements this can be expressed as $\delta = N(Z/A)e^2\lambda^2/2\pi mc^2$ where N is Avogadro's number, Z is the atomic number, A is the atomic weight, and ρ is the mass density.

Total external reflection occurs at angles $\theta < \theta_c$, where the critical angle $\theta_c = \sqrt{2\delta}$. The critical angle θ_c is approximately inversely proportional to the energy, so that high energy require smaller angles of incidence onto the mirror for reflection to occur. This allows the experimenter to choose a mirror angle (typically a few milliradians) so that the mirror reflects X-rays of the desired fundamental monochromator energy, while rejecting the undesired harmonics, which are absorbed by the mirror material. In general

[6] Synthetic metamaterials have been devised that have negative indices.

mirrors can be used as low-pass filters, so that low energies are reflected, but high energies are not.

The product of energy E and $\theta_c(E)$ is an intrinsic property of the mirror coating. Representative measured values of $E \cdot \theta_c$ (in KeV-mrad) are 31(Si), 59(Ni), 67(Rh), 62(Pd), 82(Pt), and 80(Au). For example, consider a Cu EXAFS spectrum, which extends to approximately 10 KeV. The critical angle for reflection from a palladium-coated mirror at 10 KeV is approximately 6 mrad, and the mirror angle therefore should be set to 5 mrad or less for good reflectivity.

Since the θ_c depends on the material, it is often useful to deposit multiple different metal coatings onto distinct regions of X-ray mirrors. For example, it has proven useful to apply Pt and Rh coatings in stripes to an uncoated substrate of ultralow expansion ceramic (ULE). A high atomic number coating extends the range of angles and may allow use of a shorter mirror by providing good reflectivity at larger incidence angle, but absorption edges from the mirror coating over the energy range of interest may be undesirable. A low-Z coating on top of a higher Z coating can be useful for suppressing the absorption edges of the undercoat.

The reflectivity of mirrors (and multilayers) can be calculated using Fresnel's equations from classical electromagnetic theory [19, 7]. For a sufficiently thick (at least tens of nanometers) mirror material or coating, the reflectivity (assuming zero roughness) may be written [29] in terms of the reduced angle $\phi = \theta/\theta_c$ as

$$R(\phi) = \frac{(\phi - A)^2 + B^2}{(\phi + A)^2 + B^2},$$

where

$$2A^2 = [(\phi^2 - 1)^2 + (\beta/\delta)^2]^{1/2} + (\phi^2 - 1),$$

and

$$2B^2 = [(\phi^2 - 1)^2 + (\beta/\delta)^2]^{1/2} - (\phi^2 - 1).$$

With this information the mirror reflectivity can be estimated for pure materials. A plot of this ideal reflectivity is shown in Figure 3.10. The reflectivity should be periodically measured experimentally to check against degradation of the mirror.

3.4.3 Multilayers

Multilayers are versatile and useful X-ray optics that, being literally made to order, can be tailored to experimental needs. Although multilayers are most

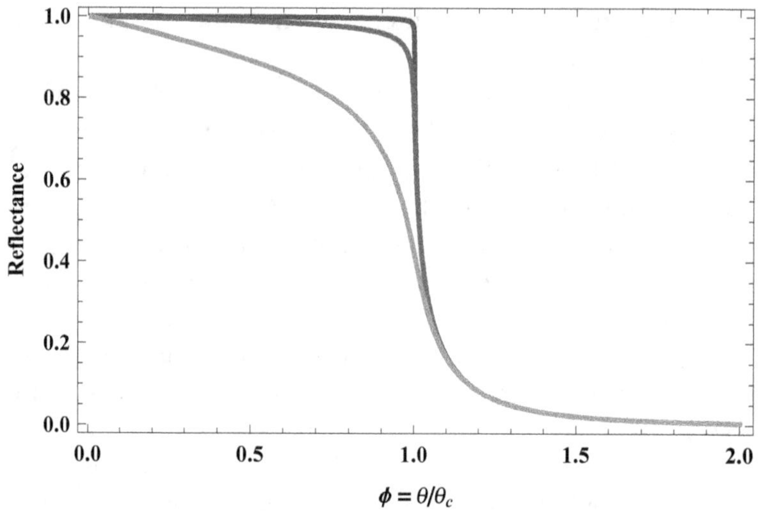

Fig. 3.10 Calculated mirror reflectivity vs. angle. The curves correspond to $\psi = \beta/\delta = 0.001, 0.01, 0.1$.

commonly used in the soft X-ray region, particularly for monochromators, they also have many uses in the hard X-ray region, such as monochromators for microfocus x-ray tubes, "supermirrors," and analyzers for X-ray fluorescence. Similar optics are used in neutron diffraction.

They consist of alternating layers of high and low atomic number materials (e.g. W/C, W/B_4C, Mo/Si), usually with uniform thicknesses on the order of 3–10 nanometers. The multilayer repeat distance (period) cannot be made much smaller than that, because of interdiffusion of atoms between layers, which reduces the reflectivity. The dependencies of the reflectivity on energy and angle can be optimized for the experiments by appropriate choice of the materials and layer spacings.

Like an oil film on water, the X-rays that are coherently scattered from each layer undergo constructive and destructive interference, depending on the angle of incidence. At a fixed incidence angle this results in high reflectivity at certain wavelengths, and low reflectivity for others. Alternatively, at a fixed X-ray energy, only X-rays incident at certain angles are diffracted. Some representative reflectivity curves are shown in Figure 3.11, as calculated using the server at http://www-cxro.lbl.gov/.

3.5 X-ray detectors

X-ray detection starts with the absorption of an X-ray photon by a material within the detector, resulting in photoionization of its atoms, and

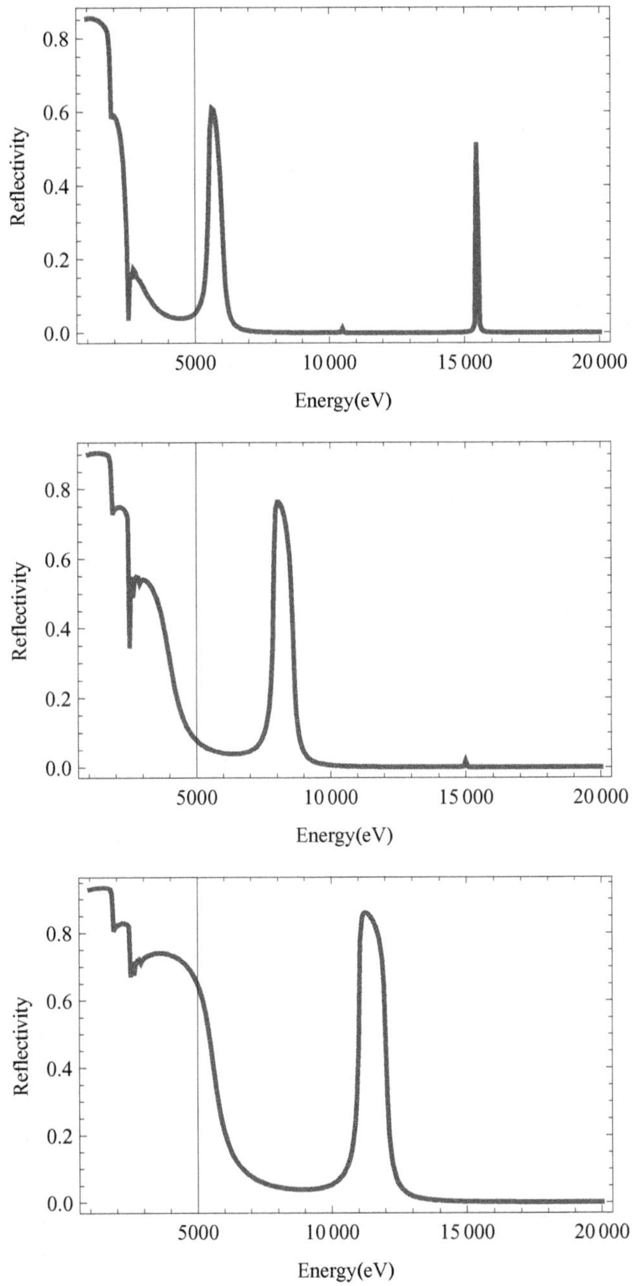

Fig. 3.11 Multilayer reflectivity vs. energy for a 7 nm period symmetric Mo/Si multilayer with a large number of periods. The plots (from top) correspond to incidence angles of $1° \approx 17.45$ mrad, $0.7° \approx 12.2$ mrad, and $0.5° \approx 8.73$ mrad. At low energies/low angles the device acts like a mirror. In the curves shown, the structure in the reflectivity curves at around 2 KeV is due to the Mo L-absorption edges.

subsequently a cascade of progressively lower energy excitations within the material, ultimately ending up as heat. It is the detector's job to convert these excitations to electrical signals in some manner. The way in which this is done has a decisive influence on the detector performance. For more detailed information on the inner workings of radiation detectors, see Knoll [13].

In general, X-ray detectors can be operated in pulse-counting mode or current integration mode. Some detector types, such as ionization chambers, effectively only operate in current mode, while others such as photodiodes and photomultipliers can be operated in both modes. Current integration is the simplest approach, but it does not allow one to distinguish photons of different energies; it merely allows one to determine the rate at which the detector absorbs energy from the X-ray beam. However, detectors operated in current integration mode do have the virtue of much higher maximum count rates than pulse-counting detectors, for reasons described below. Although the X-rays are deposited as pulses (a photon is either absorbed, or not), ionization chambers are normally operated under conditions so that many photons are absorbed during an integration period, and statistical variations in the number of absorbed photons are relatively small. However when ionization chambers are used for fluorescence detection, fluctuations in the number of absorbed photons (usually background) may be the principal source of noise.

In pulse-counting mode, each photon that is absorbed generates a pulse of charge through the separation of electrons and holes (or positive ions). In the normal counting chain (there is a separate one for each detector element), each pulse from the detector element is first passed through a preamplifier, and then a shaping amplifier, which creates an output pulse of a specified shape (e.g. gaussian, triangular) whose height is proportional to the total charge of the input pulse. Since the charge is proportional to the energy deposited by the X-ray photon, the height of the shaping amplifier output pulse is proportional to the photon energy, or at least is linearly dependent on it. The pulses are then sorted into energy bins by a multichannel analyzer (MCA) and counted, providing an energy spectrum of the detected photons. Alternatively slices of the photon energy distribution can be selected using window discriminators and counted by a device called a scaler, for example to pick out specific energy fluorescence photons. Devices such as XIA's "digital X-ray processor" carry out these functions in one compact unit. Integrated multichannel detector and electronic systems have been under continuous development at synchrotron sources such as the NSLS.

The most widely used pulse-counting detectors for XAFS are solid state detectors, which usually have 13 or 19 or more silicon or germanium elements. Such detectors are commercially available from major manufacturers such as Canberra (www.canberra.com/) and Ortec (www.ortec-online.com/). Germanium has the advantage over silicon above approximately 15 KeV because its higher atomic number allows it to better absorb the higher-energy photons. Apart from the larger active area, the main reason for using multiple detector elements is that the maximum count rate per element is typically limited to a few hundred thousand counts per second; more advanced detector systems can do several times better than this. To count millions of counts per second requires multielement detectors.

There is a tradeoff between throughput and energy resolution as well, related to the time constants in the electronics (shaping time, peaking time) and the characteristics of the detector. Longer shaping times generally result in better energy resolution, but only up to a point. The energy resolution obtainable from semiconductor detectors is sufficient for many purposes, but it is generally no better than 130 eV or so, and usually considerably worse. At present, obtaining better resolution or higher throughput requires the use of X-ray diffractive analyzers, of which there is a variety with different characteristics. These are described below.

Many detectors of this type require a high (KV) voltage to be applied across the detector. To prevent excessive current through the detector element, they must be operated at liquid nitrogen temperature. Such detectors have integrated Dewar flasks filled with LN2 that require periodic refilling, and if not protected electronically, may require constant attention to ensure that the elements do not warm up while the high voltage is on, which can severely damage or destroy the detector.

Non-cryogenic silicon detectors are also widely available. These typically use integrated thermoelectric coolers to reduce electronic noise, which improves energy resolution. Silicon Drift Detectors (SDDs) are a relatively recent and important innovation. These detectors provide higher throughput (photons/sec) than traditional detectors by optimizing charge collection. SDDs and arrays are currently available from several vendors (e.g. www.ketek.net, www.radiantdetector.com, www.amptek.com), and at the time of this writing are under active development.

3.5.1 Ionization chambers

Arguably the simplest X-ray detector is the ionization chamber, which is essentially a pair of electrodes with a gas ("fill gas") between them. The internal construction of an ionization chamber is illustrated in Figure 3.12. A

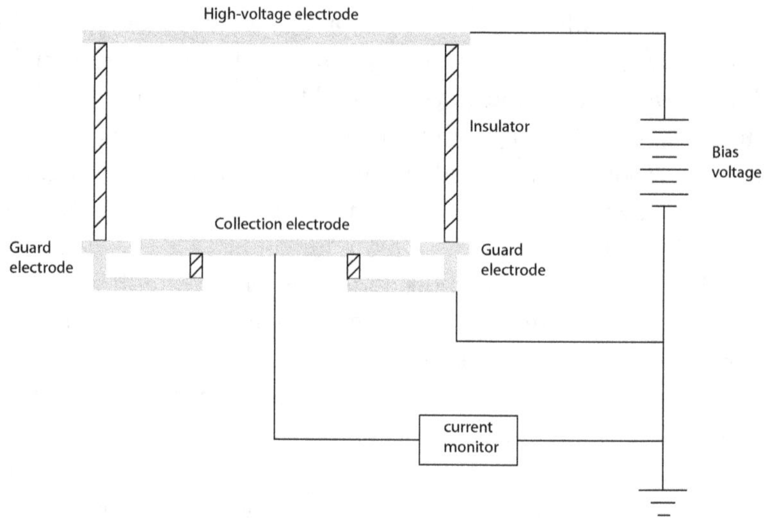

Fig. 3.12 Ion chamber internal structure. The guard electrodes reduce leakage currents due to insulator leakage. Adapted from Rossi and Staub [15].

constant high voltage, typically a few hundred volts (or as much as 1–2 KV, depending on plate separation), is placed across the electrodes, producing an electric field between the plates. With no X-rays present, the gas is a good insulator, because the gas atoms are neutral and there are virtually no free charge carriers. However if an X-ray beam is directed through the space between the plates, some X-ray photons are absorbed by the gas, generating electron/positive ion pairs. Under the influence of the electric field the electrons migrate toward the anode, and the positively charged ions migrate toward the cathode. The resulting current, usually between 10^{-7}–10^{-11} amps, is measured using a sensitive current amplifier by integrating over a time interval, typically in the range 0.1–10 seconds. This converts the small current signal to a voltage signal of magnitude several volts.

In XAFS experiments incident and transmitted beam fluxes (photons/sec) are generally measured using transmission ionization chambers. These typically consist of two parallel plates on the order 10 cm in size, separated by a gap on the order of 1 cm, inside a metal enclosure containing the fill gas. X-rays enter and exit the enclosure through thin X-ray-transparent windows made of plastic (e.g. Kapton) film a few mils (thousands of an inch) thick, beryllium, or other low atomic number materials. In addition, fluorescence experiments often use ionization chambers of a different design described below.

Most chambers are operated at atmospheric pressure, either sealed up, or more often, with gas flowing through them at a slow rate. Many beamlines have gas mixing facilities so that suitable levels of absorption can be attained. If gases are mixed, it is very important that the composition remains extremely constant over the duration of a scan, particularly if scans are slow, otherwise spurious pseudo-signals may be generated. Some ionization chambers are designed to be sealed, with sufficiently strong windows that they can support pressures higher or lower than atmospheric pressure. This offers considerable experimental flexibility, because the absorptivity of the chamber can be continuously varied, but this comes at increased cost.

It is important that the pressure inside an ion chamber remain very constant over the duration of a scan, because the signal is directly dependent on the fill gas density, and the XAFS modulations are small over much of the data range. Obstructions to the gas flow, transients in flow rates, or even slow variations in temperature can produce subtle output variations that can masquerade as signal. A temperature variation of 3 °C gives $\approx 1\%$ variation in detector output at constant beam flux; this variation is about the size of the XAFS signal. If the temperature were to vary on a similar time scale as an XAFS scan it could introduce spurious artifacts into the data.

Construction of high-quality ionization chambers requires careful attention to their internal construction, such as incorporation of guard electrodes, use of high quality insulators (to prevent leakage), and good surface finish to prevent sharp features that concentrate the local electric field, and may cause arcing.

3.5.2 Acquisition of data from ionization chambers

The currents from ionization chambers are generally in the nanoamp range, between picoamps and microamps, 10^{-12}–10^{-6} ampere. To acquire data, these small currents must be converted into something the beamline control computer can interpret as digital data. The traditional, and still widely used, approach is to use a current (transimpedance) amplifier to convert the small currents into voltages usually between 0–10 V. It is best to avoid the extremes of the range so as to avoid amplifier saturation on the high end, and noise and amplifier drift on the low end. Amplifiers also are usually more linear in the middle of their ranges. A gain setting of 10^9 volt/amp is a reasonable choice for picoamp currents.

On many amplifiers it is possible to apply a small additive offset voltage (e.g. a few hundred millivolts) to the input signal to keep it well above zero. Even if the amplifier is bipolar (it amplifies both positive and negative inputs) the subsequent electronics may not handle negative inputs. In general one must assume there is such an offset voltage (i.e. nonzero output for zero input), measure it, and subtract it out. Data acquisition programs generally support this, and it must be done regularly.

To carry out its task, the amplifier integrates the current signal over a time interval, preferably on the order of one tenth of the data acquisition interval. This time constant usually is set by a control on the amplifier, but even if it is not there is always an implicit time constant, which may depend on the gain. The time constants of the amplifiers used for I_0 and I (or I_f) should be matched, otherwise fluctuations in count rate may not divide out properly.

Ionization chambers have their own intrinsic limitations to their time response which must be recognized. Most chambers have a time response no faster than 100 msec, i.e. 10 Hz frequency response. This can be measured by periodically interrupting the X-ray beam with a variable speed chopper. This is best done by treating the detector/amplifier combination as a system. If fast time response is needed, it may be necessary to use specially designed chambers or pulse-counting detectors such as Ge or Si detectors, or Scintillator/Photomultiplier detectors, if they can handle the count rates.

In the traditional method, the voltage signal is then fed into a voltage to frequency (V/F) converter which produces a series of output pulses whose frequency is proportional to the voltage input. Typically 1 volt produces 10^5 pulses per second up to a maximum frequency of 1 MHz. These pulses are then counted for a fixed time interval in a multichannel counter (scaler) interfaced to the computer. It is important that the integration time intervals (i.e. start times, and duration) are identical for the various detectors so their signals can be compared on the same basis.

It is also important that the time intervals be sufficiently large that enough counts (at least 10^3 or 10^4) are collected during each time interval; otherwise a small number of dropped or added counts falling by chance in a time bin would cause a large relative variation in the signal.

An alternative approach to reading the data is to feed the amplified voltage signal from the current amplifier directly into an analog to digital (A/D) converter. The signal can then be computationally divided into different time slices after the fact. This approach is particularly useful for time resolved and slew-scanning ("Quick XAFS," continuous scanning) modes.

3.5.3 Plateauing ion chambers

If conditions are right, the resulting signal is proportional to the rate at which energy is deposited in the gas, which itself is proportional to the X-ray flux (photons/sec). For this ideal linear dependence to be realized, the applied electric field must be sufficiently large that it fully separates the electron–ion pairs before they recombine. The schematic dependence of ionization chamber current on bias voltage is illustrated in Figure 3.13. The electric field must not be too large, otherwise it may deliver enough kinetic energy to the electrons and ions that they will ionize other gas molecules when they collide, resulting in amplification. The total current would not be proportional to the number of photons absorbed, but would also depend on their energy. This process occurs in proportional counters and geiger counters but is undesirable in ionization chambers. The X-ray intensity also must not be so large that the cloud of electron–ion pairs ("space charge") itself screens the applied field or substantially reduces the local field experienced by the charges. This may be a concern on insertion device beamlines.

In practice it is very important to experimentally ensure that ionization chambers are operated within their plateau region, otherwise the data acquisition system will be nonlinear, and the XAFS data will be noisy and distorted. The range of plateau voltages depends on the ion chamber dimensions, the fill gases in use, total flux, and beam intensity. The plateau region can be determined by measuring the current vs. applied bias voltage under conditions of constant X-ray flux.

3.5.4 Choosing fill gases

When designing an experiment one needs to choose a fill gas or mixture of fill gases so that, given the known active length of the ionization chamber, enough of the incident beam is absorbed that it does not contribute significantly to the noise level in the data, but it allows enough of the beam through the detector onto the sample and other detectors. In most cases this implies that between 5–15% of the beam intensity should be absorbed in the incident flux monitor I_0. With a high intensity (e.g. undulator) beam, it may be necessary to use a fill gas of lesser absorption than one would ordinarily use (e.g. helium, when otherwise one might use nitrogen) to obtain linearity, because of space-charge effects.

The most common fill-gases are nitrogen and inert gases such as helium, neon, argon, krypton, xenon. Helium, nitrogen, and argon are most used because they cover a good range of absorption and are relatively inexpensive.

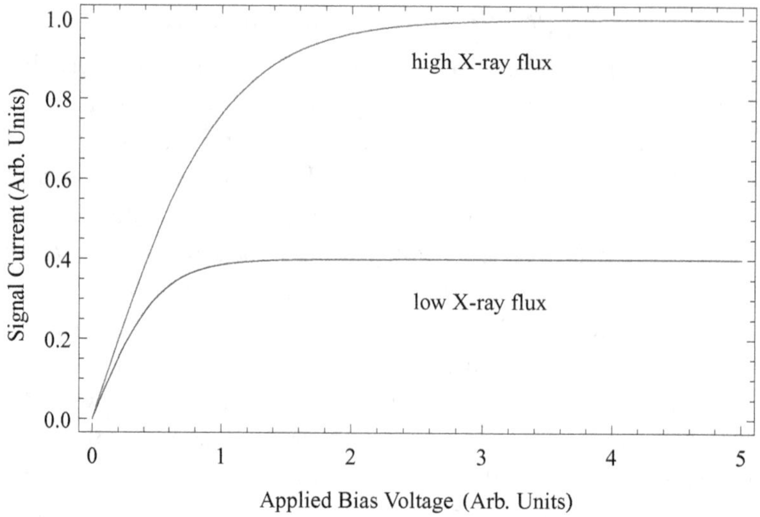

Fig. 3.13 Schematic dependence of ion chamber output vs. bias voltage for high and low X-ray fluxes. Ion chambers should be operated in the plateau (flat) region. The greater the intensity, the higher the bias voltage needed to reach the plateau.

Krypton and xenon are usually used in sealed chambers because of their cost. Mixtures of gases can be useful for tailoring the ion chamber absorption to the experiment, but some gases (e.g. air, oxygen, and water vapor) are less satisfactory, because free electrons tend to attach to the molecules, inhibiting their mobility and increasing recombination, and also producing reactive oxygen species such as ozone which chew up windows. If mixtures of gases are used, it is important that the mixture does not vary over time. Irregular gas flow can create artifacts in the data if it results in small changes to the composition of the fill gas mixture.

Calculation of a suitable choice of fill gases is usually straightforward. The probability of a photon of energy E being absorbed by the fill gas is $1 - \exp(-\mu_{fg}(E)l)$, where l is the length of the electrode along the direction of beam travel, and $\mu_{fg}(E)$ is the energy dependent fill gas absorption coefficient in the chamber. This can be calculated from the cross section. For example, using the cross-section server at http://csrri.iit.edu/mucal.html (or http://csrri.iit.edu/periodic-table.html, or NIST's FFAST server), we find the photoelectric absorption cross section for a nitrogen atom at 8 KeV is 6.853 cm^2/g, or $6.853 * 28 \approx 192$ cm^2/mol N_2. Atmospheric gases at room temperature have a number density of about 1 mol/24.5 l, giving an absorption coefficient for nitrogen gas at 8 KeV of $\mu_{fg} = 192/24.5/1000$ $\text{cm}^{-1} \approx$ 0.0078cm^{-1}. This implies that a 10 cm long ionization chamber with nitrogen gas would absorb about 8% of the incident X-rays. Figure 3.14 shows curves of fractional absorption vs. energy for a 10 cm ion chamber plate length for the gases He, N_2, Ar, Kr. The code to produce such plots is given in the following program, which can be used by the reader to generate curves for other ionization chambers, elements, and energy ranges.

Mathematica 7 program to plot fractional absorption for various fill gases and ion chamber plate lengths

```
(*note: this uses the function "GetData" defined in code listing 1.
"name" is element name as string; mw=molecular weight in g/mol of gas molecules,
l is active length of ion chamber plates in cm, e1 and e2 are plot range in KeV *)
ion[name_, mw_, l_, e1_, e2_] := (GetData[name, e1, e2]; mux = l*photo*mw/24.5/1000;
 ListLogLogPlot[Transpose[{e, (1 - Exp[-mux])}], PlotRange -> All, Joined -> True,
 PlotStyle -> Thick, GridLines -> True, Frame -> True,
  FrameLabel -> {"Energy (KeV)", "Fraction Absorbed", name}])
```

3.5.5 Estimating fluxes from ion chamber currents

It takes on average about 20–40 eV to produce an electron–ion pair in the fill gases in common use, regardless of the type of high-energy particle used to ionize the gas. This parameter is referred to as W – typical values (see e.g. http://xdb.lbl.gov/) are given in Table 3.1; the values for gases obtained from various sources vary by approximately 0.5 eV. The analogous quantity for creation of electron–hole pairs in semiconductor diode detectors is also included in the table.

These W values can be used to estimate X-ray fluxes. In most cases the ion chamber output is passed through a current amplifier (transimpedance amplifier) that is operated at a chosen gain G, e.g. 10^9 volt/amp, which means that one nanoampere of current produces one volt output. Typically these signals are displayed on a panel voltmeter at the beamline. The current can be easily determined if the gain is known (once the dark currents and amplifier offsets are subtracted out).

The current is directly related to the rate of photon absorption. Suppose that R is the number of photons per second each of energy E (eV) that enter the ionization chamber, a fraction F is absorbed (as given for example by Figure 3.10), and each requires an energy W (roughly 30 eV) to produce an electron–ion pair. Combining these factors gives for the measured voltage: $V = F{\cdot}R(\text{photon/sec}){\cdot}E(\text{eV/photon}){\cdot}(1\ e^-/W\ \text{eV}){\cdot}K\ (\text{C/e}^-){\cdot}G\ (\text{volts/amp}) = FREKG/W$, where K is the constant $1.6 \times 10^{-19}\text{C/e}^-$. This gives the photon flux $R = VW/(FGEK)$. As an example, this expression gives 3.5 million absorbed photons/volt at 10^{10} gain at the Fe K_α energy (6.4 KeV).

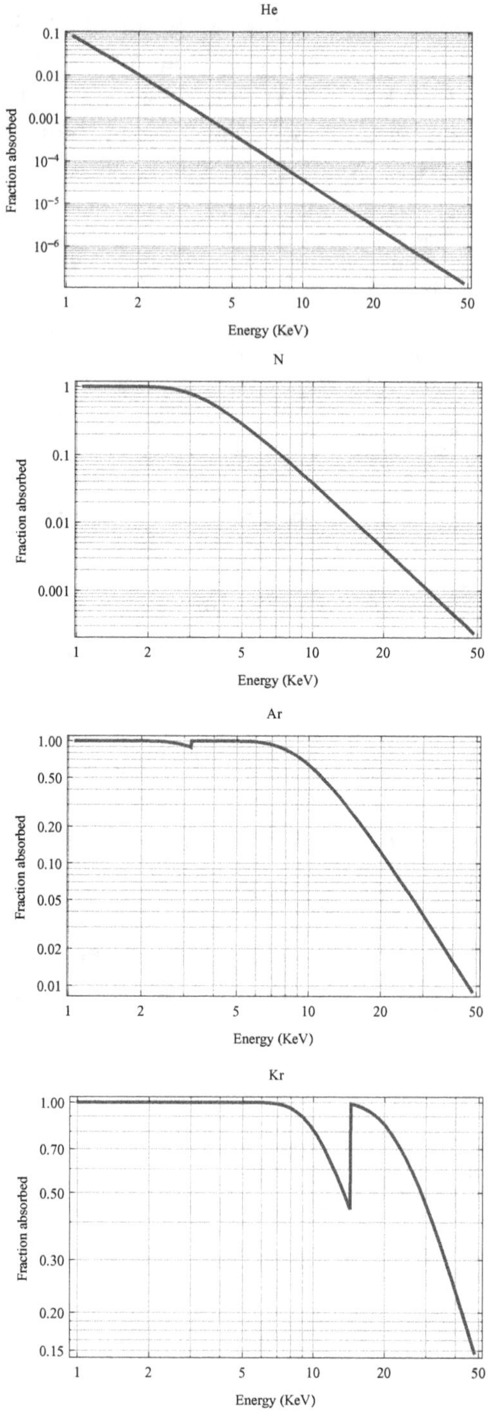

Fig. 3.14 Fraction of photons absorbed in 10 cm long ionization chamber containing fill gases (He,N_2,Ar,Kr) at room temperature.

Table 3.1 *Average energy W needed to create an electron–ion pair*

Fill gas	**Energy(eV)**
Helium	41
Neon	36
Nitrogen	36
Argon	26
Krypton	24
Xenon	22
Silicon @300K	3.6
Silicon @77K	3.7
Germanium @77K	3.0

3.5.6 Photodiode detectors for current mode

Another type of detector that is often employed in current mode is the photodiode, which acts essentially like a solid state ionization chamber. They are useful for both transmission and fluorescence detection, and have the virtue of high linearity. Common examples are PIN (Positive-Intrinsic-Negative) diodes, and PIPS (Passivated Implanted Planar Silicon) detectors, a low leakage-current photodiode design from Canberra, Inc (www.canberra.com). Instead of producing electron–ion pairs as in an ionization chamber, an X-ray photon produces a large number of electron–hole pairs in the semiconductor. When these diffuse through the material, they encounter a junction between positively doped and negatively doped regions. The electric field within the junction separates the electrons from the holes. These are collected and measured in current mode in a manner similar to an ionization chamber, and the fluxes incident on them can be calculated in a manner similar to that given above. In most current mode configurations no bias voltage is needed.

In contrast to ionization chambers, which are insensitive to visible light, photodiodes must be well shielded from visible light. This can be accomplished using multiple layers of aluminized mylar or Kapton, preferably with a black X-ray-transparent material (e.g. black paper) between the layers. When testing, light leakage can be detected as a 120 Hz signal, instead of 60 Hz, as is most low frequency electronic noise.

3.5.7 Photomultipliers

Photomultipliers (PMTs) are detectors that accept UV and visible light photons and produce pulses of current that are large enough to directly count, after they have passed through a PMT preamplifier. A major manufacturer of PMTs is Hamamatsu (www.hamamatsu.com); alas, Burle/Photonis is no more.

PMTs work in the following way: a visible photon enters the detector (which is typically housed in a sealed evacuated glass tube), hits a photocathode, which ejects electrons. A nearby electrode (dynode) is held at a potential such that the electrons are attracted to it. The electric field provides sufficient kinetic energy to the electrons that, upon collision with the dynode, they liberate more electrons from the dynode, so the current pulse is amplified several-fold. A high voltage on the order of 1 KV is distributed between approximately 10 dynodes using a passive or active voltage divider circuit. The successive collisions with the dynode chain yield an exponentially growing current pulse that is finally delivered to the anode of the PMT. PMTs depend on electron optics, and therefore should not be used for high magnetic field applications. They are often magnetically shielded using mu-metal for low fields, but they cannot tolerate large fields. Consult the manufacturer for details, or measure the response experimentally.

PMTs detect visible light photons, not X-rays, so a scintillator must be used to convert the energy of an X-ray photon into a burst of visible photons that enter the PMT. Commonly used scintillators include: doped NaI (which are bright, but produce long pulses, which restricts the maximum count rate); Plastic (anthracene), which are less bright, but produce short pulses, and are suitable for fast counting; and YAP (yttrium aluminum perovskite), which is bright and reasonably fast. One major vendor of scintillators is Saint-Gobain crystals (www.bicron.com).

PMTs can be used as current-mode detectors as well. Since the current amplification depends exponentially on the bias voltage, a very stable high voltage supply is required for this use.

Related devices called Microchannel Plates (MCPs) produce electron amplification by electrons rattling through thin channels under the influence of strong electric fields; the channels act functionally like distributed dynodes. MCPs are particularly useful for imaging and position sensitive detector applications, or where compactness is essential.

A new class of devices called Silicon PMTs recently has become available from Hamamatsu which may prove useful for X-ray work.

3.5.8 Solid angle

The solid angle Ω subtended by a detector measures its geometrical ability to collect photons. It takes account of the fact that the intensity of X-rays

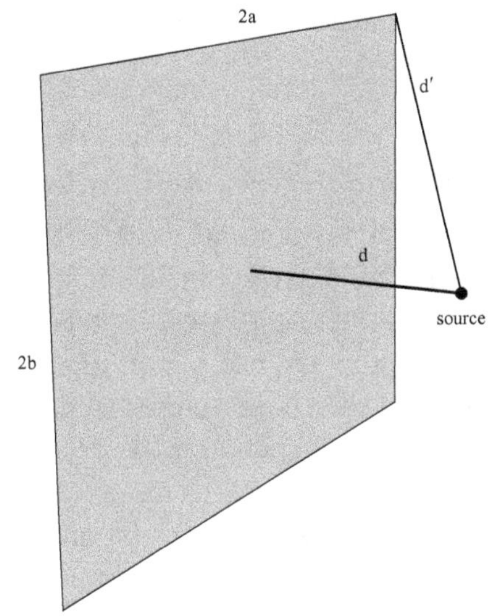

Fig. 3.15 Solid angle subtended by a rectangular detector is $\Omega = 4\arctan(ab/dd')$.

emitted from a point source decreases inversely as the square of the distance; it also accounts for the orientation and shape of the detector surface. Just as angles are measured in dimensionless units of radians, solid angle is measured in dimensionless units of steradians.

A general expression is

$$\Omega = \iint_{\text{surface}} \frac{\hat{n}\cdot\vec{r}\,dA}{r^3},$$

where $\vec{r}$ is the vector from the point source to a variable point on the detector surface, r is its length, dA is a differential element of area, $\hat{n}$ is the (variable) unit vector normal to the surface, and the integral is over the surface of the detector window. If the surface fully encloses the source, the solid angle is 4π. If the center of a large planar detector is very close to a point source it subtends nearly 2π.

The following equations are useful for practical calculations, e.g. estimating count rates. The solid angle subtended by a circular disk detector, centered on and with center normal pointing directly at the source, is $2\pi(1 - \cos\theta)$, where θ is the half-angle in radians that is subtended by the detector. Similarly, the solid angle subtended by a rectangular detector of dimensions $2a$ and $2b$, with center at distance d, and the center normal pointing directly at the source, is $\Omega = 4\arctan(ab/dd')$, (arctangent expressed in radians); d and $d' \equiv \sqrt{a^2 + b^2 + d^2}$ are respectively the minimum and maximum distances from the source to points on the detector, as shown in Figure 3.15.

3.5.9 Pulse-counting detectors and dead-time corrections

Pulse-counting detectors such as photomultipliers (PMTs), Ge and Si multi-element detectors, Silicon Drift Detectors (SDDs) all share a common characteristic: they have limited counting rates, because the physical processes in which the charge is collected are not instantaneous, and also because it takes a finite time to process the charge pulses to determine the photon energy. The maximum count rates depend on the detector characteristics and also various adjustable parameters such as amplifier pulse shaping time.

A common occurrence at high count rates is pulse pile-up. If two pulses occur too close to each other in time, they cannot be distinguished from a single pulse of twice the energy. This effect can be readily seen on a multichannel analyzer. Sophisticated pulse processing can be used to interpret and correct such events, but these capabilities are not widely available.

Pulse pile-up can be a serious problem when using pulse-counting detectors, particularly at synchrotron radiation sources, which deliver X-rays in short pulses, typically of a few hundred picoseconds duration. When a large number of photons arrive in a very short time detectors cannot resolve them – they see them as a single pulse. So the count rate of each channel of a detector is limited by the pulse repetition rate of the source. These issues are discussed and simulated in some detail by Bateman [14] and references therein.

Each electron bunch in the ring will emit a pulse of X-rays every time it passes the beam port as it circulates around the ring. This occurs at a frequency c/L, where c is the speed of light, and L is the circumference of the ring. Usually there are many bunches N; if they are equally spaced, the pulse frequency will be Nc/L. As an example, the Advanced Photon Source has a circumference of about 1100 m, so the repetition rate for a single bunch is $3 \times 10^8/1100\text{Hz} \approx 272$ KHz. If 24 bunches are equally spaced in the ring, the repetition rate is about 6.54 MHz. Therefore if the ring is operating in single bunch mode, there is a maximum count rate limit per channel of 270 KHz, which is pretty severe, and must be included in the dead-time corrections. On the other hand, if 24 bunches are used, the ceiling of 6.5 MHz/channel is not usually a problem.

Detection systems are characterized in part by their dead time. This is the finite time interval τ following the detection of a pulse, during which the detection system is at least partially blind. There are two types of idealized

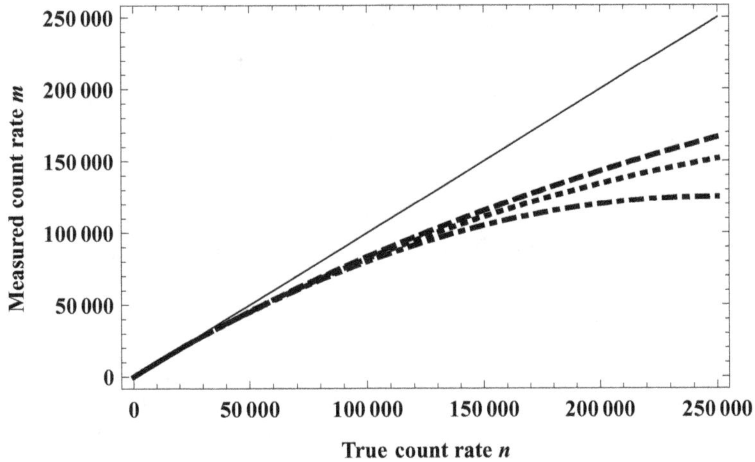

Fig. 3.16 Measured counts vs. real counts as a function of count rate: non-paralyzable model (dashed line), paralyzable model (dotted line), and the empirical model of Zhang *et al.* (dot-dashed line). The solid line is the true counts and τ is taken as 2μsec.

detection system behaviors [13]. Consider first a detection system that is described by the "paralyzable" model. Following an initial pulse, if a second pulse is detected within the time τ, the dead interval is extended by τ past the second pulse. Similarly, if a third pulse arrives within τ of the second pulse, the dead time is again extended by τ after the third pulse; and so on. The double pulse (or triple pulse, if it happens) is counted only as a single pulse. The system reverts to its initial behavior if there are no pulses within a time τ.

In contrast, a "non-paralyzable" detection system is effectively blind for a period τ after the first pulse. Any other pulses that occur within τ after the first pulse are lost. The system reverts to its initial behavior if there are no pulses within a time τ.

As a consequence of these behaviors, the measured count rate m is less than the true count rate n. A paralyzable detection system satisfies the relation [13] $m = ne^{-n\tau}$, with a maximum count rate of $1/\tau$; a non-paralyzable detection system satisfies $m = n/(1 + n\tau)$, which saturates only at infinite count rate.

There is of course no guarantee that a real detector follows these ideal behaviors. Indeed, Zhang *et al.* [40] found that the model $m = n(1-n\tau)$ best fit the experimental curve for a Canberra 13 element Ge detector system, with 4.3μ sec dead time. The maximum count rate for this model is $1/2\tau$.

Inspection of the curves in Figure 3.16 shows that significant deviations from linearity occur well below the maximum count rate. Even at 1/10 of the maximum count rate (in this example 25 KHz out of 250 KHz) the measured counts are nearly 5% lower than the true counts. These nonlinearities have important deleterious consequences, because variations in I_0 do not divide out, introducing noise, glitches, and baseline shifts. Performing dead-time corrections can clean up these problems dramatically [40, 42], however, and allow the detector to be effectively used over a broad range of count rates, although they should never be operated very close to the maximum, even with dead-time corrections.

For these reasons, dead-time corrections (or the equivalent) should always be carried out when using such detectors. It is not difficult to do so by experimentally measuring calibration curves like those shown in Figure 3.16, and to use a smooth interpolation of or best fit to the points to correct the raw data. Ideally, beamline control software will be provided to automate this process. It may be desirable to record the incident count rate (ICR) for each channel in addition to the counts within the selected energy window, if those signals are available. Keep in mind that different calibration curves will be obtained if the shaping times are changed.

3.5.10 Diffractive analyzers

The key benefit of energy-dispersive detectors such as Ge and Si multielement detectors is their energy resolution. With a multichannel analyzer or equivalent, such as XIA's Digital X-ray Processor (www.xia.com), they permit a range of energies to be monitored simultaneously, providing a picture of the relative amounts of different elements in the sample. They are useful for many Fluorescence XAFS experiments because they also allow the experimental to pick out just the fluorescence of the element of interest. Despite their utility, they have some serious limitations in maximum count rate and energy resolution. Count rates are generally several hundred thousand counts per channel, and energy resolution is typically 200–250 eV at high count rates. The best that such detectors can do is about 130 eV with present technology.

Often one is faced with measuring a dilute species in the presence of a large background due to elastically scattered radiation or fluorescence from other elements in the sample. Pulse-counting detectors spend the majority of their time counting background photons that are ultimately rejected. This is fine if the detector can handle all of the flux, as is usually the case with microfocus beamlines, but with the full beam from intense modern sources, the detectors may saturate, unless one sacrifices beam intensity, to bring it into the range the detector can handle.

It would be ideal if there were a way to reject the background *before* it got to the detector. Fortunately, there are practical ways to do exactly this, using diffractive optics. These can be used as pre-filters for ionization chambers, PIN diodes/PIPS detectors, photomultipliers, or Ge/Si detectors. Historically crystals in Bragg geometry have been used for this purpose, but the relative narrow angular acceptance creates two problems: they require very small focal spot sizes and very precise mechanical alignment to provide reasonable performance. The solid angle accepted by such analyzers generally has been impractically low.

Several useful new types of diffractive analyzer systems have been developed in the last decade or so: Pyrolytic Graphite [43], adaptive Bragg [44], Multilayer Array Analyzers Detector [45] (www.hdtechinc.com), and bent crystal laue analyzers (BCLAs) [46, 47] (www.quercustech.com).[7] Examples of the BCLA and MAAD devices are shown in Figures 3.17 and 3.18. These all offer the ability to reject the majority of the background *before* it reaches the detector, and better resolution than Ge and Si detectors. The graphite monochromator, adaptive Bragg optic, and BCLA approaches all use crystals of logarithmic spiral shape, which is the correct form to meet the Bragg condition at various points along the surface. The Multilayer analyzer in contrast uses graded synthetic multilayers as diffractive elements. The BCLA differs from the other designs in using the Laue geometry, in which the X-rays are diffracted as they pass through the log-spiral bent, asymmetric, thin crystal. All approaches require a point focus beam on the sample, and they differ significantly in their operational complexity, flexibility, and system cost. They are particularly useful for discriminating against background fluorescence lines that are close in energy to the element of interest. The MAAD and BCLA approaches intentionally have broader rocking curves to permit the use of larger source spot sizes, on the order of 100 micron, and to simplify alignment of the units by experimenters. Recent MAAD double-reflection prototypes promise excellent background rejection and good tunability. MAAD's are particularly effective at low energies ($\approx$1–10 KeV), while BCLAs are more effective at higher energies (5–30 KeV).

[7] In the interest of full disclosure, the author is a principal of Quercus X-ray Technologies, but has no financial interest in HDTech.

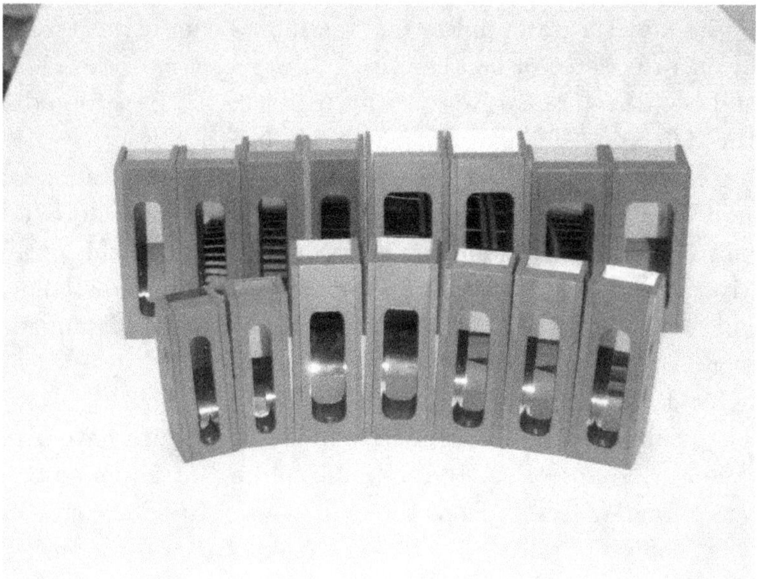

Fig. 3.17 Bent crystal laue analyzers (BCLAs) for various fluorescence energies. Each acts as an energy filter over a several KeV energy range.

3.5.11 Fluorescence ionization chambers

Fluorescence XAFS experiments require detectors of large area, or at least large solid angle, for which transmission ionization chambers are not optimal. E. Stern and S. Heald developed a detector [16] for fluorescence XAFS experiments by changing the geometry of the transmission ionization chamber so that the X-rays pass through one or more of the electrodes and are absorbed by the fill gas. The through-plate geometry, combined with use of suitable filters and slits, permits collection of photons over a large area and the rejection of elastically scattered background from the sample. The electrodes can be made of any electrically conductive but X-ray transparent material, such as tightly stretched aluminized mylar, nickel mesh, or electrically conductive polymers. It is very important that the electrodes be mechanically stable, otherwise vibrations and sound waves can introduce excessive noise.

Detectors of this type (commercially available from the EXAFS Company www.exafsco.com, and informally called "Lytle Detectors") with 3 or 5 electrodes of alternating polarity (which serve to increase the sensitive volume of fill gas) are available on many beamlines. A 17 electrode version designed for submillisecond time response has been reported [30]; the fast time

Fig. 3.18 Low-energy version of the Multilayer Array Analyzer Detector. The photodiode detectors are on the left.

response is a consequence of the increased electric field that is obtained when the plates are close together, but the bias voltage is kept the same.

In most cases the effective use of such detectors requires the use of "$Z-1$" filters and slits to reject elastically scattered background, i.e. scattered X-rays that are at the energy of the incident beam. The concept and usage of such filters is illustrated in Figures 3.19 and 3.20. The X-ray filters usually are composed of a thin support with a uniform coating consisting mostly of the element one atomic number lower than that of the element of interest in the sample. This places the absorption edge of the filter between the K_α fluorescence peak and the absorption edge of the element of interest, resulting in low absorption by the filter of the fluorescence, but high absorption for the elastically scattered background, which is at the energy of the incident beam.

For example, the K_α transition for Fe is at 6.40 KeV,[8] while the Fe

[8] The transitions are generally composed of a number of more closely spaced transitions which are usually not resolved by the detector; here we lump them together for simplicity.

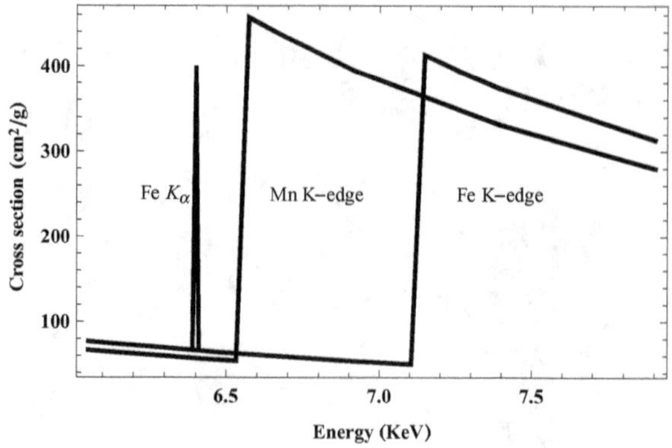

Fig. 3.19 $Z-1$ filters for Fe edge fluorescence.

absorption edge is at 7.11 KeV. The absorption edge of Mn at 6.54 KeV falls nicely between them, and is suitable as a filter for the Fe edge. For elements in the first transition series (e.g. Fe) a $Z-1$ filter works well, but for heavier elements it may be necessary to use $Z-2$ or even $Z-3$ filters.

The positions of K_α lines for the element of atomic number Z can be estimated with useful accuracy by Moseley's law $E_{K_\alpha} \approx 10.2(Z-1)^2$ eV. It should be noted that the K_β transition falls at an energy 10–15% higher energy than the K_α, and it is typically 10–15% of the intensity of the K_α transition.

Filters designed for compatibility with Lytle Detectors are available from the EXAFS Materials Company. They can also be made in a manner similar to making a good quality large area transmission XAFS sample. A detailed description regarding optimal use of filters is given in Appendix 3.

3.6 XAFS experiments

3.6.1 Linearity and noise

Noise in EXAFS experiments has a variety of sources, among them random fluctuations in the number of photons absorbed by the detectors, electronic noise, and noise that arises from different sensitivities of the detectors to fluctuations of the X-ray beam. In the typical EXAFS experiment, recognizing the fact that the incident X-ray intensity varies with time, one attempts to compensate for it by dividing the measured signal by the measured incident flux, each integrated over precisely the same time interval. A number of experimental problems can interfere with exact compensation for these beam intensity fluctuations. The problems can be minimized if care

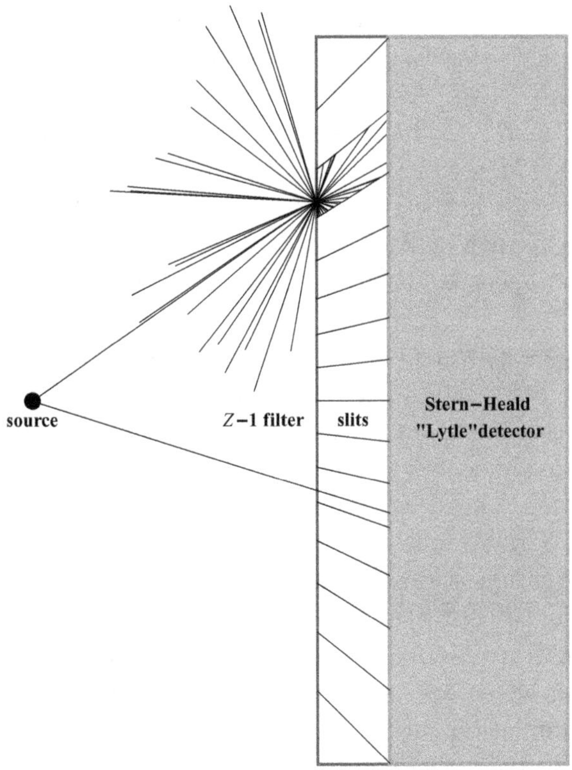

Fig. 3.20 $Z-1$ filters preferentially absorb elastic scatter above the filter's absorption edge. The slits transmit the sample fluorescence, but block most of the filter refluorescence.

is taken, but they can (and often do) cause glitches, noise, and distorted amplitudes if they are neglected. The most important of these problems can be summarized in the mnemonic HALO: Harmonics, Alignment, Linearity, Offsets. It is helpful to keep these in mind during the experiment, so they can be systematically managed.

In order to precisely compensate for the incident intensity fluctuations, it is necessary that the I_0 and I_f detectors be very linear, and that they "see" exactly the same beam, in the same way. If harmonics are present in the incident beam, the I_0 and I_f or I detectors will in general measure different proportions of the harmonic radiation and the fundamental, even *if the detectors are identical.* The reasons are quite simple. The I detector absorbs the beam that has passed through the sample. If the fill gases are chosen to be the same for I and I_0, the current I is usually at least one or two orders of magnitude lower than I_0, which breaks the symmetry between the detectors, upon which the hypothetical noise cancellation is based. On the other hand, if the I_0 chamber is chosen to have a lighter gas (e.g. N_2) than the I chamber (e.g. Ar), the symmetry between the chambers also is broken. Furthermore, residual harmonics are relatively enhanced in beam absorbed by the I detector, because they are mostly unattenuated by the sample. This increase in the proportion of high-energy X-rays conventionally is called "beam hardening": the fraction of hard (high-energy) X-rays in the beam is increased as the beam passes through the sample, windows, air, etc.

In fluorescence experiments, the I_f detector senses the fluorescence photons (which are essentially of a single energy), plus scattered background (most of which is elastic scattering at the excitation energy, but some of which is inelastically scattered radiation at lower energies), plus scattered harmonic radiation. On the other hand, if the I_0 detector is an ion chamber, it measures the X-ray intensity at the excitation energy (the fundamental), and also some fraction of the harmonic radiation.

Thus the I_0 and I_f (or I) detectors, in effect, see different beams. Even if one uses a scatterer and detector, instead of using an ion chamber for I_0, the I_f and I_0 detectors "see" beams of different spectral content. The result is that fluctuations in the incident beam intensity do not divide out perfectly. For this reason it is essential to minimize the harmonic content of the beam, either by detuning the monochromator, or by using a harmonic rejection mirror. This is the H in HALO.

The second major reason why the two detectors may see different beams is that the sample alignment may be incorrect. The X-ray beam size and shape should be defined by a slit placed before the I_0 detector, and nothing except for spatially uniform objects such as smooth windows and a homogeneous sample should intercept the beam after that point. If something does clip the beam between the front detector and the sample, whether it is a lead mask, the edge of the exit aperture of the front ion chamber, the edge of a cryostat entrance window, or even merely a wrinkled cryostat entrance window, then small fluctuations in beam position may cause noise, glitches, or other distortions. Improper alignment is one of the primary causes of glitches in experimental data.

The L in HALO stands for linearity. The intensity fluctuations will divide out only if the detectors and electronics are linear, that is, their outputs are proportional to the inputs (also a constant offset is okay, as long as it is subtracted out). It is sometimes believed that nonlinearities will somehow cancel out if identical detectors are used, but this is incorrect, because the beam intensity (and possibly spectral content) is generally different between the two detectors.

A more subtle cause of errors is a mismatch in the time constants of the amplifiers that are used. If ion chambers are used, the applied voltage must be high enough that they are in their plateau region, where the output for a given X-ray flux is independent of high voltage. The value of the plateau voltage depends on the construction of the ion chamber, the photon flux, the fill gas(es), and other variables. Typical values for the plateau voltages are 300–400 V for transmission ion chambers with incident fluxes of about 10^9–10^{10} photons/sec, and 50–100 V for fluorescence ion chambers absorbing $10^6 - 10^7$ photons/sec. If other types of detectors are used, it is crucial that their linearity also be checked. Finally, the electronics (e.g. amplifiers, voltage-to-frequency converters) should be operated over their linear ranges by suitably adjusting the gains and offsets. When dealing with small signals, it is often wise to set the amplifier offset to a few tenths of a volt or so, so the signal always is much larger than electronic noise and drift. This offset should be measured and subtracted out by the data collection computer program. For the equipment in common use (e.g. Keithley 428 amplifiers, and Analog Devices V/F converters) it is best to keep signals in the range 1–9 volt.

Finally, it is very important to periodically measure the offsets or dark currents (i.e. the signal produced when the X-ray beam is off), so that they can be subtracted out. Fluctuations in incident intensity will not divide out between I_0 and I (or I_f) if the proper offsets have not been subtracted. Sometimes offsets can drift, so it is a good idea to check them regularly. If the detectors are sensitive to visible light as well (such as PIN diodes and scintillator/PMT detectors), then the detectors should be well shielded from the light, or at a minimum, the ambient light should be kept as constant as possible. These are the four aspects of HALO. Even after these precautions are taken, some noise is, of course, still present in the data. Some noise is a consequence of residual nonlinearities, which require more advanced modulation techniques (which are beyond the scope of this brief note) to eliminate. Other problems also may potentially exist, such as differing sensitivities of the I_0 and I_f detectors to beam position. If one uses an ionization chamber to detect I_0, and the average position of the beam (i.e. centroid of the intensity) fluctuates horizontally, the I_f detector will be more sensitive to the fluctuations than will I_0, which will result in increased noise. This problem can be corrected by using a symmetrical detector configuration in which a thin scatterer or fluorescer emits X-rays or visible light into a detector at 90 degrees, or alternatively by using modulation techniques.

For dilute specimens, however, the noise is usually dominated by fluctuations in the number of detected photons. If N photons on the average are absorbed per unit time, and they are absorbed independently of each other, and N is much large than one, general statistical arguments indicate that the number of photons actually absorbed per unit time fluctuates by about $\sqrt{N}$. For example, if one million photons are absorbed on the average, one would expect (root-mean-square) fluctuations of about one thousand. Such fluctuations are a major source of noise for EXAFS studies on dilute samples.

The measured signal consists of the desired fluorescence photons (N_s) and scattered background N_b (plus refluorescence from the filter). The noise arises from fluctuations in the total number of photons: noise $\propto \sqrt{N_s + N_b}$, and thus the signal to noise ratio is given by: $S/N = N_s/\sqrt{N_s + N_b}$. It is more convenient to deal with the square of the signal to noise ratio, called the effective number of counts $N_{\text{eff}} = N_s^2/(N_s + N_b) = N_s/(1 + N_b/N_s)$. If you were to measure N_{eff} signal photons with no background, you would have the same signal to noise ratio as you obtain with N_s signal photons and N_b background photons. Calculating N_{eff} during the experiment is quite straightforward: N_s is the difference in the number of photons above (A) and below (B) the edge, and $N_s + N_b$ is just the value above the edge; thus $N_{\text{eff}} = (A - B)^2/A$. When optimizing the experiment (for example when determining the best filter thickness to use), it is quite practical to (for example) add or remove a filter, move the monochromator above and below the edge, measure A and B, and calculate $(A - B)^2/A$, and use whichever combination of filters that gives the higher number.

It is important to optimize the right expression. Although intuitively plausible, it is NOT correct to optimize the ratios A/B or $(A - B)/B$. Optimizing the wrong function will give degraded results. For example, choosing a filter by maximizing $(A - B)/B$ in general results in a filter that is too thick. Typically the amplitude of the EXAFS over the data range is only a few percent of the size of the edge step ($A - B$). If we want our EXAFS data to have a signal to noise ratio of 1–10%, then, relative to the edge step, we must obtain a signal to noise ratio of 0.01–0.1%, i.e. $S/N > 1$–3×10^3. Attaining such a signal to noise ratio requires 10^6–10^7 effective counts total per energy point. This rough criterion allows us to estimate how long it will take to obtain the data during the experiment. If the time required exceeds the time available, it may be wise to reduce the energy range of the scan, because the data at high k probably would not be useful anyway. If the ring is not operated in "top-off" mode, but rather exponentially decays, the time T available until the next fill of the electron beam is given by $T = \tau \ln(I/I_{\rm min})$, where I is the present current, τ is the beam lifetime, and $I_{\rm min}$ is the current at which the ring is refilled.

Example: Suppose you are using a fluorescence ion chamber at the iron K edge, the amplifier is set to 10^{11} gain, and you are getting 260 000 V/F output pulses in 2 seconds integration time below the edge, and 520 000 above the edge. Many V/F converters in popular use put out 100 000 pulses/sec for one volt input, so below the edge we have 1.30 V, and above the edge we have 2.60 V coming from the amplifier. Since the signal and background are equal (1.30 V), the effective voltage $(A-B)^2/A$ is 0.65 V. Using the conversion factor of 3.5×10^6 photon/sec for 1 volt at 10^{10} volt/amp gain at the Fe K_α fluorescence energy, and accounting for the difference in gain, we obtain $N_{\rm eff} \approx 3 \times 10^5 \times .65 \approx 200\,000$ effective photons/sec. To obtain statistics of 2 million effective counts it would take 10 seconds integration time per energy point. If you are integrating for 2 seconds/point in a scan, and each scan takes 30 minutes, it would take roughly 5 scans and 2.5 hours to obtain the required data. If the electron beam lifetime is short, you may wish to include the effect of reduced beam current as well. If the beam lifetime is t minutes, and the present beam current is I, the average current for the next T minutes is: $I(\tau/T)(1-\exp(-T/\tau))$. The average current until the next fill is: $(I - I_{\rm min})/\ln(I/I_{\rm min})$, where $I_{\rm min}$ is the current at which the next fill will occur.

In summary, it is useful to commit to memory the simple equation

$$N_{\rm eff} = \frac{(A-B)^2}{A},$$

where A and B are the counts above and below the edge, and to use it when optimizing experiments. It may also be useful to remember the mnemonic HALO: Harmonics, Alignment, Linearity, Offsets. In most cases one should obtain enough fluorescence scans that $N_{\rm above} > 1 - 3 \times 10^6$ for the sum of all scans. In combination, these simple rules can be a great help in diagnosing problems, in planning experiments, and in making effective and efficient use of beam time.

3.6.2 Energy shifts

Monochromators are not perfectly stable devices. Thermal drifts in particular can cause minute changes in the dimensions of components within the monochromator, resulting in small angle changes, which can cause energy shifts. Alterations to the orbits of the electron beam in the storage ring also may shift the position of the X-ray source, resulting in energy shifts because the angle of incidence onto the crystal changes. This occurs without the knowledge of the beamline control program, so the effect is a miscalibration of the energy.

Both XANES and EXAFS are sensitive to the energy zero, and all shifts more than ≈ 0.1 eV should be compensated when data are analyzed. It is a good idea to monitor the energy calibration by recording the spectrum of a reference sample that remains unchanged for the duration of the experiment.

A simple way to accomplish this in transmission is to use a third reference ionization chamber. After passing through the I_0 chamber, the sample, and the I_1 chamber, a fraction of the X-rays then pass through a reference sample and the third, I_2 chamber. The spectrum of the reference is then obtained from $\ln(I_1/I_2)$. If the monochromator calibration shifts, this will be detected as a shift in the position of the edge of the reference sample, and it can be used to compensate for such shifts. An alternate approach is to periodically measure the reference spectrum to monitor drifts.

3.6.3 Beam intensity leveling

It has been observed that if the beam intensity is kept constant during a scan, for example by dynamically detuning the monochromator, the apparent noise level in the data gets better. As appealing as this approach might seem, I think it is not a good idea, because it merely covers up the basic problem, which is a nonlinear data acquisition system. Even if the data look smoother, they will suffer from systematic amplitude errors, distortion of edge features, and other problems. The preferred approach would be to linearize the detection system, or at least develop procedures to compensate for the nonlinearity, such as dead-time corrections.

In fact, in my experience, the opposite approach is very useful: using beamline software tools (or additional gadgets placed into the beam), intentionally modulate the beam intensity to monitor and debug nonlinear elements in the system. The overall data quality will be much improved.

3.6.4 Bragg glitches

Monochromators have energies at which more than one Bragg plane will diffract. At these energies the monochromator may produce a beam that quickly sweeps across the slit, producing a monochromator glitch. These can be calculated and their locations mapped and predicted. Depending on the monochromator design it may or may not be possible to adjust the alignment of the crystals to prevent the glitches. The simplest approach is just to use a crystal set or monochromator that does not produce any glitches in the energy range of interest.

Other sources of Bragg glitches are diffraction from crystallites in the sample, ice crystals, or from a crystalline substrate. A beam is momentarily diffracted into the detector, or away from the sample, affecting the apparent absorption or fluorescence. These artifacts are best prevented during the experiment, for example by using diffractive analyzers that greatly reduce the Bragg diffracted background.

If the sample substrate is diffracting, it may be possible to choose a range of incident angles in which it cannot. Alternatively, since the Bragg spots are directional, masking off parts of the detector may help. Spinning the sample to average the effect has been reported to be of use. Finally, when spinning, Bragg spots should diffract only at certain times in the cycle, and synchronous blanking of the detector or scalers at specific time intervals could be of use, although to my knowledge this has not been proposed elsewhere or tried.

3.6.5 Thickness effects

In transmission experiments it has been observed that, if samples are made too thick, systematic distortions will occur in the data. These thickness effects are particularly serious in the near-edge region where large amplitude, high-contrast structure may be found. A symptom of thickness effects is that the XAFS oscillations are compressed – they have reduced amplitudes, which has the effect of erroneously reducing the estimated coordination numbers that are obtained in data analysis. It is best to keep the sample edge step $\Delta\mu x$ to about 1.0 or less, and the total sample thickness μx to less than 3 or so, preferably less.

As described by Parratt *et al.* [48] the tails of the monochromator resolution function can cause thickness effects, even for homogeneous samples. In principle this effect could be removed by a deconvolution procedure, although this is seldom done. Sample inhomogeneity and harmonics are other, equally important factors [49, 51]. Here we present a simple model [52] to computationally explore the effects of sample inhomogeneities.

For simplicity, let us suppose the incident beam is uniform over the sample surface, and but the sample may be of nonuniform thickness. We define the thickness distribution $P(x)$ so that $P(x)dx$ is the fraction of the sample area that is between thickness x and $x + dx$, with $\int_0^\infty P(x)dx = 1$. Then the effective attenuation

$$(\mu x)_{\text{eff}} = -\ln\Big(\int_0^\infty P(x)e^{-\mu x}dx\Big),$$

where $\mu = \mu(E)$ is the total attenuation coefficient (including both photoelectric absorption and scattering) of the material of which the sample is composed. This can be expanded in powers of μ as

$$(\mu x)_{\text{eff}} = -\sum_{n=1}^{\infty} \frac{C_n}{n!}(-\mu)^n = C_1\mu - \frac{C_2}{2}\mu^2 + \frac{C_3}{6}\mu^3 \cdots,$$

where the C_n are the cumulants of the thickness distribution. The first cumulant, C_1 is the average thickness $\bar{x} = C_1 = \int_0^\infty P(x)x\, dx$, and C_2 is the mean square variation of the thickness $\int_0^\infty P(x)(x-\bar{x})^2 dx$. The second cumulant cannot be negative, which shows that the net attenuation is reduced by the thickness variation.

For a Gaussian thickness variation of width σ[9]

$$P(x) = \frac{1}{\sqrt{2\pi\sigma^2}} \exp\left(\frac{(x-\bar{x})^2}{2\sigma^2}\right),$$

we have $(\mu x)_{\text{eff}} = \mu x_0 - (\mu\sigma)^2/2$. If there are pinholes (regions of zero thickness) in the sample of fractional area a, the thickness distribution is $P(x) = 2a\delta(x) + (1-a)\delta(x-x_0)$, where $\delta(x)$ is the Dirac delta function. The effective attenuation in that case is

$$(\mu x)_{\text{eff}} = -\ln\left(a + (1-a)\exp\left(-\mu x_0\right)\right),$$

while the average thickness is $\bar{x} = (1-a)x_0$, and the second cumulant is $a(1-a)x_0^2$.

As a simple example of a quasi-realistic thickness distribution, consider a layer of spherical particles of unit radius in a square array, with another layer placed on top, so that the spheres in the second layer cover the holes in the first layer; subsequent layers alternate. The contributions to the thickness from a minimum cell and the combined thickness are shown in Figures 3.21 and 3.22. Finally, the thickness distribution, which was obtained by randomly sampling 10^6 points over the minimum cell and constructing a histogram is shown in Figure 3.23. The mean thickness is 1.047 and the standard deviation (second cumulant) is 0.207. The thickness distribution by itself demonstrates even simple configurations of particles can result in a broad distribution of thicknesses; if the total sample thickness is too large, distortions in the data will result.

[9] For simplicity we assume that the width $(\sigma/\bar{x})^2 << 1$, so that the low-x tail of the gaussian is negligible for $x < 0$.

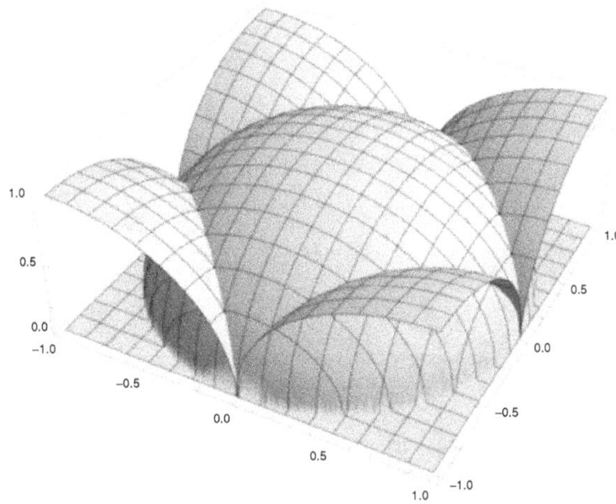

Fig. 3.21 Contributions to thickness from alternating layers of spherical particles.

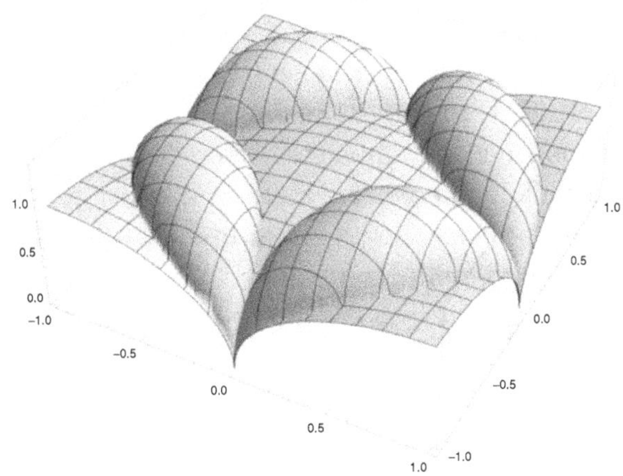

Fig. 3.22 Combined thickness of alternating layers of spherical particles.

3.6.6 Planning experiments

A number of factors must be considered when planning experiments. Foremost among them are: what kind of information is sought during the experiment? Which elements are of interest in the sample, and what are their concentrations and distribution within the sample? What is the physical state of the sample (solid, liquid, gas; lump, thin film, powder, oriented single crystal)? Do the measurements need to be made under cryogenic, high

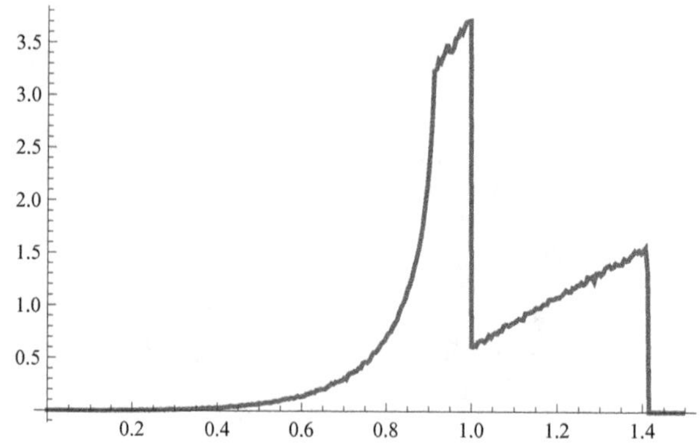

Fig. 3.23 Thickness distribution of alternating layers of spherical particles of unit radius.

temperature, high pressure, or other special conditions? All of these factors shape the design of the experiment.

In transmission experiments clearly the key thing is to get enough X-rays through the sample that they can be measured accurately, while still minimizing "leakage," i.e. residual high-energy harmonics getting through the sample, X-rays getting through thinner areas of the sample (samples are never perfectly homogeneous), ambient scattered and fluorescence radiation from air and materials in the environment, and scattered and fluorescence radiation from the sample itself. Steps can be taken to reduce these effects, such as shielding of I detectors from ambient radiation (e.g. air scatter, fluorescence from materials nearby), and physically separating/shielding the I detector so that fluorescence from the sample isn't collected. If the transmitted beam is sufficiently intense, such contributions will be relatively small, but if the transmitted beam is attenuated a great deal by a thick sample, these contributions can be large enough to serious distort the XAFS spectra.

To prevent thickness effects, as a rule of thumb the total sample thickness (+ substrate if there is one) should be kept to less than $\approx$ 4 absorption lengths, while the edge step $(\mu(E_a)x - \mu(E_b)x)$ of the element of interest should be about 1.0 (between 0.2 and 2.0). It should be mentioned that these values are only rough guidelines.

An attempt in Figure 3.24 has been made to produce a flow-chart to help choose an appropriate measurement mode. The numbers are approximate and are just a rough guide. To make good decisions about the measurement mode, several quantities must be calculated or measured. First, one must determine the energy of the absorption edge of interest. Then there are several absorption lengths that must be computed right above (say 50 eV) the selected energy, such as the absorption μ_i of the element of interest within the sample; the total absorption μ_T of the sample excluding the substrate; and the absorption μ_s of the sample substrate (e.g. if it is supported or deposited on some material, or there are windows the beam must go through). Here, x is the thickness of the sample (excluding substrate) and y is the thickness of the substrate alone. The flow chart should be traversed for each absorption edge of interest because there is no guarantee that a sample or measurement mode that is appropriate at one energy will be so for another.

The absorption lengths $1/\mu(E)$, which are strongly energy dependent (roughly as E^3), determine the relevant length scales. For example, suppose you are studying Ni^{+2} and Ca^{+2} ions in aqueous solution at the same concentration. One absorption length for water is about 1mm at 8 KeV, but it is only roughly 125 microns at 4 KeV. This limits the X-ray penetration depth into the sample, and correspondingly reduces the fluorescence signal, by nearly an order of magnitude.

In the following we distinguish three different absorption lengths: $\mu_i(E)$ is the absorption coefficient of the element of interest within the sample; $\mu_T(E)$ is the total absorption coefficient of the sample material; and $\mu_s(E)$ is the absorption coefficient of the substrate (if any) upon which the sample is deposited. Furthermore, we take x as the sample thickness, and y as the substrate thickness.

As a concrete example, suppose one wants to look at the Cu edge in a film of (BCCO) Ba-Ca-Cu-O superconductor that is on a $SrTiO_3$ substrate. The $\mu_i(E)$ would be that of the Cu in the BCCO material; $\mu_T(E)$ would be that of BCCO; and $\mu_s(E)$ would be that of $SrTiO_3$.

A recent review of grazing incidence is given by Waychunas [34].

3.6.7 Fluorescence and self-absorption

To understand fluorescence detection we must consider what happens when an X-ray beam at energy E is incident on a slab of a uniform sample.[10] Suppose that it enters the sample at angle α as shown in Figure 3.27. Then the probability that it will propagate to depth x' without being absorbed is evidently $\exp(-\mu_T(E)x' \csc(\alpha))$, where μ_T is the absorption coefficient of the sample; the probability the photon is absorbed by the element of interest between x' and $x' + dx'$ is $\mu_i(E)\csc(\alpha)dx'$, where μ_i is the absorption

[10] The more general case of distributions in particle size, taking into account particle shape, packing fraction, and nonfluorescent particles has been investigated by Tannazi [106] using analytical and Monte Carlo methods.

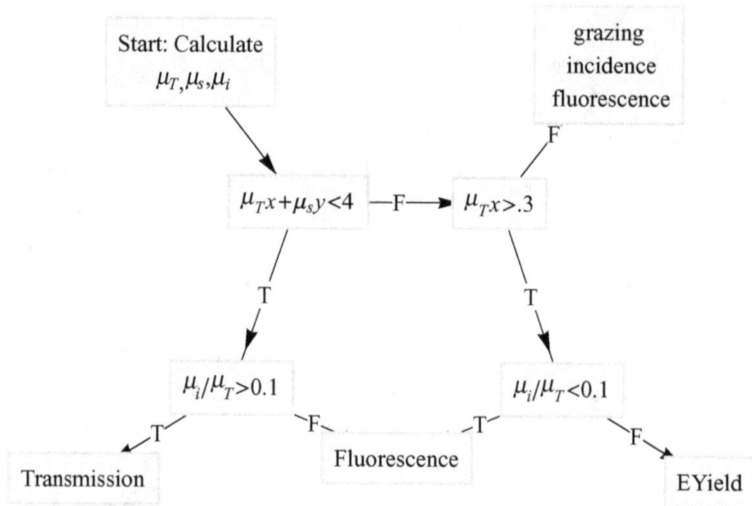

Fig. 3.24 Flow chart for choosing an experimental mode. The "T" and "F" labels on the arrows designate True and False outcomes for each test.

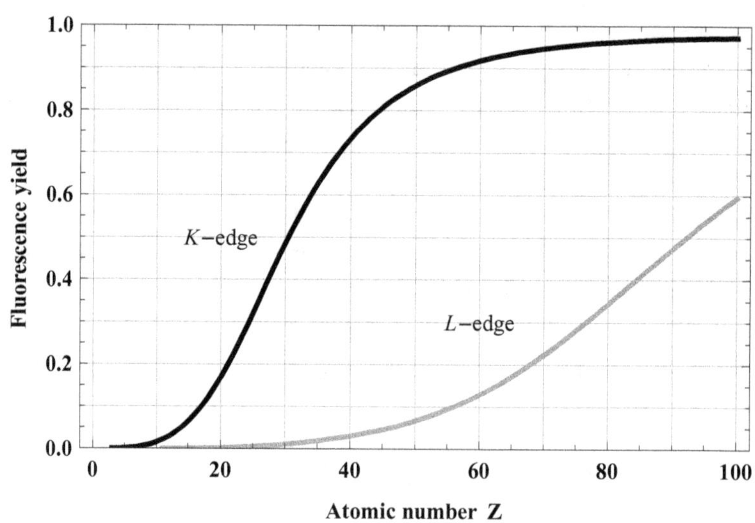

Fig. 3.25 Fluorescence yield vs. atomic number for K-edges and L-edges (average).

coefficient of the element of interest; the probability it will emit a fluorescence photon into differential solid angle $d\Omega$ is the fluorescence yield ϵ times $d\Omega/4\pi$; and the probability that the fluorescence photon propagates through the sample at angle β and escapes is $\exp\left(-\mu_T(E_f)x' \csc(\beta)\right)$. The fluorescence yield depends on the atomic number of the element and the edge under consideration, as plotted in Figures 3.25 and 3.26.

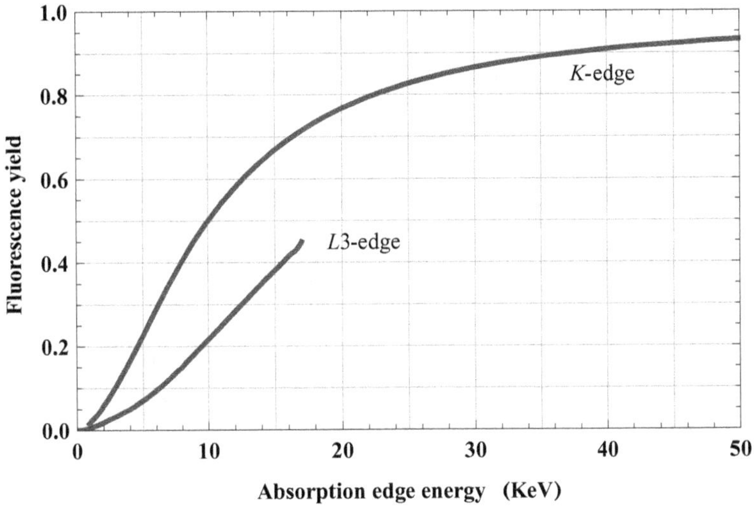

Fig. 3.26 Fluorescence yield vs. absorption edge energy for K-edges and L_3-edges.

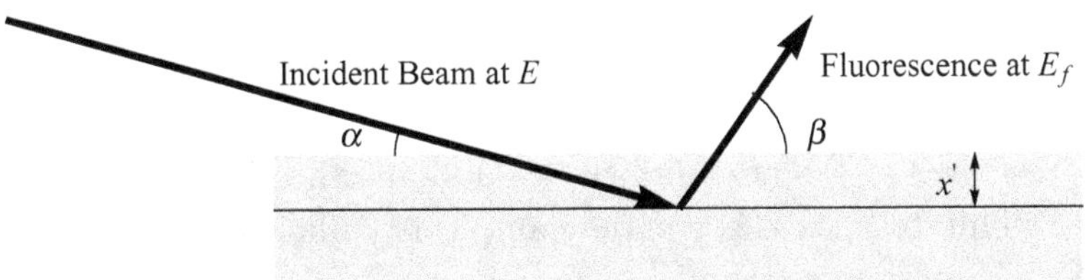

Fig. 3.27 Fluorescence from a homogeneous slab of material. The incident X-ray beam hits the sample at angle α and energy E; the fluorescence X-ray leaves at angle β and energy E_f. An average over β should be performed for a particular detector size and location.

Combining these probabilities and integrating over depth from $x^{'} = 0$ to $x^{'} = x$ (the sample thickness) we have:

$$\left(\frac{\epsilon\ \mu_i(E)\csc(\alpha)d\Omega}{4\pi}\right)\int_0^x \exp\left(-f(E)x^{'}\right)\,dx^{'},$$

with $f(E) = f(E, E_f, \alpha, \beta) = \mu_T(E)\csc(\alpha) + \mu_T(E_f)\csc(\beta)$ or, integrating

$$\left(\frac{\epsilon\ \mu_i(E)\csc(\alpha)d\Omega}{4\pi}\right)\left(\frac{1-\exp(-f(E)x)}{f(E)}\right).$$

Integrating this expression over the detector solid angle Ω, we find that the measured signal is the desired signal $\mu_i(E)$ times an energy-dependent function $g(E)$, where

$$g(E) = \epsilon/4\pi \int_\Omega \csc(\alpha) \left(\frac{1 - \exp(-f(E)x)}{f(E)} \right) d\Omega.$$

For very thick samples $f(E)x \gg 1$ the exponential can be neglected, and the expression for $g(E)$ simplifies somewhat.

$$g(E) = \epsilon/4\pi \int_\Omega \left(\frac{\csc(\alpha)}{f(E)} \right) d\Omega.$$

We see that in fluorescence mode we measure $\mu_i(E)g(E)$, not $\mu_i(E)$, as we would like. If $g(E)$ varies smoothly with energy this is easily corrected. This will be the case in the dilute case, in which $\mu_i(E)$ is much smaller than $\mu_T(E)$. So, extraction of $\mu_i(E)$ from the measured data is simple in the thick dilute limit.

It is even simpler in the thin, concentrated limit: for thin samples $f(E)x \ll 1$ the exponential can be expanded to first order, giving

$$g(E) = x\epsilon/4\pi \int_\Omega \csc(\alpha) d\Omega,$$

which is just an energy-independent scale factor that is proportional to the thickness. This tells us that, if the sample is sufficiently thin, we directly measure the function we want, i.e. $\mu_i(E)$.

However, if the sample is thick, but not dilute, it gets more complicated. From its definition, $g(E)$ depends on $\mu_T(E)$ which itself contains contributions from $\mu_i(E)$. In fact, for a pure element, $\mu_i(E) = \mu_T(E)$. Since the measured signal is $\mu_i(E)g(E)$, this implies the measured signal depends nonlinearly on the desired function $\mu_i(E)$. The general tendency is to suppress structure in the X-ray absorption. When $\mu_i(E)$ increases, the term $\mu_i(E)$ (which is buried in $\mu_T(E)$) in the denominator also increases, which tends to partially cancel the increase in the measured absorption. Physically what happens is that as the absorption coefficient increases (say, over an absorption edge), the penetration depth ($\approx 1/f$) decreases, so the number of atoms contributing to the fluorescence decreases. In this case the fluorescence signal no longer mirrors the absorption coefficient.

There are a number of recipes for disentangling these nonlinearities (under certain assumptions) to generate a correction for the isolated XAFS oscillations $\chi(k)$. A particularly simple and effective one (although approximate) is given by Booth and Bridges [31].

It is possible to reduce the importance of the term proportional to $\mu_T(E)$ in the function f by choosing β as small as possible while keeping $\alpha \approx$

$\pi/2$. Limiting the detector angle in this way reduces the solid angle of data collection substantially, but it does work [56].

Arguably the best approach is to avoid those situations in which such self-absorption corrections would be necessary. One basic problem is that unless one can generate error bars on the correction function, systematic errors of unknown magnitude can be introduced. In other words, if possible, try to work in the thin concentrated or the thick dilute limits. If it is necessary to measure thick concentrated samples, consider electron yield detection, or total external reflection fluorescence.

Finally, if a sample is made of particles, as many are, it is *not* sufficient that the average concentration is low, e.g. by diluting the particles in something else. In addition the particle sizes of the component(s) containing the element of interest should be considerably smaller than one absorption length *of that subcomponent.* Otherwise the spectra will suffer the types of distortions described above, and the final parameters determined in data analysis likely will be incorrect.

3.6.8 Grazing incidence/glancing angle and reflectivity/ReflEXAFS

There are many cases in which one wishes to probe only the surface of a sample. It is possible to do so with flat samples using grazing incidence and total external reflection. For example, one may wish to measure films coated on substrates, or films adsorbed onto mineral surfaces. Elastic scattering or fluorescence from the substrate may produce unacceptable background in the standard 45° fluorescence geometry in such cases.

It is clear from the description of fluorescence above that the effective penetration depth is $1/f = 1/(\mu_T(E)\csc(\alpha) + \mu_T(E_f)\csc(\beta))$. This implies that the penetration depth can be substantially reduced by setting α to a few degrees, or a few tens of milliradians. In this case the fluorescence detector is placed in the standard position so to collect as much fluorescence as possible.

If α is set below the critical angle for the coating or the substrate materials, which typically happens at angles less than a few milliradians, total external reflection will occur (see the description of X-ray mirrors). This confines the X-ray wave-field even more strongly to the surface, typically with a penetration depth of a few nanometers. In some conditions it may be possible to choose an angle that will transmit through the coating but be reflected totally from the substrate.

If the coating consists of multiple layers of different materials, or a film deposited on a crystal, and one wishes to study them as a function of angle, it is advisable to use the Fresnel equations of electromagnetic theory to calculate the spatial distribution of the wave-field in the material [7]. The square of the wave-field will be proportional to the probability density of photons interacting with atoms in the material. Information can be obtained about the spatial distribution of atoms normal to the surface by varying the angle and measuring the fluorescence emitted as a function of angle. This phenomenon/technique is called X-ray Standing Waves (XSW).

Several reviews are available [34, 32, 33] on grazing incidence and total external XAFS, XSW, and related applications as applied to mineral surfaces.

Analysis of the XAFS data under conditions of total external reflection or interference between multilayers can be subtle. As the energy is varied, the real and imaginary parts of the X-ray scattering amplitude vary substantially with energy. Because the X-ray wave field near the surface is strongly affected by the complex index of refraction, which itself depends on the $\mu_i(E)$, one can get nonlinear dependencies of the signal that may not be trivial to untangle.

Another method of obtaining spectra (sometimes called ReflEXAFS) is to monitor the intensity of the totally externally reflected beam, using an ionization chamber or PIN diode, as a function of E at fixed angle [35]. Within certain approximations, simple procedures have been devised for obtaining $\mu_i(E)$ from the reflectivity curve $R(E)$ [36, 37]. It can offer greatly enhanced surface sensitivity relative to bulk measurements, and some control over the penetration depth into the sample.

3.7 Samples

One of the virtues of XAFS is that it can be applied to nearly any material under a variety of conditions. Samples can be solids, liquids, or gases, and they can be measured under a variety of environmental conditions, using cryostats, furnaces, high pressure cells, strong magnetic fields, etc. Depending on the nature of the material and the regulatory environment of the facility, safety concerns and sample containment may require some creative experimental design.

3.7.1 Sample cells

The construction of sample cells can range from virtually nonexistent to quite elaborate. It is not possible to describe all sample cell variants here, but a few examples may be helpful.

Cells for many solutions and powder samples at room temperature and below can be rather simply constructed by drilling a hole or milling a slot several mm wide in a sheet of lucite or other appropriate plastic. Thin ($\approx$0.5–1 mil) windows of Kapton (which is fairly X-ray tolerant) can then be glued onto the faces with epoxy or other glue. Small diameter holes can be drilled through the plastic for the purpose of washing the cell, filling it, and letting out air. Powder samples generally will require filling before the second window is attached. These can be plugged with wax or by any number of other means.

If the cells have to support vacuum, for example within a cryostat, it becomes more complicated. In this case threaded, sealable plug-holes are needed. Reactive specimens may dictate the use of Teflon cells and windows, requiring the use of additional hardware to hold the windows. If the sample must be cooled to cryogenic temperatures, there must be good thermal contact with the cold finger. In this case the sample cell preferably should be made of a good thermal conductor at the temperatures of interest and low heat capacity. The ductile metal indium is suitable for fabricating seals at such temperatures.

Biological and other samples are often measured using a helium cryostat (e.g. Oxford XAS cryostat) with quick-change features, which do not require breaking vacuum to change samples. The heat exchange medium is cold He gas. Helium is effective for this purpose in comparison to other gases because of its high average speed, a consequence of its low mass.

3.7.2 Fluid samples

Fluid samples are simple in that particle size is not a concern – usually solutions are naturally homogeneous. They can make good transmission samples if the element of interest is sufficiently concentrated. Solutions also often make good fluorescence samples. Concentrations of 1 millimolar, or 100 part per million are routine, and smaller concentrations are routinely measured. However, solutions can become inhomogeneous during an experiment because of phase separation, and radiation damage/denaturation. In the presence of intense beams, e.g. those from a focussed undulator beam, photolysis of water by the beam can actually produce holes in aqueous samples. Often radiation damage is suppressed by freezing below the glass transition temperature of the solution. In biological (e.g. protein) solutions, it is often helpful to add glycerol to facilitate formation of a glassy sample, to inhibit Bragg diffraction from ice crystallites. Diffractive analyzers can be helpful in reducing artifacts from Bragg peaks, because they reject elastically scattered X-rays.

3.7.3 Particulate samples

It should be clear from the previous discussions on transmission and fluorescence that if a sample is made of particles, they should be considerably smaller than one absorption length in the material at the energy of interest. This is true in both transmission and fluorescence measurements. Electron yield is more forgiving in this respect, because only a thin layer at the surface of the particles is probed.

Before preparing samples one must calculate the absorption length of the material at an energy right above the absorption edge of interest. Then one must prepare particles that are at least several times smaller than one absorption length ($1/\mu$) in the material. Many materials require particles of micron size, which are not always easy to prepare. The particles must then be distributed uniformly over the sample cross-sectional area, by coating onto a substrate or by making a suspension or solid solution.

There are many ways to prepare fine particles. The best method to choose depends very much on the nature of the material. During sample synthesis it may be possible to choose conditions to induce the formation of small particles. It may be possible to procure some samples already in the form of sufficiently fine particles. If not, it is often possible to grind the material into smaller particles, and then to separate them using sieves, sedimentation, electrostatic precipitation, or other techniques.

This processing is generally best done in a fume hood or with a HEPA dust mask so that small particles (and possibly fumes) are not inhaled. Often the particles that are needed are as fine or finer than ordinary airborne dust. One must ensure that the sample material does not become chemically modified during grinding, due to heating. It is always wise to independently confirm (e.g. by spectroscopy or diffraction) that the samples have not been significantly altered by processing. With dilute samples one also must take care to prevent cross contamination of samples.

Samples can be ground by hand in a porcelain or agate (aluminum oxide) mortar and pestle. Inexpensive small volume ball mills (e.g. "Wig-L-Bug") with agate vials also can be used to grind samples. Care should be taken to ensure that, if steel balls are used for grinding, iron or other components of the steel do not contaminate the sample detrimentally. Frisch mills (e.g. MiniMill 2 Panalytical, Gilson Micromill) are available for several thousand dollars US.

To separate particles of different sizes, the simplest approach is to sieve the samples. One passes the ground material in series through progressively smaller sieves, obtaining the smallest particles at the end. Stainless steel mesh ASTM sieves are readily available with different diameters (A set of 3 inch diameter sieves in 100, 200, 325, 400, 500, 600 mesh sizes is generally useful). Sieve shakers and sonic sifters are also available as well. The mesh size indicates the number of wires per inch, and it is inversely related to the maximum particle size passed by the sieve. The particle size in microns is approximately $15\,000/M$ where M is the mesh size for sieves from 100–450 mesh. For 500–600 mesh the particle size is about 5 micron smaller than given by the $15\,000/M$ rule.

3.7.4 Sedimentation

Even a fine sieve such as 600 mesh only guarantees a maximum particle size of 20 microns, which often is not fine enough, although they are a very fine dust. When one needs smaller particles, it may be possible to use sedimentation. This works by suspending a distribution of different particle sizes in a fluid, and isolating small particles by using the fact that large particles fall through the fluid more quickly, leaving smaller ones in suspension. The suspended particles can then be decanted, and the fluid removed (e.g. by evaporation or filtration), or immobilized (e.g. by freezing or cross-linking). This approach assumes that no collective flow occurs among the particles in solution, so it is best to start with already sieved fine particles, and to avoid sedimenting too much material at once.

It is of course essential to ensure the fluid does not react with the particles, or dissolve them, both for reasons of science and safety. For example, sedimenting a strong oxidizer in acetone would be a very bad idea. The Material Safety Data Sheet (MSDS) for the substance should be consulted.

The drag force on a spherical particle of radius R moving at a velocity v in a fluid of viscosity η is $F = 6\pi\eta Rv$. Particles of density ρ will fall through the fluid of density ρ_0 at a speed at which the particle's weight, less the buoyancy force, equals the drag force:

$$(\rho - \rho_0)\frac{4}{3}\pi R^3 g = 6\pi\eta Rv.$$

If the height of the fluid in the container is h, the time that it would take all particles of radius R to fall to the bottom would then be:

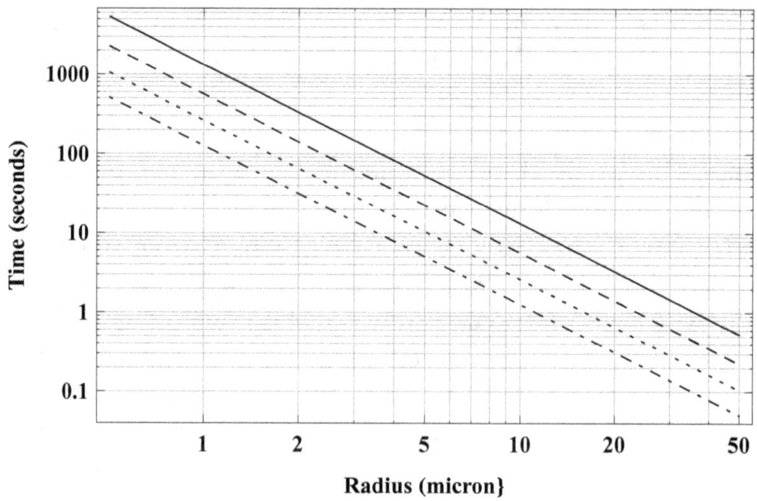

Fig. 3.28 Sedimentation time in acetone at 20 °C vs. radius and particle density, 5 cm drop height. Solid line: $\rho = 3\text{g/cm}^3$; Dashed line: $\rho = 6\text{g/cm}^3$; Dotted line: $\rho = 12\text{g/cm}^3$; Dot-Dashed line: $\rho = 24\text{g/cm}^3$.

$$t = \frac{9}{5}\frac{\eta h}{(\rho - \rho_0)gR^2}$$

To use these equations, from the known densities of the material and liquid (e.g. acetone), and the viscosity, fluid height, and required particle radius, calculate the time t. Then mix in the (pre-sieved) particles, wait the required time, decant the supernatant with a pipette, and dry the particles in a dish in a fume hood.

For example, MnO in acetone at 20 °C the viscosity of acetone is 0.0032 Poise = 0.0032 g/(cm · s). The density of acetone is $\rho_0 = 0.79$ g/cm^3; the density of MnO is $\rho = 5.4$ g/cm^3; $g = 980$ cm/s^2. If our beaker height $h = 5$ cm and we want $R = 1\mu\text{m} = 3 \cdot 10^{-4}$ cm, then $t \approx 640$ seconds, or about 11 minutes. Since $t \propto 1/R^2$, a factor of 3 smaller particle takes about an order of magnitude longer.

The sedimentation time required to obtain a chosen particle size for materials of various densities can be read off the graph in Figure 3.28. This approach can be easily generalized to plate-like or rod-like particles, assuming that one knows their orientations as they fall through the solution. It may be appropriate under some circumstances to make use of a centrifuge if the particle density is close to that of the fluid, or the viscosity of the fluid is large.

3.7.5 Assembling and checking samples

There are a number of ways to assemble samples. The goal is to maximally disperse the material in a uniform manner. In transmission mode one additionally requires that the areal density of material presented to the beam be uniform, and especially that thin regions (gaps, pinholes) be minimized. The following approaches are suitable for transmission and fluorescence.

One useful approach is to apply a uniform coating of fine particles (e.g. using a small disposable brush) to commercial adhesive tape. Try to avoid brush marks and striations. Scotch Magic Transparent tape is a very uniform product, mostly clean of transition metals, and is of low absorption at most energies of interest. Commercial Kapton tape is less suitable because of absorptive elements in the adhesive. One must always check such materials to make sure they do not have trace quantities of the elements of interest.

Often particles can be made to form something close to a packed monolayer on tape, but that is not sufficient in transmission: one should use multiple layers (typically between 2–8, by folding/stacking the coated tape) to cover gaps between particles, and to obtain the desired absorption. However, if the particles are too fine they will coat the tape's adhesive, but not provide much absorption. In that case so many layers may be needed to get the required edge step that the total absorption will be too high. In such a case, use particles that are not quite so small.

Another approach is to uniformly mix the sample with a filler/binder material. This can be done efficiently with a Wig-L-Bug or equivalent, perhaps using disposable cells to prevent cross-contamination. The filler material must be chemically nonreactive with the sample, devoid of the element of interest, non-hygroscopic, and its X-ray absorption coefficient must be small at the relevant energy range. Typical materials include boron nitride powder, activated carbon powder, polyvinyl alcohol, corn starch, or sucrose. The latter carbon-based compounds would not be suitable for oxidizers.

The well-mixed sample can then be placed into a sample cell with X-ray transparent windows (e.g. Kapton, polypropylene), or made into a pellet using a press.

A third approach is to make a paint by mixing particles with a suitable polymer binder, and applying it to substrate such as Kapton plastic film. Duco cement thinned with acetone is a suitable carrier, provided that acetone does not dissolve the substrate. It needs to be viscous enough that particles do not sediment out.

It is wise to verify that samples were not altered during their preparation using spectroscopy, diffraction, or other means. They should also be checked visually for homogeneity. An inexpensive digital microscope (e.g. Olympus Mic-D) can be useful for particle size analysis using open source software (e.g. Image/J). More sophisticated instruments such as laser scattering particle size analyzers may also be available to the experimenter.

There is one caveat to be aware of in visual inspection: the optical absorption lengths and the X-ray absorption lengths are not directly related, and may be vastly different in order of magnitude. This implies that a sample that appears inhomogeneous by visible inspection of the absorption may or may not be practically inhomogeneous for X-rays. Refraction and scattering still make particle size analysis by microscopy feasible however.

During experiments, it is always wise to measure a blank sample – one in which there should be negligible amounts of the element of interest. Many common materials have unexpected elements in them. For example, many experimenters have measured iron fluorescence spectra from aluminum sample holders without knowing it, believing it was their sample. Often there is fluorescence (induced by air-scattered X-rays) from elements in materials near the sample. Measuring a blank will determine if this is a problem.

3.7.6 Polarized XAFS and magic angle spinning

If particles naturally grow in plates or rods, there is potential for preferred orientation (or *texture*), i.e. nonrandom orientation of the particles when deposited on a substrate such as tape, or compressed into a pellet. The X-ray beams from synchrotron sources are nearly always plane-polarized, and if the sample's grains are not randomly oriented in space systematic errors can arise from having a partially oriented sample in the polarized X-ray beam.

X-ray linear dichroism (Polarized XAFS) is quite beneficial if systematically applied to oriented samples [49, 50], but the dichroism can be a nuisance or worse if it is unintentional. Fortunately magic angle spinning [57] can be used to average out the dichroism on a partially oriented sample. Perhaps surprisingly it is possible to do this at the same time as grazing incidence fluorescence.

The sample is placed on a motorized stage that is made to spin about an axis that is normal to the sample surface, as shown in Figure 3.29. This axis is oriented so that it makes an angle of $\theta^* = \arccos(1/\sqrt{3}) \approx 54.7^\circ$ with the X-ray polarization vector. The angles $\pm\theta^*$ or $\pm(180^\circ - \theta^*)$ are also suitable, as they satisfy the basic equation $3\cos^2\theta^* = 1$.

The explanation for this method can be expressed simply. The X-ray absorption transforms as a second rank tensor (in the dipole approximation), which implies that it can be written in the form $\hat{\epsilon} \cdot M \cdot \hat{\epsilon}$, where $\hat{\epsilon} = (\sin\theta\cos\phi, \sin\theta\sin\phi, \cos\theta)$ is the X-ray polarization vector expressed in spherical coordinates θ, ϕ, and M is the absorption tensor (see Chapter 4). Expanding out the sum, this becomes

$$M_{11}\sin^2\theta\cos^2\phi + M_{22}\sin^2\theta\sin^2\phi + M_{33}\cos^2\theta$$
$$+\,2(M_{12}\sin^2\theta\sin\phi\cos\phi + M_{31}\sin\theta\cos\theta\cos\phi + M_{23}\sin\theta\cos\theta\sin\phi).$$

A spherical average over 4π solid angle is accomplished by multiplying this expression by $\sin\theta$ and integrating θ from 0 to π, and ϕ from 0 to 2π, yielding a quantity proportional to $M_{11} + M_{22} + M_{33}$, the trace of M. This corresponds to taking the isotropic spherical averaged spectrum, as one would get in a solution, or a randomly oriented polycrystalline sample.

We now choose our coordinate system so that the sample normal is the polar axis $\theta = 0$. Then θ is the angle between $\hat{\epsilon}$ and the polar axis. If we rotate the sample about the polar axis (sample normal) and rotationally average, it is equivalent to averaging over the azimuthal angle ϕ of the polarization vector.[11] Integrating over ϕ then gives (within a constant factor)

$$(M_{11} + M_{22})(\sin^2\theta) + M_{33}(2\cos^2\theta).$$

We wish to eliminate the effect of orientation so that the result is proportional to the isotropic average $M_{11} + M_{22} + M_{33}$. This will occur at the value of θ at which the factors $\sin^2\theta$ and $2\cos^2\theta$ become equal. This is the "magic angle" θ^*, which satisfies the condition $3\cos^2\theta^* = 1$.

3.8 Data acquisition

3.8.1 Scan setup

Experimental XAFS data are often obtained on a nonuniform energy grid in step-scan mode or continuous scan mode. A typical scan setup would be 10 eV spacing from −100 eV to −20 eV (to give a good pre-edge background trend line for normalization); 0.5 eV from −20 to 20 (to adequately sample near-edge structure); uniform in k $\delta k = 0.05$ Å^{-1} from that point up to the $k = 14$ (to adequately sample the data in k-space). The time per point also typically will be increased with k to improve signal to noise ratio at high k.

The energy grid in the region spanning the edge should be sufficient to resolve the features there. Core-hole broadening and monochromator resolution conspire to broaden the features. It is not worthwhile to choose

[11] The same result is obtained if one averages the expression $\hat{\epsilon} \cdot (R(\alpha) \cdot M \cdot R^{-1}(\alpha)) \cdot \hat{\epsilon}$ over α from 0 to 2π, where R is the rotation matrix about the polar axis (sample normal).

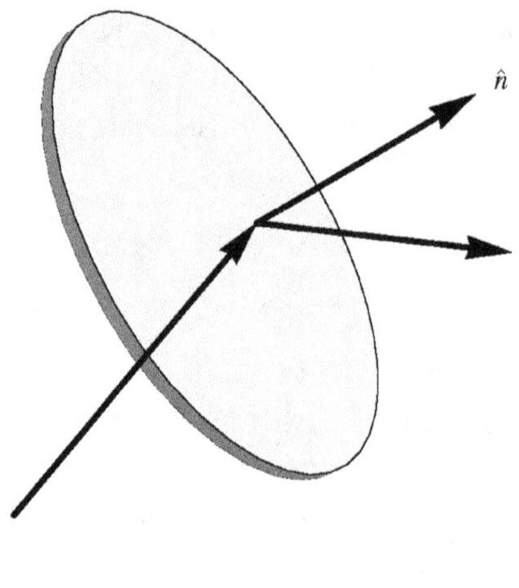

Fig. 3.29 Magic angle spinning geometry: $\vec{k}$ is the incident beam direction; $\hat{\epsilon}$ is the x-ray polarization; $\hat{n}$ is the normal vector about which the sample spins. The angle between $\hat{\epsilon}$ and $\hat{n}$ should be $\theta^* \approx 54.7°$ to average out the dichroism.

an energy spacing δE that is smaller than about 1/3 of the core-hole level width. An exception would be if one intends to remove the core hole or instrumental broadening via a deconvolution procedure, which is feasible if the noise level is low enough.

3.8.2 k and E sampling

By the Nyquist criterion of Fourier analysis, the data must be sampled (at a bare minimum) twice in every oscillation of the highest frequency contained in the data. If this isn't done, high-frequency data will appear at a lower frequency than it really is. When acquiring EXAFS data it is wise to sample at several times that minimum frequency, so that high-frequency noise does not alias into the frequency range of the data: $2\Delta k R_{\max} = 2\pi$. If $R_{\max} \approx 10$ Å, this gives a minimum $\delta k \approx 0.3\text{Å}^{-1}$. Oversampling in the EXAFS part of the spectrum by a factor of 4 to 6 is reasonable. Experience has shown that a k grid of 0.05 Å^{-1} is generally sufficient.

If the data acquisition program does not allow uniform sampling in k-space, it must be approximated by the experimenter using a series of regions in E-space, each of which has a uniform δE. Since $E - E_0 = 0.2625k^2$ in eV Å units, or $(E - E_0) = 3.81k^2$, we see $\delta E = 3.81 \cdot 2 \cdot k\delta k$. If we require $\delta k = 0.05\text{Å}^{-1}$ this implies $\delta E \approx 0.2\sqrt{E - E_0}$. The maximum energy grid

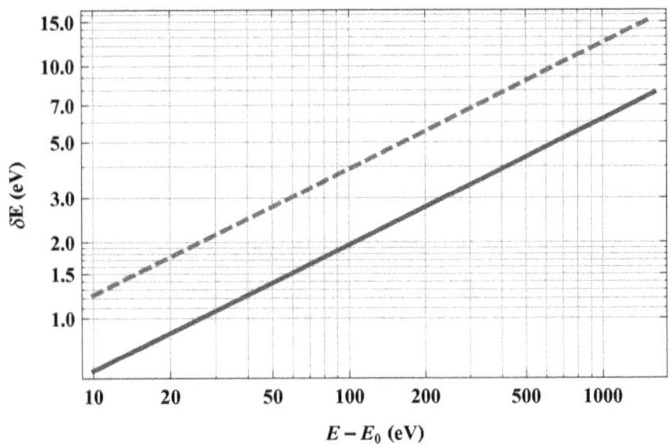

Fig. 3.30 Maximum energy grid to use for EXAFS data acquisition, as a function of energy above the edge. The solid line is for a desired k grid $\delta k = 0.05\text{Å}^{-1}$; the dashed line is for $\delta k = 0.1\text{Å}^{-1}$.

spacing that should be used for data acquisition is plotted as a function of energy above the edge in Figure 3.30.

In some cases data acquisition systems (e.g. continuous scanning) may record a series of energy points on a more or less even spacing in energy or monochromator angle. This is generally fine, but in the early stages of data reduction, the data will need to be re-binned into k-space to get the proper statistical weight from the data. This is described in Chapter 5.

4
Theory

4.1 Introduction

XAFS theory has advanced significantly over the last several decades, driven by creative theorists, experimental demand, and Moore's law. It continues to evolve at a rapid rate. XAFS theory has benefited from and contributed to advances in many-body physics, quantum field theory, and scattering theory. Some current areas of focus include work to improve the accuracy of computing non-spherical molecular potentials, vibrational effects, and multilelectron excitations. Efforts to improve parallelization of codes to take better advantage of multiple processor cores, multiple processors, graphics processing units, computational grids, and cloud computing are also under development.

Theoretical advances have transformed the practice of data interpretation and analysis. Computer programs are now readily available for the calculation of X-ray Absorption spectra with very good accuracy in the EXAFS region, and useful accuracy in the XANES. A number of complementary theoretical approaches and the computer programs that implement them are widely disseminated and used, among them EXCURV, FEFF, GNXAS, MXAN, and FDMNES.

There is little fundamental disagreement about the correct basic physics underlying all these approaches. They differ principally in the sets of approximations that are used to make possible efficient calculation of the spectra, and in the computational algorithms that are employed. Recent progress has shown that one theoretical approach, specifically the Real Space Multiple Scattering (RSMS) formalism, is useful also for calculating X-ray emission, X-ray magnetic circular dichroism, X-ray elastic and inelastic scattering/X-ray Raman Scattering, and dielectric response functions.

Although, in the near-edge region, it is often still intuitive and useful to think about spectra in terms of wavefunctions, it has become clear that the complementary description in terms of scattering theory provides the most general unified framework for bound and continuum states, particularly because of its ability to describe many-electron states in terms of quasi-particles. Other complementary approaches, such as multiplet theory and molecular orbital theory, continue to prove useful for describing near-edge features.

This chapter is intended to give a brief practical introduction for experimentalists to the theory of XAFS, from several complementary points of view. XAFS theory is advancing rapidly, and a highly detailed and mathematically sophisticated development would be beyond the scope of this book, both in length and prerequisite knowledge. For further reading the following review articles are recommended, as are the references cited by them: the review article by Rehr and Albers [64]; the paper by Fujikawa [67]; the recent book by de Groot and Kotani [68]; review articles in the Coordination Chemistry Reviews issue on Synchrotron Radiation, in particular the one by Rehr and Ankudinov [62]. Earlier review articles remain quite useful: the Handbook article by Stern and Heald [61]; and the reviews by Stern (on EXAFS), Durham (on XANES), and Stöhr (on SEXAFS) in the book edited by Koningsberger and Prins [10].

4.2 Transition rate

XAFS is intrinsically a quantum mechanical phenomenon. For simplicity in the following treatment we will approximate the X-ray electromagnetic wave as a classical field, which will be sufficient for our purposes, although in reality it is quantized. We will also use the nonrelativistic approximation unless stated otherwise, despite the fact that relativistic effects are important, especially for heavy atoms. The electronic structure can be determined by solving the Schrödinger equation, or its counterpart in the relativistic quantum theory of the electron, the Dirac equation. A modern approach to XAFS theory has made it mostly unnecessary to calculate the wavefunctions, however – the spectrum is calculated directly from the Green's function in the real space representation.

We are interested in the transition rate between a core level (e.g. $1s$) and final states that is induced by a weak time-dependent perturbation such as an electromagnetic wave. Time-dependent perturbation theory [59] tells us that the transition rate, to first order in the perturbation, is proportional

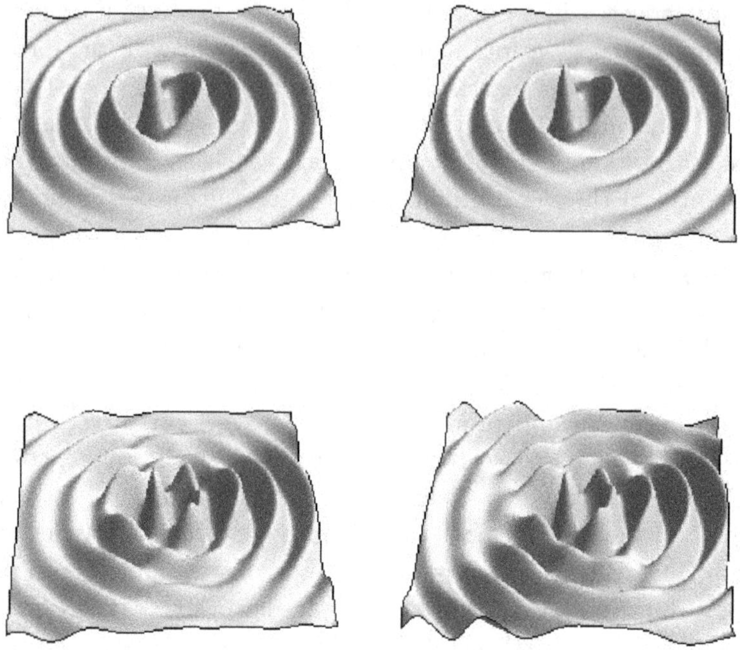

Fig. 4.1 Stereo pairs of p-symmetry final state wavefunction, polarized in x-direction, with and without backscatterers. Top: No backscattering atoms. Bottom: with two backscattering atoms in the x direction.

to the squared modulus of the transition amplitude (matrix element)[1]

$$\frac{2\pi}{\hbar}| < \psi_f|H^{'}|\psi_i > |^2\rho, \tag{4.1}$$

where $|\psi_i>$ and $|\psi_f>$ are respectively the initial and final electron states, of energy E_f and E_i, ρ is the density of states, and H' is the interaction Hamiltonian between the electromagnetic field and the electrons.

In the case of excitation by electromagnetic waves, to first order in the field (which, in quantized field picture, corresponds to single photon interactions) $H' \propto \vec{A} \cdot \vec{p}$, where $\vec{A}$ is the vector potential of the wave and $\vec{p}$ is the momentum operator of the electron. Using commutation relations with the Hamiltonian this can be written as

$$\mu \propto \left| < \psi_f|\hat{\epsilon} \cdot \vec{r}\, e^{i\vec{\kappa}\cdot\vec{r}}|\psi_i > \right|^2 \rho(E_f - E_i - \hbar\omega), \tag{4.2}$$

and $\hat{\epsilon}$, $\hbar\vec{\kappa}$, and $\hbar\omega$ are the X-ray's electric polarization vector, momentum, and energy.

[1] Enrico Fermi found Dirac's expression so useful that he called it [60] his "golden rule #2". Now nearly everyone calls it "Fermi's golden rule."

4.2.1 Digression: optical spectra

The expression for the transition rate (Equation (4.2)) is the same as what is used to calculate optical spectra, and it is reasonable to ask why the interpretation of XAFS spectra appears qualitatively so different from optical spectroscopy. First, it must be said that interpretation of the localized bound states in the XANES of molecular systems often can be done in a manner similar to optical spectra of molecules. At higher energies this approach breaks down.

A key difference lies in the nature of the initial and final states. First, while the initial state in XAFS is a bound state, above the absorption edge the final states in XAFS are continuum states of large kinetic energy (as illustrated in Figure 4.1), and there remains a core hole (core-level vacancy) with which the photoelectron interacts. Below the absorption edge there may be allowed dipole or quadrupole transitions into unfilled states of appropriate symmetry, which gives localized peaks; often there are not. In contrast, in the optical spectroscopy of molecular systems, the initial and final states usually can be well described by bound state molecular orbitals, which are classified according to the manner in which they behave when subjected to the symmetry operations of the molecule. The initial and final state wavefunctions in optical spectra generally are spread out over the molecule, and they have complex shapes. In contrast, in X-ray absorption spectra, the initial state ($1s$ state for K-edges, $2s$ or $2p$ states for L-edges) is very simple in symmetry, and it is highly localized near the center of the absorbing atom. Therefore the XAFS spectra only depend on the value of the final state wavefunction in the small region where the initial state is localized.

4.2.2 Dipole approximation

The exponential in the matrix element in Equation (4.2) can be expanded as

$$< \psi_f|\hat{\epsilon}\cdot\vec{r}e^{i\vec{\kappa}\cdot\vec{r}}|\psi_i > \approx < \psi_f|\hat{\epsilon}\cdot\vec{r}|\psi_i > +i < \psi_f|(\hat{\epsilon}\cdot\vec{r})(\vec{\kappa}\cdot\vec{r})|\psi_i > +\cdots. \tag{4.3}$$

The first term is the dipole term; the second term is the quadrupole term. We wish to estimate their ratio, which is of order of $\langle \kappa r \rangle$, where the bracket indicates the average over the size of the initial state wavefunction. The initial $1S$ state sees the (mostly) unscreened nuclear charge; we will roughly approximate the core states of an atom of atomic number Z as that of a hydrogenic atom of nuclear charge Ze but with a single electron. In that case we simply replace $e^2 \to Ze^2$ in the expressions for hydrogen, which implies that the energies scale as Z^2 and the characteristic size of the wavefunctions scale as $1/Z$. The ground-state energy of hydrogen is $E = -(1/2)\alpha^2 mc^2 = -(1/2)me^4/\hbar^2$ where $\alpha = e^2/\hbar c \approx 1/137$ is the fine structure constant. The X-ray wave vector then is related to the K-edge energy $\hbar\kappa c \approx (1/2)Z^2(e^2/a_0)$ or $< \kappa r > \approx ka_0/Z \approx Ze^2/2\hbar c \approx Z\alpha/2 \approx Z/(2\cdot 137)$. Therefore the ratio of the quadrupole to dipole matrix elements for Fe, $Z = 26$, is about 10%, and the ratio of the transition probability is about 1%, for K-edges.

This shows that dipole approximation is good, but that quadrupole transitions are not always negligible. They can be readily observed [76] in some situations when not masked by allowed dipole transitions, for example in the pre-edge region of transition metal complexes. Our simple estimate above also indicates the quadrupole/dipole ratio grows quadratically with Z, so that it can be expected to be substantial for heavier atoms, and for L-edges because of size of the initial state wavefunction. Multipoles of higher order than quadrupole are usually negligible.

4.2.3 Orientation dependence

Summing over final states of different energies, in the dipole approximation we can rewrite the expression for the absorption probability as $\mu \propto \hat{\epsilon}\cdot M\cdot\hat{\epsilon} = \sum_{j,k} M_{jk}\epsilon_j\epsilon_k$, where the tensor M has components

$$M_{jk} \propto \sum_f < \psi_i|r_j|\psi_f >< \psi_f|r_k|\psi_i >, \tag{4.4}$$

and $\hat{\epsilon} = (\sin\theta\cos\phi, \sin\theta\sin\phi, \cos\theta)$ is the X-ray polarization vector expressed in spherical coordinates θ, ϕ. Expanding, this is

$$\begin{aligned} M_{11}\sin^2\theta\cos^2\phi + M_{22}\sin^2\theta\sin^2\phi + M_{33}\cos^2\theta \\ + 2(M_{12}\sin^2\theta\sin\phi\cos\phi + M_{31}\sin\theta\cos\theta\cos\phi + M_{23}\sin\theta\cos\theta\sin\phi), \end{aligned}$$

where we may take $M_{jk} = M_{kj}$ without loss of generality, because the double sum is symmetric.

Although the tensor can be diagonalized by choosing an appropriate coordinate system at one energy, it is not generally possible to do this at all energies. However, when the sample itself has certain symmetries the form simplifies considerably. An important special case is if the sample is isotropic (e.g. solution or randomly oriented polycrystalline, or is of cubic symmetry) the absorption tensor is proportional to the identity matrix – it may as well be taken as a scalar.

For example, suppose the sample has an N-fold rotational axis along the z

direction, i.e. it is unchanged by rotations of $2\pi/N$. The matrix for rotation by an angle α about the z axis is

$$R(\alpha) = \begin{pmatrix} \cos(\alpha) & -\sin(\alpha) & 0 \\ \sin(\alpha) & \cos(\alpha) & 0 \\ 0 & 0 & 1 \end{pmatrix}.$$

A vector $\vec{v}$ is rotated by applying the rotation matrix to it: $\vec{v}' = R \cdot \vec{v}$. A second rank tensor $\mathbf{T}$ is rotated by applying R and its inverse R^{-1}: $\mathbf{T}' = R \cdot \mathbf{T} \cdot R^{-1}$; the inverse matrix equals the transpose $R^{-1} = R^{\mathrm{T}}$.[2]

Rotating the tensor (which corresponds to rotating the sample) through an angle $2\pi/N$ and equating it to the unrotated tensor $M = R \cdot M \cdot R^{-1}$ shows that for $N > 2$ the tensor is diagonal, and $M_{11} = M_{22}$. This implies that $\mu \propto M_{11} \sin^2 \theta + M_{33} \cos^2 \theta$: the ϕ dependence vanishes. A two-fold rotation axis gives the conditions $M_{23} = M_{32} = 0$, $M_{13} = M_{31} = 0$. If the sample has cubic symmetry, so that x, y, and z are all equivalent (so that rotations by $\pi/2$ about each axis make no difference), these considerations show that the absorption tensor is proportional to the identity matrix: the absorption is isotropic, independent of orientation.

Now suppose that we have a pair of atoms aligned along the z axis, and we excite them with X-rays. Evidently we can expect both $\cos^2 \theta$ and $\sin^2 \theta$ angle dependencies, where θ is the angle between the polarization vector and the vector connecting the atoms. Calculation shows that in K-edges the $\sin^2 \theta$ contribution dies off rapidly with energy above the edge, leaving only the $\cos^2 \theta$ dependence. However, it must be remembered that the $\sin^2 \theta$ term is generally present in the near-edge region, as it is for L-edges [49, 61].

For an analysis of the orientation dependence of XAFS spectra in terms of spherical tensors see the paper by Brouder [78]. Additional information and references on polarized XAFS can be found in Chapter 3.

[2] The tensor transformation rule can be inferred from the requirement that the scalar product be invariant under rotations $v' \cdot \mathbf{T}' \cdot v' = (R \cdot v)^{\mathrm{T}} \cdot (R \cdot \mathbf{T} \cdot R^{-1}) \cdot R \cdot v = v \cdot R^{-1} \cdot R \cdot \mathbf{T} \cdot R^{-1} \cdot R \cdot v = v \cdot \mathbf{T} \cdot v$.

The following simple Mathematica 7 program calculates the form of the absorption tensor when the system has n-fold rotational symmetry about the z axis. It should be easy for the reader to modify it for rotations about other axes, or to include other symmetry operations.

Mathematica 7 code to calculate the form of μ tensor with n-fold axis

```
RotMu[n_]:=(R[alpha_]:=RotationMatrix[alpha,{0,0,1}];m=.;M=.;
m=Array[M,{3,3}]/.{M[2,1]->M[1,2],M[3,2]->M[2,3],M[1,3]->M[3,1]};
soln=Solve[m==R[2 Pi/n].m.Transpose[R[2 Pi/n]]][[1]];
eps={Sin[\[Theta]] Cos[\[Phi]],Sin[\[Theta]] Sin[\[Phi]],Cos[\[Theta]]};
nice={M_[i_,j_]->Subscript[M,i,j]}; (*let nice={ } to disable subscripts *)
Return[Column[{"n="<>ToString[n],(eps.(m/.soln/.nice).eps)//Simplify,
(m/.soln/.nice)//MatrixForm," "}]])
(*the following evaluates the function Rotmu for 1,2,3,4,5,and 6-fold z-axis*)
ColumnForm[RotMu/@{1,2,3,4,5,6}]
```

4.2.4 Selection rules

The core state from which the electron is removed is deeply bound, and the dominant interaction the electron has is with the nucleus and electrons of the central (photoabsorbing) atom. As a consequence, the initial state is well approximated as an atomic-like state of angular momentum quantum numbers l and m. The transition operator in the dipole approximation is proportional to the vector $\vec{r}$. So far, things are simple. In contrast, the spatial dependence of the final-state wavefunction is complicated, because it reflects the environment of the central atom.

The final-state wavefunction can be expressed as a linear combination of functions with specific l, m, i.e. spherical harmonics *centered on the absorbing atom*, which are eigenfunctions of the operators L^2 and L_z. Because of the orthogonality of spherical harmonics, the transition matrix element projects out only the part of the final-state wavefunction that has the appropriate symmetry for an allowed transition. Electric dipole selection rules tell us that $\Delta l = \pm 1$, and quadrupole selection rules $\Delta l = \pm 2, 0$. For example, excitation of the $1s$ level ($l = 0$) only has nonzero dipole transition strength for transitions to the p-symmetry ($l = 1$) part of the final state wavefunction. The transition operator $x \pm iy$ causes transitions from $l = 0$, $m = 0$ to $l = 1$, $m = \pm 1$, while the transition operator z causes transitions to $l = 1$, $m = 0$.

Dipole selection rules therefore imply that for K- and L_1-edges, allowed transitions are to states containing p symmetry. For L_2- and L_3-edges, allowed transitions are to final states containing d and s symmetry. In applying these rules, it is important to realize that angular momenta are always referred to a specific origin. A p-state centered on a neighboring atom is not a p-state with respect to the central atom, although it can be expressed as a linear combination of different angular momentum states with respect to the central atom.

4.2.5 Symmetry

Group theory offers a general framework to describe symmetries and selection rules. Much of spectroscopy uses the methods and terminology of group theory, and some familiarity with it is helpful when reading the XAFS literature. A number of accessible texts on group theory and spectroscopy are available [80, 79].

The central atom's environment generally will have some set of symmetries, i.e. operations that leave the environment invariant. For example, in a square planar environment, rotations by $N\pi/2$ with integer N about the normal vector leave the system unchanged. These symmetry operations also are symmetries of the Hamiltonian. The set of symmetry operations will be closed, in the sense that carrying out any pair of them is equivalent to performing another one of the operations in the set. Each operation also can be inverted (reversed); they are associative; and there is an operation that does nothing at all (identity). These properties of the set of symmetry operations constitute the formal properties of a mathematical group.

The theory of group representations is used for the systematic study of symmetry, and group-theoretical notation and language permeates much of spectroscopy and diffraction. Spectroscopy tends to be concerned with operations about a point: rotations, inversions, reflections. Diffraction, being all about spatial periodicity, is concerned with space groups, which has translational symmetries in addition to point group operations.

The wavefunctions can be classified according to their behavior under the operations of the Hamiltonian's symmetry group. In general, when a symmetry operation is carried out, a wavefunction will be transformed into linear combination of other wavefunctions. For example, in an octahedral or tetrahedral environment, the d_{xy}, d_{yz}, d_{zx} orbitals of a transition metal transform amongst themselves under all of the operations in the group, but they do not mix (within an eigenstate) with the $d_{\mathrm{x}^2-\mathrm{y}^2}$ and d_{z^2} orbitals. They are said to transform according to (or "belong to", or more correctly "form a basis for") different irreducible representations. Another example is the fact that, in a spherically symmetric potential, the orbitals of different l do not mix with each other under rotation of the coordinate system. However, the various wavefunctions with the same l but different m do mix.

A p_z orbital when rotated may be expressed as a linear combination of p_x and p_y.

The product of two or more quantities, each of which belong to particular irreducible representations, will in general not belong to a single irreducible representation. However, it will generally be possible to reduce the product to a linear combination of quantities, each of which transform according to different irreducible representations. For example, a second rank tensor such as the absorption tensor can be constructed as the outer product of two vectors $\vec{a}$ and $\vec{b}$ with components a_i and b_j: $a_i b_j$. This tensor can be decomposed into a sum of three terms, each of which transform differently under rotations:

$$a_i b_j = \frac{\vec{a} \cdot \vec{b}}{3} \delta_{ij} + \frac{a_i b_j - a_j b_i}{2} + \left(\frac{a_i b_j + a_j b_i}{2} - \frac{\vec{a} \cdot \vec{b}}{3} \delta_{ij} \right).$$

The first term is independent of orientation; it transforms like a scalar. The antisymmetric second term has three independent elements that transform like the components of a vector. The third term transforms as a traceless symmetric tensor, with five components. These behave under rotations respectively like s (1 component), p (3 components), and d (5 components) orbitals, i.e. $l = 0$, $l = 1$, and $l = 2$. Altogether there are $1 + 3 + 5 = 9$ elements, as there must be for a 3×3 matrix.

These spherical tensors have important applications. The Wigner–Eckart theorem [72, 73] states that the matrix elements of spherical tensor operators of a given l depend on the azimuthal quantum number m in exactly the same way: they are proportional to a Clebsch Gordan coefficient, which are related to integrals over three spherical harmonics. This implies that for a given l, matrix elements only need to be calculated for one value of m; all the rest are proportional to it. A general theorem tells us that the matrix element will be zero unless the product is reducible to a quantity that contains the totally symmetric representation of the group.

Consider the case of K-edge X-ray absorption. The initial and final state wavefunctions and the dipole transition operator (essentially $\vec{r}$, in the dipole approximation) contained in the transition matrix element each transform in specific ways under the operations of the Hamiltonian's symmetry group. The transition dipole transforms like an object with angular momentum $l = 1$, so it can couple the initial s ($l = 0$) state to final p ($l = 1$) states, but no others.

4.2.6 Multiplets

A standard and very useful way to find stationary states (eigenstates) of well-defined energy is to use variational methods. A simple theorem in quantum mechanics states that the expectation value of the energy $< \psi|H|\psi >$ in a state is minimum if that state is an eigenstate of the Hamiltonian H. By using the true Hamiltonian, one can put in trial wavefunctions for ψ with undetermined parameters, and by minimizing the energy with respect to the parameters, find the best approximation to the energy eigenvalue, and the eigenfunction of H. The greater the flexibility of the function, the better one can approximate the true values. This is a standard way to find wavefunctions in chemistry, by expanding molecular wavefunctions as a linear combination of atomic orbitals centered on the various atoms in the molecule. The trial wavefunctions should build in antisymmetry with respect to interchange of electrons, for example by using Slater determinants.

In quantum mechanics anything that might happen affects what does happen. We tend to describe atomic configurations within atoms as if we knew the precise occupancy of each orbital, but this is only an approximation. In our variational wavefunction we really ought to include different atomic configurations (i.e. different patterns of occupancy of the various atomic orbitals) to get the most accurate eigenstates. The various configurations turn out to be important when the central atom suddenly finds itself with a core hole after an X-ray absorption event. As you might guess, "multiplet effects" involve calculating matrix elements between states on the same atom, and some of these are large. Generally when several interacting orbitals are highly localized in the same region of space it may be necessary to consider using the multiplet approach. This is often the case when considering pre-edge transitions in XAFS. This approach is described in the references [63, 68].

4.3 Band structure, molecular orbital, clusters

There are many different complementary approaches to calculating electronic structure, and a number of good books are available on the subject [69]. Here we briefly describe some that have been used to calculate XAFS.

When individual atoms are placed in close proximity to each other, their orbitals interact and energy levels split apart. If there are many atoms, the levels spread out to form a quasi-continuous energy band. The more strongly the orbitals interact – which depends on the orbitals' spatial overlap and the closeness of their energies – the greater the bandwidth. If there are only a few atoms, and only low energies near the Fermi level are considered, the electronic structure can be described in terms of molecular orbitals using linear combinations of localized states, such as atomic orbitals (LCAO) that are centered on different atoms. Each such molecular orbital must transform according to a particular irreducible representation of the symmetry group of the Hamiltonian, and the MOs are classified on that basis.

The older and simpler approaches of crystal field and ligand field theory still can be useful for understanding the effects of symmetry on the energy states. In these approaches, one views the central atom as subject to a perturbing potential with the symmetry of the atom's environment. This perturbing potential removes the degeneracy found in the spherically symmetric isolated atom. The energy splittings are parameterized and used for fitting experimental data.

Description in terms of electron multiplets [63, 68] has proven to be another very useful approach when states are highly localized, partially filled, and electron–electron interactions are strong. None of these localized orbital approaches work well at high photoelectron energies, however, because the electron final state wavefunction extends over many atoms, and it would require very diffuse wavefunctions in the basis set.

Yet another useful approach [71] is to use the recursion method of Haydock [69] to successively build in interactions of the central atom with increasingly remote atoms. This efficient method is particularly useful for modeling clusters.

An alternative to the localized cluster approach is to treat the wavefunctions as periodic functions that are modulated by the periodic potentials of the atoms. In a crystal the potential experienced by an electron is periodic because the atoms are situated in regular arrays. The periodicity imposes conditions on the nature of the energy levels, and they can be parameterized in terms of a wave vector $\vec{k}$, so that the energy bands depend on direction in k-space. These methods of which there are far too many variants to describe here are collectively called band-structure theory. An important version of band-structure theory is the KKR [69, 70] method, which treats the interactions between the electrons and atoms as a multiple scattering process.

KKR is an ancestor to current multiple scattering methods, which, like KKR, use an angular momentum expansion and Green's functions, but they do not impose translational symmetry. A great virtue of the scattering approach is that it handles extended continuum states at large positive energies as well as localized states (even bound states at negative energy), and scattering resonances, all on the same unified theoretical basis.

As described in Chapter 2, a useful approximation is to treat the behavior of electrons in a many-electron system as that of a single electron moving in the average field of the other electrons. Because of their fermionic quantum statistics, which are the origin of the Pauli Exclusion principle, the electrons have an effective interaction that is distinct from their ordinary coulomb repulsion. It can be accounted for by including a term in the ground state Hamiltonian called an exchange-correlation potential. The Kohn–Sham theorem tells us that the ground-state energy of any system is given by a universal functional[3] of the particle density, but the exact form of this functional is unknown. In materials various approximations are commonly used that depend on the electron density and its gradients. With this approach (Density Functional Theory) the ground-state energy levels can be calculated with good accuracy in many cases.

4.3.1 Electron mean free path

Most of these methods are used primarily for calculating the ground state electronic structure, and perhaps low-lying excited states. However, one essential requirement for XAFS calculations is to account for the interaction of the photoelectron with the core hole that is produced in the photoexcitation process. This makes the central atom potential special, breaking the translational symmetry within a crystal, and therefore methods that assume strict translational symmetry in the system are not suitable for such calculations.

Another key ingredient for accurate calculations at high energies is to account for the fact that the photoelectron interacts dynamically with the gas of other electrons in the material, and its behavior is substantially modified by these many-body effects. The photoelectron together with the others it interacts with collectively act as a "quasiparticle," with behavior that differs from free electron behavior. These interactions are accounted for using an energy-dependent complex function called the self-energy. The use of a complex potential implies that the probability is not conserved [72]. Such a potential is sometimes called an "optical potential," in analogy to the way the propagation of light in absorbing media is modeled using a complex index of refraction. The "leaking away" of probability from the wavefunction effectively introduces a mean free path into the formalism, which means a loss of coherence as the electron propagates through the material. Probability is lost from the main single electron (quasiparticle) channel into a large number of multielectron channels that need not be explicitly calculated.

[3] A function*al* is a quantity that depends on the global properties of a function, not just its value at each point. For example, the integral of the squared derivative of a function is a functional.

The electron mean free path $\lambda(k)$ is a key ingredient in the success of XAFS theory. The effect limits the range of the photoelectron, so the spectra are sensitive only to the short range order in the sample. The electron mean free path is energy dependent, of magnitude $\approx$ 5–10 Å, which limits the sensitivity of XAFS to regions of about nanometer scale or less.

The electron mean free path emerges naturally from the many-body nonequilibrium Green's function formalism, and the real space multiple scattering expansion.

4.4 Multiple scattering expansion

The golden rule expression Equation (4.2) for the transition rate in the dipole approximation can be recast as

$$\mu(E) \propto \sum_f \langle i|\hat{\epsilon}\cdot\vec{r}'|f\rangle\, \delta(E-E_f)\, \langle f|\hat{\epsilon}\cdot\vec{r}|i\rangle.$$

The sum over final states and the energy-conserving delta function together can be compactly written in terms of the Green's operator $G = (H - E + i\eta)^{-1}$,[4] and $(-1/\pi)\mathrm{Im}\, G = \sum_f |f > \delta(E - E_f) < f|$. From this it follows that the density of states $n(E) = (-1/\pi)\mathrm{Im}\,\mathrm{Tr}\,(G)$, where Tr is the trace, which is independent of representation. Inserting this form into the matrix element we have

$$\mu(E) \propto (-1/\pi)\mathrm{Im}\,\langle i|\hat{\epsilon}\cdot\vec{r}'\; G\; \hat{\epsilon}\cdot\vec{r}|i\rangle.$$

In the position representation G can be expressed in terms of the wavefunctions ψ_n of energy E_n,

$$G(\vec{r},\vec{r}';E) = \sum_n \frac{\psi_n(\vec{r}')\psi_n^*(\vec{r})}{E - E_n + i\eta}.$$

The wavefunctions ψ_n are solutions to the equation

$$h'\psi_n = \left(\frac{p^2}{2m} + V'_{\mathrm{coul}} + \Sigma(E)\right)\psi_n = E\psi_n.$$

Although this equation resembles a time-independent Schrödinger equation, it is actually a Dyson equation that allows one to solve for the quasiparticle states at energy E. The operator h' is the non-Hermitian final

[4] η is a small positive quantity; the limit is to be taken as it approaches zero.

state Hamiltonian. The non-hermiticity implies the eigenvalues are complex, which is responsible for the finite lifetime of the final state. The many-body effects are approximated via the complex self-energy term $\Sigma(E)$, and a screened central atom core hole is included in Coulomb term $V_{\text{coul}}^{'}$.

Multiple scattering theory separates the potential into localized contributions from each atom at positions $\vec{R}$:

$$V_{\text{coul}}^{'} + \Sigma(E) = \sum_{\vec{R}} v_{\vec{R}}(\vec{r} - \vec{R}).$$

Most multiple scattering approaches currently employ the "muffin-tin" approximation, in which the atomic potentials are approximated as spherically symmetric out to a finite radius, and the interstitial potential between the atomic sphere is approximated as a constant. This approximation tends to work well for close-packed structures, but is less successful with open structures. A muffin-tin potential is sketched in Figure 4.2. Typically the muffin-tin radii are chosen so that the charge inside equals that of the atom (Norman criterion). In practice the muffin-tin radii often are scaled up so that they overlap slightly, which experience has shown partially makes up for deficiencies in the muffin tin approximation.

In the EXAFS region, the kinetic energy of the electron is large enough that small variations in the interstitial potential do not make much difference, and the muffin-tin approximation works well. In the near-edge (XANES) region the muffin-tin approximation is less satisfactory for quantitative analysis. Alternate approaches ("full potential" methods) have been developed for the XANES region that avoid the muffin-tin approximation altogether, such as the finite difference approach embodied in FDMNES [75]. Extending real-space multiple-scattering approaches beyond the muffin-tin approximation currently is an area of active development.

The Green's function separates into contributions from the central atom and multiple-scattering contributions from the neighboring atoms, so that we can write $\mu = \mu_0(1 + \chi)$, where μ_0 is the contribution from the central atom, and the scattering effects from other atoms are encoded in χ.

The scattering portion of the Green's function is proportional to $(1 - G^0T)^{-1}G^0$ where G^0 is the free particle Green's function, and the T is the scattering operator, which accounts for all the serial scattering processes off each atom. Writing it in this way permits a decomposition into scattering processes from different atoms. In an angular momentum basis (which is what the muffin-tin potential does for us), this equation can be solved

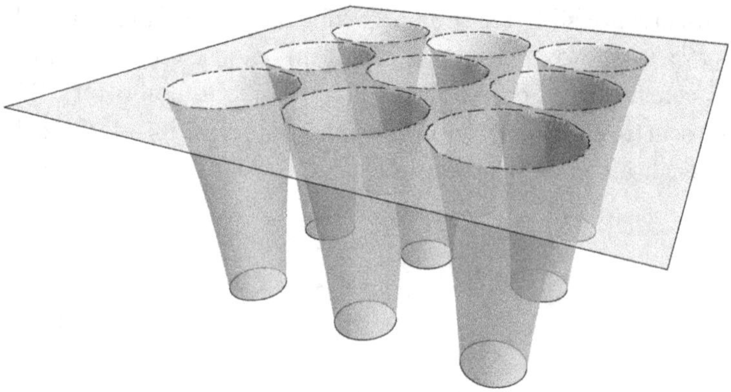

Fig. 4.2 Sketch of a "muffin-tin" potential.

by matrix inversion, or it can be expanded[5] in a multiple scattering series $G = G_0 + G_0TG_0 + G_0TG_0TG_0 \cdots$, iff the series converges. Theory and experiment have shown that the series is generally convergent in the EXAFS region, but it may or may not be convergent in the near-edge region, in which case direct matrix inversion must be done.

The form of this Green's function expansion implies that χ can be expanded as a sum of terms, each of which corresponds to a scattering path Γ that begins and ends at the central atom, and involves scattering one or more times from the atoms in the vicinity. The repeated contribution of the free particle propagator G^0 (which has the form of a spherical wave e^{ikr}/r), in the expansion implies that the longer scattering paths oscillate more rapidly vs. k, and they decrease as $1/r$ along each leg of the path.

4.5 Outline of theoretical calculation

In this section we briefly describe the calculations that are carried in a theoretical multiple-scattering XAFS program, using FEFF8.4 as an example. Most but not all of these steps are used in other scattering-based XAFS theory programs.

The starting point of the calculation is to calculate the electronic structure

[5] This can be inferred from the operator identity $A^{-1} = B^{-1} + B^{-1}(B - A)A^{-1}$ which can be easily verified by multiplying both sides by A from the left and from the right.

of the atoms of the material, as isolated free atoms. FEFF8.4 does this by solving the Dirac–Fock equation, which is a relativistic cousin of the Schrödinger equation as applied to atoms. The atomic charge densities are superimposed in the desired geometry, and averaged to create a muffin-tin potential with a prescribed degree of overlap. Hedin–Lundqvist self-energy terms are then added. Self-consistent field iterations are optional. The absolute energy of the transition is calculated from the difference in the total energy with the core hole, and without it. The mean free path is calculated from the imaginary part of the interstitial potential between the atoms and the core-hole lifetime.

Once the spherically symmetric muffin-tin potentials are generated, the partial wave scattering phase shifts can be calculated by integration within the atomic potential outward to the radius, and matching the boundary conditions to free spherical waves between the atoms. The set of partial wave phase shifts for various angular momenta l as a function of k completely describe the scattering from each scatterer. Relativistic transition matrix elements between the initial state and the wavefunction within the central atom as a function of k are calculated to generate μ_0.

Full Multiple Scattering (FMS) can then optionally be used to (in effect) sum the multiple scattering expansion to all orders, but only for a limited range in k (below $k \approx 4$ Å^{-1}). This involves numerical inversion of large matrices.

To calculate out to higher values of k, a multiple scattering expansion is done. A huge number of paths enter into it, but many of them are equivalent (or nearly so) to each other, and many others are weak and do not contribute significantly to the signal. For this reason a path filter is implemented in FEFF that efficiently (and approximately) estimates the importance of each path. The path is rejected if it is less than a minimum threshold. In fact, several different levels of path filters with increasing degrees of accuracy are used at different levels of calculation.

Once the paths are enumerated, the effective scattering amplitudes for each path are calculated using the algorithm of Rehr and Albers [66], and the spectrum is generated by summing the paths, including Debye–Waller factors that are calculated internally or externally input. FEFF8.4 allows the input also of force constants and calculations of Debye–Waller Factors by equation of motion and recursion methods.

4.5.1 Example: MnO

As an example, consider Mn K-edge excitation of manganosite, MnO, which has the NaCl structure in which Mn and O atoms occupy alternate positions on a simple cubic lattice. A model cluster of atoms was generated[6] (using the Mathematica code shown) out to a distance of about 7 Å and the multiple scattering was computed using FEFF 8.4. The shortest path has the electron leaving the central atom, scattering once off a nearest neighbor oxygen, and returning to the central atom. There are six equivalent paths, because there are six symmetrical oxygen atoms around the central Mn atoms; the "path degeneracy" is six. The total path length is twice the nearest neighbor distance. Since paths are visualized using Fourier transforms, in which the distance scale is calibrated for single scattering paths, the "effective distance" r_{eff} for each path is defined as one half its path length.

Mathematica 7 code to make FEFF 8 input file (MnO) & Plot in 3D.

```
(*generate points, select within radius, assign characteristics, write file, plot *)
SetDirectory[NotebookDirectory[ ]]; rpath = 7.0;rbond=2.22;
pts = Select[Tuples[Range[-2, 2], 3], Norm[#] <= rpath/rbond &]; (*identify Mn*)
MnQ[x_] := EvenQ[Total[x]]; Mnpts = Pick[pts, Map[MnQ, pts]];
Mnpts = Sort[Mnpts, Norm[#1] < Norm[#2] &]; Opts = Complement[pts, Mnpts];
Opts = Sort[Opts, Norm[#1] < Norm[#2] &]; Mntag = Table[2, {Length[Mnpts]}];
Mntag[[1]] = 0; Otag = Table[1, {Length[Opts]}];
hdr = {"TITLE MnO 80K", "CONTROL 1 1 1 1 1 1", "PRINT 0 0 0 1 0 3",
"*SCF 4.5", "EDGE K 0.0", "POTENTIALS", "0 25 Mn", "1 8 O", "2 25 Mn",
"RPATH "<> ToString[rpath], "CRITERIA 6.0 3.0", "RMULTIPLIER "<>ToString[rbond],
"DEBYE 80 400", "EXAFS 14","ATOMS \n"}; (*add ipot tags to coords*)
coords = Union[Transpose[Append[Transpose[Mnpts], Mntag]],
Transpose[Append[Transpose[Opts], Otag]]];(*sort lines by R*)
coords = Sort[coords, Norm@Take[#1,3] < Norm@Take[#2,3]&];
outstring =ExportString[hdr, "Table"] <> ExportString[coords, "Table"];
Export["feff.inp", outstring, "Lines"];
Mnatom[x_] := {Blue, Sphere[x, .3]}; Oatom[x_] := {Red, Sphere[x, .3]};
Graphics3D[{Mnatom/@Mnpts, Oatom/@Opts}]
```

Mathematica 7 code to import FEFF8 scattering path data, make chi

```
SetDirectory[NotebookDirectory[ ]];Print[fnames=FileNames["feff0*"]];
readfeffpath[m_]:=(dat=Import[fnames[[m]],"Table"];
nskip=Position[dat,"real[2*phc]"][[1,1]]; n=dat[[Position[dat[[2;;]],"Path"]
[[1,1]]+1,2]];{nleg[n],deg[n],reff[n]}=Take[dat[[11]],3];path[n]=Take[dat,{13,12
+nleg[n]}];{k,twophc,amp[n],phase[n],redfact,lambda,realp}=Transpose[Drop[dat,
nskip+1]];phifun[n_]:=Interpolation[Transpose[{k,2 k reff[n]+twophc+phase[n]}]];
ampfun[n_]:=Interpolation[Transpose[{k,deg[n]*redfact*amp[n]*
Exp[-2 reff[n]/lambda]/(k*reff[n]^2)}]];);kgrid=Range[.1,14,.05];
npaths=Length[fnames];Do[readfeffpath[m],{m,1,npaths}];
chi[n_]:=(ampfun[n]/@kgrid)*(Sin[phifun[n]/@kgrid]);totalchi= Sum[chi[n],
{n,1,npaths}];plt[n_]:=ListPlot[Transpose[{kgrid,chi[n]}],PlotRange->All]; plt[1]
```

[6] The program ATOMS (which is part of the horae/ifeffit suite of programs by Bruce Ravel and Matt Newville) can be used to generate feff.inp files for known crystal structures.

Depending on path filtering criteria, FEFF 8.4 estimates that on the order of 100 paths are important in MnO. Two scattering paths with three, four, and five legs are shown in Figures 4.3–4.5. Stereo pairs are shown to enable the reader to visualize the 3D structure. With a little practice these can be viewed by most without special glasses.[7]

It is important to realize the multiple scattering expansion is basically just a useful way to visualize and compute a perturbation series. Some of the terms may seem counterintuitive, but they all need to be present. For example, there are terms corresponding to the electron leaving the central atom, scattering from a second neighbor atom, and returning, without having scattered from the nearest neighbor directly on the line between them. In Figures 4.4–4.5 the similar appearance between one of the five-leg paths and one of the four-leg scattering paths is due to this skipping-over of a nearest neighbor in the five-leg path.

The strongest scattering processes are single scattering, and linear (or near-linear) scattering processes. Large angle scattering processes tend to be weaker. If two neighboring atoms are nearly in a line, the attractive potential of the intervening atom literally focusses the electron wave on the more distant atom, enhancing the contribution from it. The same effect occurs for more distant atoms that are in a line, as many atoms are in MnO. This forward scattering effect is strong in molecular complexes such as hexacyanides, imidazole complexes such as histidine residues in proteins, and heme groups. Examining the paths in MnO confirms this observation.

However, in a highly symmetrical structure such as MnO the path degeneracy drastically changes the importance of different paths. Even if some multiple scattering paths are intrinsically weak, they may have a high degeneracy. For example, a number of three-leg paths have degeneracy 48; one in the cluster shown even has a degeneracy 144. Weighting the intrinsic scattering contribution by the degeneracy factor makes a large number of the paths in MnO non-negligible.

The first eight scattering paths for MnO are shown in Figure 4.6. They are weighted by k^2 to emphasize their high k behavior. The paths are dominated by single scattering and small angle multiple scattering (focussing effect), but because of the high degeneracy factors, many of the weak large angle scattering paths are non-negligible. Path #8 is large because it consists of the linear scattering (focussing effect) Mn-O-Mn. Its amplitude is large at high k because the atomic number of Mn is substantial ($Z = 25$). In

[7] Primers on viewing stereo pairs without glasses can be found at www.edge-3d.com/how_to.htm and http://spdbv.vital-it.ch/TheMolecularLevel/0Help/StereoView.html

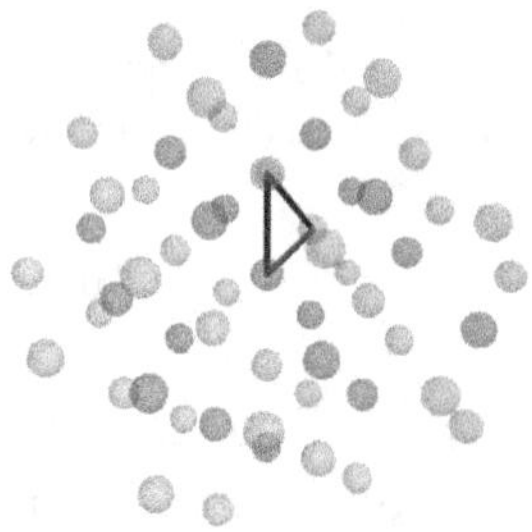

Fig. 4.3 Three-leg multiple scattering path in MnO (stereo pair).

Fig. 4.4 Two distinct four-leg multiple scattering paths in MnO (stereo pair). The path including the triangle goes directly from the center atom to the second neighbor Mn.

contrast path #7 is lower in amplitude because it consists of two consecutive backscatterings from the oxygens on either side of the central atom.

Inspection of the 80 or so remaining paths (not shown) follows the same pattern. In addition to the single scattering and obvious focussing multiple scattering paths, there are other MS paths in diagonal directions that are important, e.g. paths #58 and #60. In these cases several Mn atoms at coordinates such as (1,1,0) and (2,2,0) are aligned with each other – they are focussing MS paths like the type described above.

4.5.2 Alternatives to path expansions

At sufficiently low energy, in the XANES region close to the Fermi level, the path expansion may not converge, and a full matrix inversion must be done,

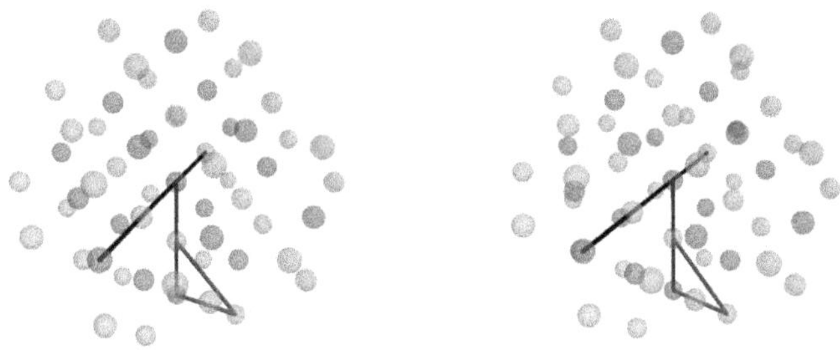

Fig. 4.5 Two distinct five-leg multiple scattering paths in MnO (stereo pair). Many of the 100 or so important scattering paths in MnO are linear.

or an alternative used, such as Haydock's recursion method [69] or the finite difference method. One approach (embodied in GNXAS [98, 99]) is to sum together all terms of scattering order N and describe them by correlation functions, that are characterized by fitting to experimental data.

4.6 EXAFS equation

Because the absorption coefficient μ depends linearly on the Green's function, $\chi(k)$ can similarly be expressed as a sum of contributions corresponding to increasing numbers of scattering events. Experiment and theory have shown that above the edge an expansion in terms of single and multiple scattering processes is convergent. At lower energies the expansion may not converge, and in effect it must be summed to all orders by performing a matrix inversion.

It is useful to express the total absorption coefficient $\mu(E)$ as the isolated atom absorption $\mu_0(E)$ times a correction factor χ: $\mu = \mu_0(1 + \chi)$, or $\chi = (\mu - \mu_0)/\mu_0$. This says that χ is the fractional change in absorption coefficient that is induced by the presence of neighboring atoms. Within the context of the single scattering approximation (and others discussed below), Stern, Sayers and Lytle [1, 2, 3, 4] derived a simple and extremely useful expression for χ, which has come to be called the standard EXAFS equation. According to this theory, for K-edge (and L_1) excitations,[8] an atom (index

[8] Because $L_{2,3}$ edge excitations couple p states to either d and s states, the EXAFS equation has a more complicated form for those edges, with three types of terms like those in Equation (4.5): one for d final states, one for s, and cross terms between them. The phase terms include a term $(-1)^l$, which implies that the $s \rightarrow p$ and $p \rightarrow d$ signals will be π out of phase (neglecting

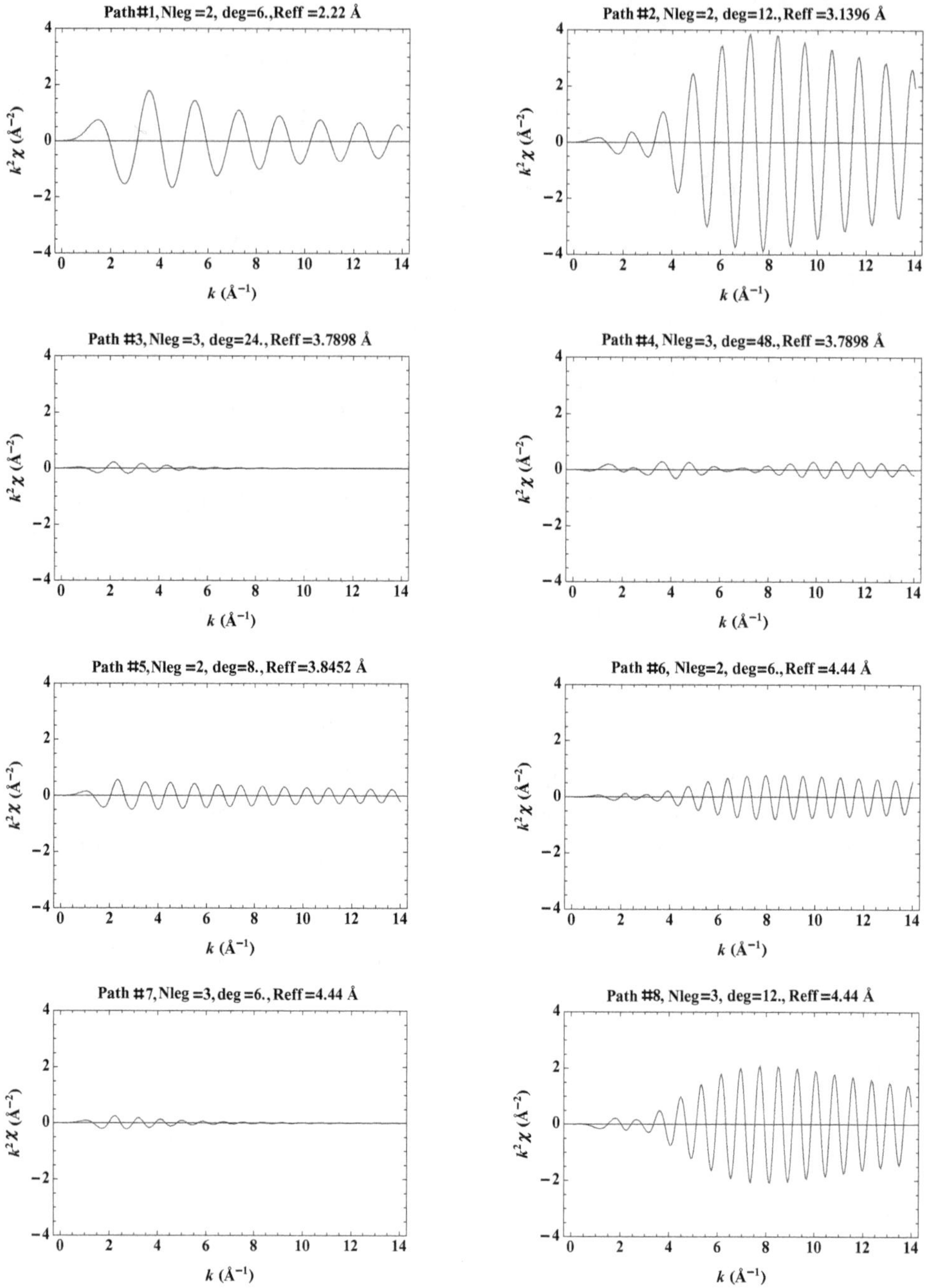

Fig. 4.6 First eight paths for MnO, calculated using FEFF 8.4. These are weighted by the degeneracy factor, e.g. the intrinsic strength of the large angle MS path #4 is only 1/48 of that shown. Path #8 is a linear scattering path Mn-O-Mn.

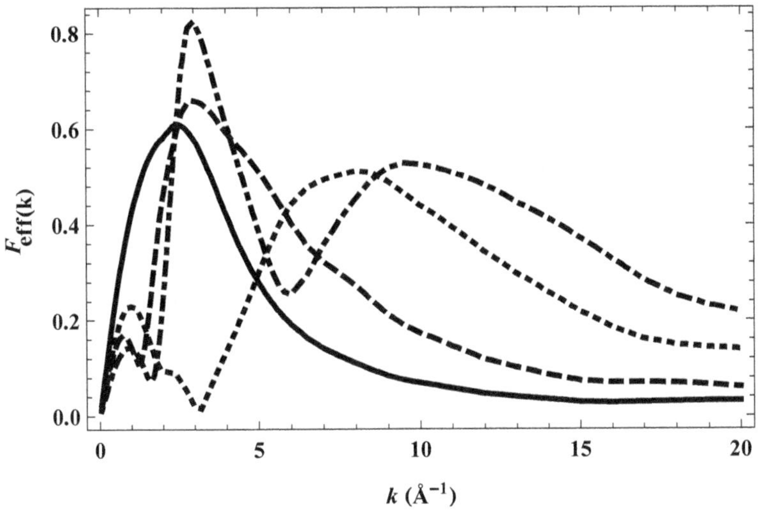

Fig. 4.7 The scattering amplitude times the mean free path is shown for Fe central atom and O ($Z = 8$), S ($Z = 16$), Se ($Z = 34$), Te ($Z = 52$) backscatterers as calculated by FEFF 8.4 at $R = 2.0$ Å. Shown are the single scattering amplitude contribution multiplied by kR^2/N. The different k dependencies are used to determine the approximate atomic number of the backscatterer.

i) at relative distance r_i makes a contribution to the EXAFS

$$\chi(k) = -S_0^2 \sum_i \frac{3\cos^2\theta_i}{kr_i^2} |f_i(k)| \, e^{-2r_i/\lambda_e} \sin(2kr_i + 2\delta_1(k) + \arg(f_i(k)), \quad (4.5)$$

where $k = \sqrt{2m(E - E_0)/\hbar^2}$ ($\approx \sqrt{.2625(E - E_0)}$ in eV, Å units), θ is the angle between the X-ray polarization vector $\hat{\epsilon}$ and the vector $\vec{r_i}$ connecting the central atom with the atom in question. Here, $|f_i(k)|$ and $\arg(f_i(k))$ are the modulus and phase of the complex electron scattering amplitude for each atom (as graphed in Figures 4.7 and 4.8); $\delta_1(k)$ is the $l = 1$ partial wave phase shift[9] central absorbing atom; and S_0^2 and the photoelectron mean free path λ_e account for losses of coherence due to multielectron excitations. The $1/r^2$ factor accounts for the $1/r$ decrease in intensity of the photoelectron wave propagating out to the scatterer and then back to the central atom.

spherical wave effects). These edges also show the "non-$\cos^2\theta$" dependence described in Section 4.2.3.

[9] Partial wave phases $\delta_l(k)$ describe the effect on the phase of waves of specific angular momenta l when they scatter from a spherical potential such as a muffin-tin potential. The scattering process is fully characterized by their k dependence. The plane wave scattering amplitude is given by $f(\theta) = \frac{1}{k}\sum_{l=0}^{\infty}(2l+1)e^{i\delta_l}\sin\delta_l P_l(\cos\theta)$.

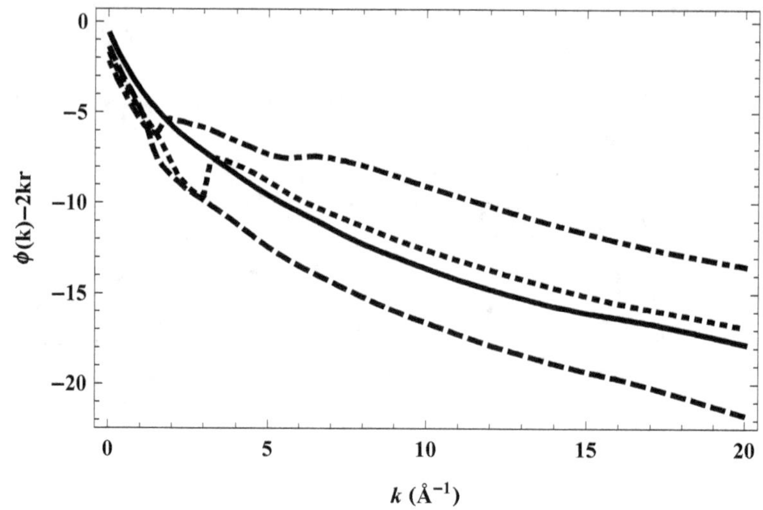

Fig. 4.8 Single scattering phases ($2kR$ subtracted) as calculated by FEFF 8.4 for Fe central atom and O ($Z = 8$), S ($Z = 16$), Se ($Z = 34$), Te ($Z = 52$) backscatterers. (O solid line; S dashed; Se dotted; Te dash-dotted). Calculated at $R = 2.0$ Å. The average phase decreases with k because of the central atom phase shift. Note the roughly π (modulo 2π) shift between elements in adjacent rows of the period table.

Several approximations underpin the simple theory: the potential energy of the photoelectron propagating through the solid is approximated as that of spherically symmetric atoms with a flat region between them (the "muffin-tin approximation" described above); only a single electron is directly excited, which interacts with the electron gas in the material, giving rise to a mean free path; only backscattering from each neighboring atom is included, i.e. multiple scattering is neglected; and in the scattering process the photoelectron is approximated as a plane wave.

The plane wave and single scattering approximations are known to be inadequate, and the modern multiple scattering theory properly takes account of them.[10] It is remarkable that, when spherical wave effects, many body interactions, etc. are included, the basic structure of the original EXAFS equation is preserved as long as the plane wave scattering amplitude $f(k)$ is replaced by an *effective* scattering amplitude $f_{\text{eff}}(k;r)$: the scattering amplitude acquires a weak r dependence. Also, as described earlier, in the case

[10] Historically spherical wave effects were compensated for by using empirical standards with similar bond distances, so that the spherical wave corrections and the mean free path approximately cancelled out.

of oriented samples, additional terms proportional to $\sin^2 \theta$ appear at low k, which may not always be negligible.

The neglect of multiple scattering in the simple equation implies that the total $\chi(k)$ is a simple linear sum of contributions from backscattering atoms. This is a useful first approximation, but it is well known that multiple scattering is often important in EXAFS, particularly when the absorbing atom and scatterers are collinear. Closer to the edge multiple scattering is generally important. The overall structure of the simple equation is that of a sum of damped sine waves: a k-dependent amplitude prefactor times the sine of a phase which is approximately linear in k. In other words, each atom contributes a sinusoidal signal which, when plotted vs. k, oscillates more rapidly the larger the distance – rapid oscillations in the EXAFS mean long distances. The stronger the scattering amplitude, the larger the signal. In an oriented sample, the signal from a given atom is largest when its radius vector lies along the direction of the X-ray polarization vector.

The contributions from multiple scattering also oscillate more rapidly the longer their path length. For this reason single scattering contributions from higher shells may oscillate at about the same frequency as multiple scattering contributions of similar path length. This complicates interpretation of higher shell data, but this problem has been solved by recent theoretical advances.

There are a number of exact expressions [18] for the full multiple scattering $\chi(k)$ using spherical waves, and most of them involve nested sums over angular momenta up to high order. Rehr and Albers [66, 65] showed that the form of the simple path expansion is maintained if the scattering amplitudes at each vertex of a scattering path are replaced by matrices F; six by six matrices seem to provide essentially exact results. Their expression for a path Γ (either single or multiple scattering) is

$$\chi_\Gamma(p) = \mathrm{Im}\, \mathrm{S}_0^2 \frac{\mathrm{e}^{\mathrm{i}(\rho_1+\rho_2+\cdots+\rho_\mathrm{N}+2\delta_1)}}{\rho_1\rho_2\cdots\rho_\mathrm{N}} \mathrm{e}^{-\sigma_\Gamma^2 \mathrm{p}^2/2\hbar^2} \times \mathrm{Tr}\, \mathrm{MF}^\mathrm{N} \cdots \mathrm{F}^2\mathrm{F}^1,$$

where M is a termination matrix, S_0^2 is a parameter slightly less than unity that accounts for the less than overlap between the wavefunctions of the passive electrons in the initial and final states, $\vec{\rho}_i = p/\hbar(\vec{R}_i - \vec{R}_{i-1})$, $p^2/2m = E - V_{\mathrm{MT}}$, V_{MT} is the zero energy of the muffin-tin potential, and Tr represents the trace. The effects of disorder are approximated by the Debye–Waller-like factor, where σ_Γ is the mean square variation in total length of path Γ. This term comes from configurationally averaging the complex exponential term and taking the lowest order terms as described below. This expression clearly shows that even including multiple scattering and spherical wave effects, χ is essentially represented by a product of spherical waves propagating between the scattering centers, decreasing in amplitude inversely as product of the distances between them, and the rapidity of the oscillations of $\chi(k)$ increases in proportion to the total path length.

4.6.1 Configurational Average

In an experiment one effectively averages over many sites in the sample, and the instantaneous positions of atoms may vary because of thermal and quantum zero point motion, and structural heterogeneity. XAFS essentially takes a snapshot of the instantaneous configurations of atoms, because the lifetime of the excited state is limited by the lifetime of the core hole (i.e. the vacancy in the initial ($1S$) state), and core-hole level widths ΔE usually are 1 eV or greater, corresponding to time scales of $\tau = \hbar/\Delta E < 10^{-15}$ sec, approximately 10^3 times shorter than periods of interatomic vibration.

Because of the rapid spatial variation of the phase factor $\sin(2kr + \delta)$ compared to other r-dependent factors, often the variation of the latter is neglected. The configurational disorder is manifested in the spectra mainly through variations in total path length.[11] The average of the phase factor $\langle e^{ikr} \rangle$ is 1 at $k = 0$ (if the distribution is normalized to 1), so evidently we can expand $\ln \langle e^{ikr} \rangle$ in a Taylor series about $k = 0$ as $\ln \langle e^{ikr} \rangle = ika + (ik)^2 b + \cdots$ where a and b are some constants. Below these will be shown to be related to the first and second central moments of the distribution. Exponentiating both sides we have $\langle e^{ikr} \rangle \approx e^{ika} e^{-bk^2}$. The first term gives an oscillatory contribution while the second provides a Gaussian damping factor of the sort generally found in the averaged EXAFS equations. The damping occurs because the EXAFS is a superposition of oscillations of different wavelength. At low k the wavelengths are long, so all the contributions add together in phase. As k increases, however, the wavelengths become shorter, and the signals of slightly different path length destructively interfere with each other. These damping factors are called EXAFS Debye–Waller factors in analogy with similar quantities in diffraction theory.

For randomly oriented polycrystalline or solution samples an isotropic average over angles must be performed. In this case the contributions from atoms of the same atomic number and at similar distances from the absorbing atom may not be resolvable from each other, and the atoms must be conceptually grouped together into "coordination shells." For small

[11] Variations in bond angles can affect the scattering as well, and therefore have an effect when configurationally averaged, but normally this is much smaller than that due to path length variation.

variations in distance within a shell,[12] the equation becomes:

$$\chi(k) = -\sum_j \frac{N_j}{kR_j^2}|f_j(k)|e^{-2R_j/\lambda_e(k)}e^{-2k^2\sigma_j^2}\sin(2kR_j + 2\delta_1(k) + \arg(f_j(k))),$$

where N_j is the number of atoms in the coordination shell, and σ_j^2 is the mean square variation of distances about the average R_j to atoms in the jth shell. This is the classic EXAFS equation of Stern, Sayers, and Lytle [2].

It is not obvious a priori whether the Gaussian approximation in the Debye–Waller factor in the simple EXAFS equation is adequate, so a general expansion is desired. The cumulant expansion described in Appendix 2 offers a general way to describe distributions of atoms. If the disorder is too large for the Gaussian Debye–Waller factor to suffice, additional terms must be included in the exponential, and new terms are introduced into the phase. If these terms are substantial, but are ignored, serious errors in the interatomic distance and other parameters can occur. Apart from its conceptual utility, the cumulant expansion is particularly useful for describing anharmonic systems [89, 56] and thermal expansion.

4.6.2 Thermal motion

Increasing the sample temperature increases thermal motion, which increases the mean square variation of each scattering path length, i.e. σ_j^2 of the jth path. These can be calculated from the vibrational modes and frequencies, which themselves depend on the molecular/crystal structure and the forces between the atoms. These quantities can be calculated for molecular systems and for periodic systems (crystals). The normal mode eigenvectors and frequencies can then be used to calculate the Debye–Waller factors for each path [81]. Alternatively the equation of motion method [82] or other methods can be used.

Single scattering DWFs were studied in the early paper of Sevillano *et al.* [81], in which they performed Monte Carlo calculations of σ_j^2 using several force-constant-based lattice dynamical models. For a monatomic lattice they showed

$$\sigma_R^2 = \frac{\hbar}{M}\int \rho_R(\omega)\frac{\coth(\beta\hbar\omega/2)}{\omega}d\omega,$$

where M is the atomic mass, ω is vibrational frequency, $\beta = 1/k_BT$. The

12 "Small" disorder in this case means $k_{\max}\sigma << 1$ where $k_{\max}$ is the maximum k used in fitting. When this condition fails marginally, additional terms involving higher-order moments (or cumulants – see appendix 2) of the distribution can be added to the basic EXAFS equation. For large disorder either model-dependent or regularized inverse methods should be used.

projected density of vibrational states (PVDOS)

$$\rho_R(\omega) = \sum_{\vec{k},\lambda} (\hat{\epsilon}_{\vec{k},\lambda} \cdot \hat{R})^2 (1 - \cos \vec{k} \cdot \vec{R}) \delta(\omega - \omega_{\vec{k},\lambda})$$

projects out only the part of the mode that contributes to the relative motion of the central and backscattering atoms. Extension to multiple scattering paths can be put into the same form [64], where the atomic mass is replaced by twice the reduced mass parameter for the path.

The temperature dependence of σ^2 is not very sensitive to the details of the PVDOS. Generally it approaches a constant value corresponding to quantum zero point motion as the temperature approaches 0 K, linear behavior with positive slope at high temperature, and a "knee" between, as shown in Figure 4.9. Calculations of the projected densities of vibrational modes in crystals typically give a narrow peak at high frequency with a lower-frequency tail. Two simplified approximations to the PVDOS are accurate enough to be useful. First, the Einstein model, approximates the PVDOS as a delta function – a spike at a single frequency ω_E. This amounts to approximating the central atom/scattering atom pair as that of a quantum harmonic oscillator and a mass equal to the reduced mass of the atomic pair. In this case $\rho_R(\omega) = \delta(\omega - \omega_E(R))$, and

$$\sigma_R^2 = \frac{\hbar}{M\omega_E} \coth(\beta\hbar\omega_E/2).$$

This expression can be used to fit experimental data using only a single parameter, the Einstein frequency ω_E.

The second simple model that has been found useful is the correlated Debye Model [81]. This models the vibrational dynamics in a material as those of sound waves in a homogeneous isotropic medium. The correlated model projects out the part that contributes to changes in absorber–scatterer distance. This simple model gives

$$\rho_R(\omega) = \frac{3\omega^2}{\omega_D^3} \left[1 - \frac{\sin(\omega R/c)}{(\omega R/c)}\right].$$

One practical use of the Einstein and Debye models is experimentally fitting the temperature dependence $\sigma^2(T)$. In doing so it is helpful to use the Einstein and Debye temperatures $\Theta_E = \hbar\omega_E/k_B$, $\Theta_D = \hbar\omega_D/k_B$ as parameters. In this case the constants numerically evaluate to ≈ 25 where σ^2 is in Å^2, $\mu_r = m_1 m_2/(m_1+m_2)$ is the reduced mass of the absorber/scatterer pair in g/mol. and temperatures are in Kelvins. The equation then takes

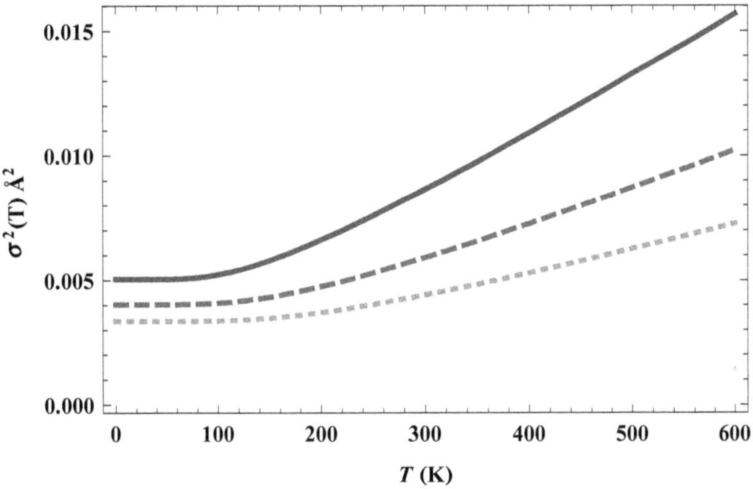

Fig. 4.9 Einstein model $\sigma^2(T)$ for Mn and O pairs (reduced mass 12.4) for $\Theta_E =$ 400K (solid line), 500K (dashed line), and 600K (dotted line).

the form

$$\sigma^2(T) \approx \frac{25^\circ}{\mu_r \Theta_E} \coth(\Theta_E/(2T)).$$

This is plotted for Mn and O pairs for several values of Θ_E.

If one wishes to carry out more detailed calculations of σ^2 it is necessary to calculate interatomic forces as a function of position. These can be approximated using tabulated force constants [81, 84, 82], calculated from quantum theoretical models [84, 85, 86], inferred from other data such as inelastic neutron scattering [83], or if there are not too many of them, used as floating parameters. A comparison of different approximations [84] to calculating of interatomic forces in molecular systems showed that accurate results for the $\sigma^2(T)$ could be obtained in molecular systems if density functional theory (DFT) (with suitable nonlocal corrections) is used. This has the advantage of accounting for changes in interatomic forces as a function of bond distance and angles, but it is time consuming. The multiple scattering DWFs have been mapped out and parameterized as a function of geometry in complexes of transition metals with amino acid residues of interest in metalloprotein research, and used for ab initio fitting [86, 87, 88]. In the future we expect such calculations to be carried out in an integrated manner with the electron scattering calculations.

5
Data analysis

In the process of data analysis one creates a hypothetical model of the data containing unknown parameters or functions that are to be determined through comparison with experimental data. Once the model is established, the goal is to identify and then describe the regions within the space of parameters that are consistent with the data, within the experimental and theoretical uncertainties.

The "forward" problem – calculating spectra for a hypothetical set of structural parameters – has been the principal activity in XAFS theory in recent decades. The inverse problem – determining structural parameters from a set of data – usually is handled through fitting and some kind of statistical analysis to generate confidence limits on the determined parameters. Nevertheless a variety of methods have been developed for solving the inverse problem in a direct manner. Some of these methods are briefly described at the end of this chapter.

XAFS data depend on the types and dispositions of atoms and the variation of those structural parameters over the sample volume. Modeling the structural variation in a sufficiently accurate manner without introducing too many parameters is often the trickiest part of formulating a fitting model. Near-edge spectra furthermore can be sensitive to oxidation states, either directly by introducing vacancies in bound state orbitals or affecting the charge distribution, or indirectly by inducing structural changes.

As described in Chapter 4, EXAFS has a standard parameterization in terms of multiple scattering paths that is descended from the original formulation by Stern, Sayers, and Lytle. There are other useful ways to parameterize the data, such as the description in terms of correlation functions, which is embodied in GNXAS [98, 99]. It is also possible to fit the data directly in energy space, bypassing most of the steps in conventional analysis described in this chapter. This approach is described briefly in the section on fitting data.

A number of capable analysis packages are available from a multiple of sources. The European Synchrotron Radiation Facility ESRF keeps an updated listing of many of them (a web search on "ESRF XAFS software catalog" should get you there). Most are free/open source; a few are sold as commercial packages; others such as recent FEFF8 [95] programs are licensed (FEFF6 is available without restrictions) for a modest fee. Some, such as GNXAS [94], consist of modular suites of Fortran programs that are controlled through command line interfaces and command files. DLEXCURV, which supports integrated 3D graphics, also is available from online [97]. The open source program IFEFFIT [96] is a flexible data reduction/fitting engine that can be used directly in command line mode, or called by external programs. Athena/Horae and SixPack have nice integrated graphical interfaces that use IFEFFIT for most of the numerical calculations. A variety of other programs have various unique capabilities.[1] It is safe to say that there is no one single program package that offers all the features of the others, and there is no reason not to use the best tool for a particular job. It is probably advisable for most people to use one of the standard analysis packages, however.

All data analysis tools require an investment in time and brain cells to learn to use well. The goal in this chapter is to provide the basic background needed to understand and learn to use various data analysis programs. Each program will have extensive documentation that must be consulted to use them intelligently. Some concise examples written in Mathematica are included here for purposes of illustration.

5.1 XANES

The analyses of XANES and EXAFS are usually separate procedures, for a couple of reasons. The first is that the quantitative accuracy of XANES calculations has not been as good as EXAFS. This difference is decreasing over time, and we can anticipate more quantitative structure determination using XANES as theory continues to improve. Another reason is that, although modern theoretical programs are capable of computing both XANES and EXAFS, they may not be able to do it well at the same time. At low

[1] I was one of the authors of the original family of UW programs, but now I mostly use a combination of Mathematica 7 and FEFF8 because of their flexibility and power – new ideas can be tested and bugs can be fixed quickly if you do it yourself. Unfortunately Mathematica is not free. Other programs, such as those mentioned above, are gladly used if needed. The authors of such open source programs perform a great service to the XAFS community.

enough k, close to threshold, the multiple scattering path expansion may not converge, in which case the spectrum must be calculated by direct matrix inversion (if one is using scattering theory), or an alternative approach such as the finite-difference method, or recursion method. This matrix inversion approach is impractical in the EXAFS region, because it requires angular momentum expansions to order $l \approx ka$; for $k \approx 15$, and $a \approx 1$ matrices get large, and this becomes computationally impractical.

Fitting discrete transitions to edge features is a conventional nonlinear fitting process that can be carried out with any number of different programs that are unrelated to XAFS, as well as XAFS-oriented programs. If peak areas are of interest, careful attention should be paid to errors associated with how the continuum part of the edge and the background are modeled. Discrete transitions in principle should be fit with a Lorentzian lineshape[2] (which is the Fourier transform of an exponential decay in time corresponding to the state's lifetime) that is broadened by the resolution function of the monochromator. The combined lineshape in practice is often approximated by the convolution of the Lorentzian lineshape with a Gaussian broadening function; this is called a Voigt function. The edge step is often approximated as an arctangent function, which is the integral of a Lorentzian. This choice can be rationalized if the density of empty dipole allowed states is taken as a step-function at the absorption edge.

Given a set of hypothetical atomic positions, with modern theoretical programs and fast computers it is usually straightforward to calculate a multiple scattering spectrum in seconds to hours (for a large cluster, self-consistent potential). Such calculations can be embedded in a fitting loop, by generating a hypothetical structure, calculating the spectrum, comparing it to the experimental data to generate an error value, and then minimizing the error by repeating through use of a suitable function minimization algorithm. Usually the time to calculate a spectrum is sufficiently large that one wishes to minimize the number of calculated spectra. One successful approach [101] is to use high-dimensional interpolation of the spectra as a function of the parameters.

[2] This has the form $(\Gamma/2\pi)/((E - E_f)^2 + (\Gamma/2)^2)$ where Γ is the full width at half maximum and E_f is the center energy of the peak. It is normalized to unity.

5.1.1 Qualitative interpretation of XANES

The final states of K and L_1 edges are p-states, and the final states in L_2 and L_3 edges are a mixture of d and s character (when projected onto the central atom[3]).

The p final states of K and L_1 edges are more diffuse than the localized d final state of L_2 and L_3 edges. All of these edges may show strong peaks at the top of the edge (the "principal maximum"). Historically these were called "white lines," because that was how they appeared on photographic plates.

The existence of a sharp intense absorption band indicates the final state electron is highly confined by the potential that it sees. This can be due to empty bound atomic states of the right symmetry for a dipole allowed transition. Suppose you have a free atom that has a bound but empty p symmetry state. This will show up as an intense peak at the absorption edge. If that atom were placed into a material, the tails of its attractive potential would be truncated by the attractive potential of the neighboring atoms, allowing the final state electron wave to leak out of the confining central atom potential, which would broaden the bound state into a resonance. The sharp peak at the edge would be smoothed out and suppressed, at least as far as the central atom absorption μ_0 is concerned.

An example calculation [52] of μ_0 (what is now often called "embedded atom" absorption) is shown in Figure 5.1. The truncation of the central atom potential broadly affects the shape of the edge.

Holland *et al.* [91] have shown that the shape of the central atom potential itself can cause low-frequency modulations in the absorption coefficient, in a manner analogous to EXAFS. This effect was later labeled "Atomic XAFS," or "AXAFS," and shown to have potential use as a probe of chemical state of the central atom. Progress along these lines and references to earlier work can be found in a recent communication [92].

Confinement of the electron to produce sharp, intense absorption features also can be due to multiple scattering effects. If the neighboring atoms are negatively charged the repulsive potential will tend to confine the electron. If the geometry is highly regular, as in MnO, interference effects between many paths of different path length also can conspire to produce a large white line peak, such as that seen in the MnO spectrum of Figure 1.2.

The strength of a transition is also related to the occupancy of the state.

[3] It is important to be clear on this point. Angular momenta are defined in reference to an origin. Orbitals of any symmetry that are centered on neighboring atoms can be expanded as linear combinations of states of various l located on the central X-ray absorbing atom. For example, an s state that centered on a neighboring atom will contain some p character when expanded with respect to the center atom.

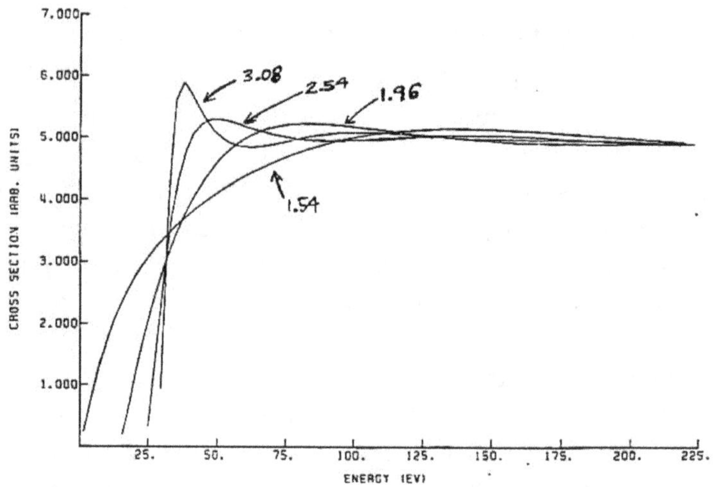

Fig. 5.1 The effect of truncation of the free Mn atom potential as a function of truncation radius [52]. Notice that μ_0 becomes less peaked when the attractive tails of the potential are truncated, which happens when a free atom is placed into a material. The curves correspond to truncation radii of 1.54, 1.96, 2.54, and 3.08 Bohr (1 Bohr = 0.529 Å). Since $\mu(E) = \mu_0(E_0)(1 + \chi(E))$ this effect on μ_0 offers another way that interatomic distance can affect the shape of the edge. Image from reference [52].

Because of the Pauli exclusion principle there can be no transitions to filled states. Often L-edges have partially empty states at the edge which can produce large white lines, which are superimposed on a rising absorption due to the continuum of final states that is available for $k > 0$. Attempts to determine the occupancy of such levels from the white line areas has been a minor industry, but there are a number of factors that may complicate the interpretation. The L_2 and L_3 edges of Pt have rather different shapes, which is attributed (like the energy splitting) to relativistic spin-orbit interactions [93].

5.1.2 Pre-edge transitions

Pre-edge transitions are often found in transition metal XANES spectra. At present their main utility has been as an indicator of local site symmetry and orbital occupancy. It is useful to describe them in terms of molecular orbital theory, although in some cases a quantitative analysis may require more detailed treatment in terms of multiplets [68].

From an MO point of view, the bound states in the pre-edge region consist of linear combinations of orbitals that have specific symmetry properties.

In transition metals such as Mn there are $3d$ orbitals with energies in the pre-edge region that mix with ligand (and other metal) orbitals to make molecular orbitals.

Suppose you were to apply a symmetry transformation to the cluster of atoms, such as a rotation or inversion. Different orbitals will transform in different ways under the symmetry operation. For example, an s orbital will be unchanged by inversion through the central atom, but a p orbital will change sign. If you invert the system twice, you get back to the original configuration, so any state must either be unchanged (even parity) or change sign (odd parity) under inversion.

Now, it is not difficult to show that if the Hamiltonian is symmetric under inversion, states of opposite parity cannot mix to make an eigenstate of the Hamiltonian. This is a special case of a general theorem in group theory (see Chapter 4 and references cited there). This implies that p states (odd parity) can only mix with d or s states (located at the central atom) if the Hamiltonian (and therefore the local environment) is not totally symmetrical under inversion. Since the final state in K- and L_1-edges are of p symmetry (by dipole selection rules), any strong transition near the energy of the metal $3d$ orbitals indicates a breaking of inversion symmetry. A spectacular example of this is $KMnO_4$ (Fig. 1.3). In $KMnO_4$ the Mn atom is surrounded by four oxygen atoms in tetrahedral arrangement, which is very asymmetrical under inversion. This permanganate ion MnO_4^- forms a unit, and the K^+ ion is further away and very loosely bound – it basically just balances out the charge. There are many tetrahedral oxyanions of this and similar types, e.g. TiO_4^{2-}, VO_4^{2-}, CrO_4^{2-}, MoO_4^{2-} and related complex ions. They all show large pre-edge peaks that proclaim a breakdown of inversion symmetry.

The reason that the pre-edge peak is so strong and sharp in MnO_4^- is its short bond length (only 1.63Å) and low occupancy (formally it's empty – $3d^0$). The compound K_2MnO_4 has an extra electron, which is forced into an antibonding orbital. This increases the bond length, and reduces and broadens the peak.

This argument would suggest that the pre-edge transition strength should be identically zero for inversion symmetric sites. This isn't quite true however. Quadrupole transitions are weakly present because the transition Hamiltonian contains small terms that are non dipolar (see Chapter 4). Such terms can be observed in the pre-edge of most transition metal complexes.

Some materials, such as the spinel and inverse spinel structures have metal ions in inequivalent sites: 2/3 of the sites are octahedral, and 1/3 are tetrahedral. Despite the dilution due to octahedral sites, the $3d$ pre-edge peaks are noticeably larger than the pure octahedral case.

In some cases the states associated with pre-edge transitions are full ($3d^{10}$ configuration) in one charge state, but have a hole ($3d^9$) in another charge state. For example the presence or absence of the $3d$ pre-edge transition can be used to detect the difference between Cu^+ and Cu^{2+}.

5.1.3 Edge shifts

There is a long practice of correlating formal charge state with absorption edge shifts. It is clear from simple physical arguments that shifts of a few eV can indeed occur by transferring charge, but the amount of charge transfer that actually occurs in chemical binding is much less than the formal charge would suggest. Atoms remain nearly neutral even when chemical bonds are formed.

The K-edge position is the difference between the energy of the $1s$ level and the final p-state. The final states in the pre-edge are fairly localized. As a simple model of chemical charge transfer between atoms, imagine the central atom is surrounded by a spherical shell of charge Q and radius R. When charge is chemically transferred because of chemical bonding, the effective radius of this shell of charge changes slightly. From electrostatics we know that the energy levels of everything inside the sphere then is shifted by $\Delta(q/R)$. The initial $1s$ state and any final state orbital that is entirely contained within the spherical shell would be shifted the same amount by the change $\Delta(q/R)$, so the transition energy in the spectrum would not be affected. The more diffuse orbitals (such as the $4p$ orbital in $3d$ transition elements) would be more strongly affected, however, because some of the final state orbital would be contained within the sphere, and some would not.

It also has long been observed (since the 1930s) that there is inverse relationship between the extent in energy of the edge and the average bond length. Hartree *et al.* suggested that the principal maximum occurs when the electron wavelength equals the interatomic distance. In eV-Å units, this gives $E - E_0 \approx 150/r^2$. From a scattering point of view the charge transfer would enter into the calculation only through alterations to the partial wave scattering phase shifts and μ_0.

I have argued [52] that although small shifts of a few eV may be directly due to such effects, larger shifts (10–20 eV) are more often due to chemically induced changes in bond length. An aphoristic way to state this is the chemistry (charge transfer) tells the atoms where to go, and physics (scattering) tells us where they are. Since there is, at least in some materials such as Mn oxides, a strong correlation between the formal charge state and the average bond length, it is difficult to experimentally distinguish between the direct correlation (formal-charge → edge-shift) and the indirect one (formal-charge → bond distance → edge-shift). An example of this correlation is shown in Figure 5.2 which shows a log-log plot of the average formal charge in Mn oxides versus average nearest neighbor distance. The linearity of the curve shows there is a power law relation between them: formal charge $\propto r^{-4}$. This sort of correlation can be very useful, but I do not mean to suggest this is a general relation; such correlations in fact may be misleading.

With this strong a correlation it is difficult (although not impossible) to disentangle the effects experimentally. It was straightforward to do so with theoretical calculations however [52]. In this study, the charge densities, muffin-tin potentials, and scattering phase shifts were first constructed, then effectively frozen; the geometry was then varied in later stages of the calculation. Figure 5.3 shows a series of spectra calculated for different interatomic distances, with frozen charge densities. There is a clear edge shift that is driven solely by bond length as suggested by Hartree *et al.* in 1934. When the energy is rescaled by $1/r^2$ the edges line up reasonably well (Figure 5.4).

Because of the primitive way in which the potentials were constructed (and discontinuities were eliminated), the muffin-tin zero was artificially low.[4] This had the fortuitous effect of exposing some filled states; it also broadened the sharp pre-edge bound state into a resonance (this was connected with Mn 3d states/T_2 molecular orbital by examination of the d-symmetry partial wave phase shifts). The approximate position of the Fermi level is shown by the short line; no transitions occur into states below that line; and the state below it would be invisible in an experimental absorption spectrum.

Although the calculations shown were done in the pre-FEFF era, similar results can be obtained by FEFF8.4 in full multiple scattering with self-consistent potentials, as shown in Figure 5.5.

It is illuminating to see what features in the edge are due to single scattering, and which are multiple scattering effects. Single scattering calculations were performed using code written by the author, and multiple scattering calculations were done with a version of the full multiple scattering code DLXANES [104] from Daresbury Laboratory that had been converted to

[4] This has come to be called the "extended continuum" approximation.

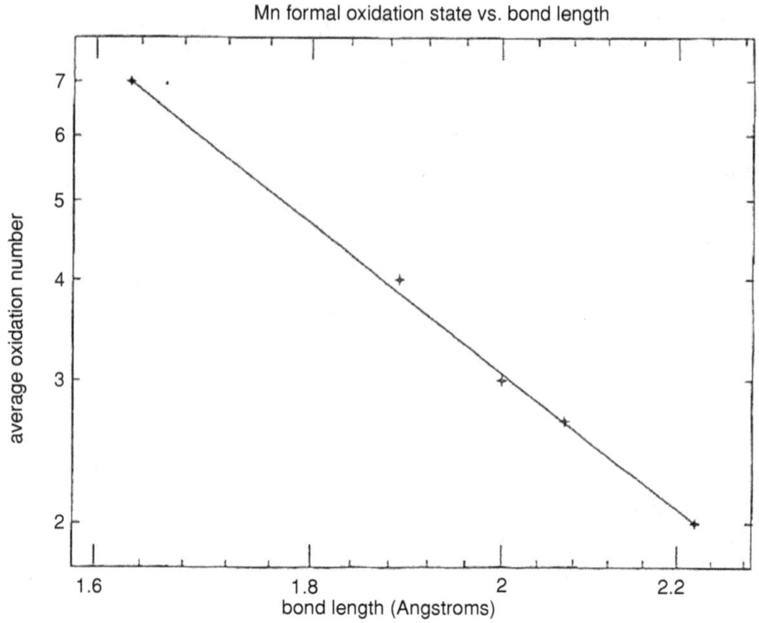

Fig. 5.2 Log-Log plot of formal oxidation state in Mn oxides vs. average bond length. Image from [52].

Fortran 77 by B. A. Bunker. The same partial wave phase shifts and μ_0 were used for both.

Results of these calculations are shown in Figures 5.6 and 5.7. The characteristic shape of the edge is due to multiple scattering, as is the pre-edge transition, which, from a molecular orbital point of view, is a T_2 orbital associated with Mn $3d$ orbitals, as described under pre-edge transitions.

5.1.4 Linear combinations and PCA

It is often useful to fit an unknown spectrum to a linear combination of known spectra. When the sample consists of a mixture of absorbing atoms in inequivalent sites, one measures a weighted average of the $\mu(E)$ or $\chi(k)$ corresponding to each site. The data can then be fit on that basis. If the measured data are nonlinearly distorted, as for example when thickness effects or fluorescence self-absorption effects occur, the situation is more complicated because the distortion must be corrected first. See self-absorption in Chapter 3.

If such effects do not occur, i.e. things are linear, the process of fitting to data is straightforward, in concept at least. Fitting EXAFS is most

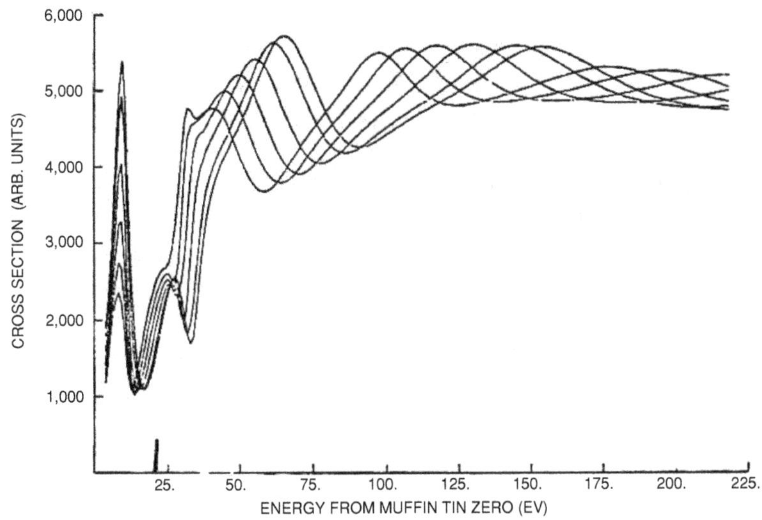

Fig. 5.3 Calculated MnO4 cluster edges vs. distance [52], calculated using DLXANES [104]. Mn-O distances shown are 1.63, 1.67, 1.76, 1.84, 1.93, 2.01 Å. Notice the strong edge shift as a function of interatomic distance. Image from [52].

robust because the basis functions (EXAFS spectra) are usually quite distinct from each other. The different $\chi(k)$ spectra can be viewed as vectors (see Appendix 1), and a collection of them is a basis set. If the scalar product

$$\frac{\int \chi_i(k)\chi_j(k)dk}{\sqrt{(\int \chi_i^2(k)dk)(\int \chi_j^2(k)dk)}}$$

between all pairs of spectra $\chi_i(k)$ in the set are appreciably different from ± 1 the fits are well conditioned. If the scalar products were all zero, the basis set would be orthogonal, but this would never occur by chance. However, if (as frequently occurs in XANES fitting), the basis functions (now $\mu(E)$) have very similar shapes, the scalar products between basis functions are close to 1, and it takes very little experimental error to rotate one of the vectors (spectra) into some linear combination of the other spectra. This makes linear combination fitting to XANES (as usually practiced) risky business, especially if fluorescence self-absorption effects or thickness effects are important [106]. The fitting can be done in a numerically stable manner using singular value decomposition (SVD) or pseudoinverse methods [111], but there is no guarantee that the derived answers will obey sensible physical constraints such as positivity. If constraints are applied in a linear fitting problem by imposing a penalty function, the problem becomes nonlinear, and then multiple minima may be encountered. Tannazi [107] developed a

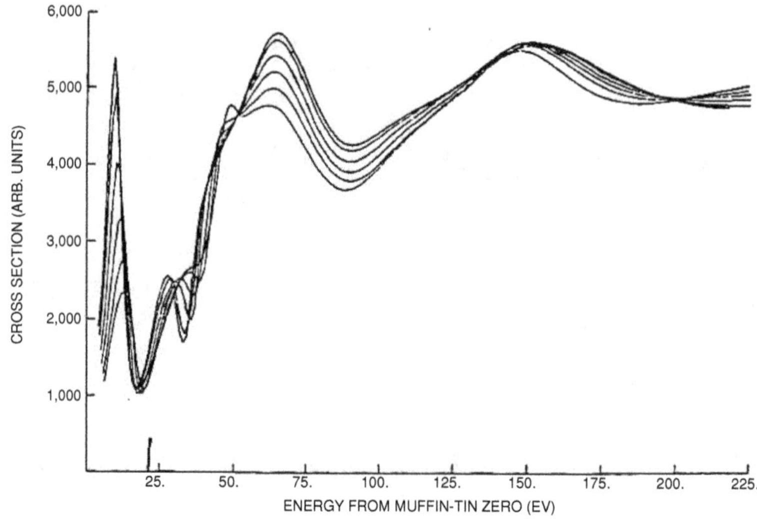

Fig. 5.4 Calculated MnO4 cluster edges vs. distance [52], with the energy axis of the spectra rescaled by $1/r^2$; calculated using DLXANES. Notice that the energy rescaling largely compensates the edge shift. Mn-O distances shown are 1.63, 1.67, 1.76, 1.84, 1.93, 2.01 Å. Image from [52].

constrained fitting method using a linear programming algorithm that has performed well in difficult fitting situations. An attractive feature of this approach is that the fitting model remains linear, even when constraints are applied, so there is a unique minimum.

It is often of interest to see if there is a subset of spectra that can be eliminated from the basis set without significantly degrading the quality of the fit. With small basis sets this reasonably can be done by fitting with all combinations, but since there are $2^n - 1$ fits to do if there are n spectra in the basis, this grows quickly out of hand for $n > 12$ or so. This approach does work, but it is impractical for larger basis sets. If a best fit says that the fraction of a particular component is small, it may indicate that it actually isn't present at all, and removal from the basis set may be indicated.

There is another class of procedures (Factor Analysis, Principal Components Analysis (PCA) [108], Singular Value Decomposition(SVD) [111]) that find linear combinations of basis spectra and produce new sets of basis vectors, so as to minimize the number of basis vectors needed to reproduce the spectra adequately. This can be viewed as a form of data compression. Unfortunately there is no guarantee that the new basis vectors actually represent any physical spectra of interest. In general the new basis vectors are not a simple subset of the original basis spectra (although they form a

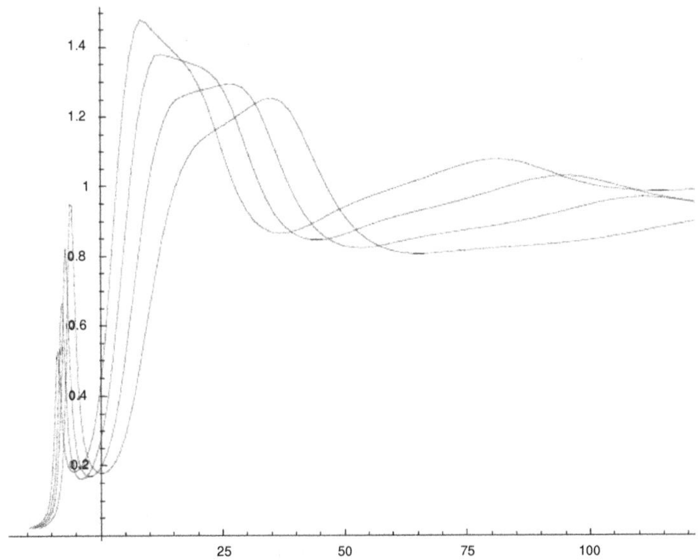

Fig. 5.5 MnO4 cluster calculated by FEFF8.4 self-consistent field, full multiple scattering as function of bond length (1.63, 1.73, 1.84, 1.94Å).

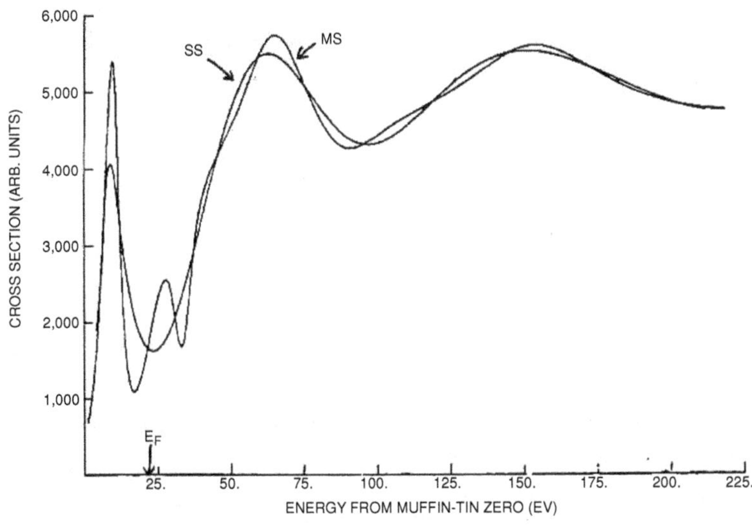

Fig. 5.6 Calculated MnO4 cluster edges [52], in multiple scattering (DLXANES) and single scattering approximations (program written by author), using the same partial wave phase shifts. The broad features are due to μ_0 and single scattering, while the more rapidly oscillating structure is given by multiple scattering. In particular the pre-edge transition is a multiple scattering effect. Image from [52].

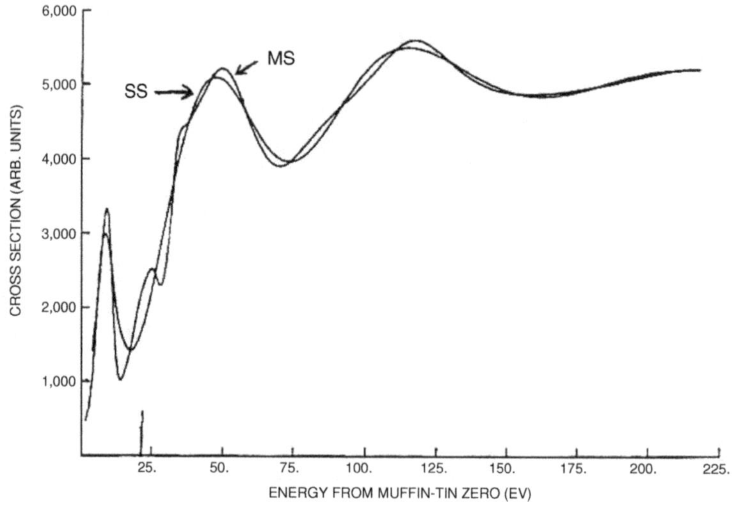

Fig. 5.7 Same as Figure 5.5, bond length 1.84Å. Calculated MnO4 cluster edges, in multiple scattering and single scattering approximations. The broad features are due to μ_0 and single scattering, while the more rapidly oscillating structure is given by multiple scattering. In particular the pre-edge transition is a multiple scattering effect. Image from [52].

subspace) which is usually what one wants. Nevertheless PCA can be useful for understanding the intrinsic complexity of a spectrum.

5.2 EXAFS

Usually in EXAFS data analysis the data are first reduced to a $\chi(k)$ function. The data can be fit in k-space (i.e. as a function of k), or they can be Fourier transformed into an r-space (half path length) representation. In this representation the faster oscillations in $\chi(k)$ (which correspond to long path lengths) are represented as r-space peaks that are located at high r. Larger amplitude oscillations in $\chi(k)$ give larger peaks in r-space. Different contributions can be isolated by Fourier filtering and fit in k-space.

It is feasible to fit the data directly in energy-space without preprocessing; this was done by Pettifer and Cox [100], and is currently supported by GNXAS [94]. This approach has the virtue that it requires only minimal processing of the data, giving fewer opportunities to make mistakes, or to inadvertently bias the results. It also makes it possible to avoid interpolation of the data, which statistically correlates nearby points, reducing their independence, an important characteristic for error estimation by standard methods. On the other hand, deconstructing the data by systematically dismantling it can be very illuminating. The two approaches are not mutually exclusive – the best procedure would be to do both: take the data apart, and put it back together again.

In most cases the data reduction process is fairly straightforward and can be done with automated or semi-automated methods. In some cases background subtraction can be tricky and manual intervention is required. It is always wise to graphically check the output of the automated programs at each stage of data reduction and analysis. If data analysis becomes very time consuming, it is usually because the modeling problem is intrinsically complicated, or there may not be sufficient information in the data to determine all the parameters in the model. Often one has to be creative to minimize the number of fitting parameters.

5.2.1 Qualitative interpretation of EXAFS

The $\chi(k)$ spectrum consists of a sum of sine waves with amplitudes that depend on the type of atoms, and how they are distributed. Rapid oscillations in the data correspond to long path lengths. A complex wave pattern indicates a well-ordered structure. Usually the more distant atoms are more highly disordered (if only because of thermal vibrations). If the amplitudes of the data decay rapidly above the edge, the backscatterers are probably of low atomic number. In most cases the contributions from nearby atoms $R < 3$ Å is considerably stronger than more distant atoms $R > 3$ Å, so slower oscillations are normally stronger than the fast oscillations. The scattering amplitudes of light (low atomic number Z) atoms decay more quickly above the edge than higher Z atoms. For a high Z atom, depending on location in k of the minima in the amplitudes, the amplitude may actually increase and then decrease. If the atoms of a particular type have a large spread in distances the EXAFS amplitude will decay more rapidly above the edge.

Atoms that differ in atomic number by about a row in the periodic table tend to have scattering phases that differ by about $\approx \pi$ radians, and if they occur at about the same distance they tend to interfere destructively and cancel out. The same effect occurs to a lesser degree if there is any phase difference between the scatterers. If two well defined groups of atoms are a few tenths of an Å apart they will produce "beats," in which the combined amplitude oscillates. This beat pattern can be used to determine the difference in distance. (See below in Subsection 5.2.15 on beats).

Although the interference patterns in the data may be complex, the data are always basically periodic. If a spectrum starts out in the normal manner at low k (e.g. slow oscillations have larger amplitude than the fast ones), but the phase starts to deviate from periodic behavior at high k (e.g. the oscillations get much slower or faster or irregular), or the amplitude is sensible at low k but then increases at high k, you should be suspicious of the data. There should not be any steps in the data, except possibly some very small secondary edges associated with multielectron transitions.

5.2.2 χ and μ

The "χ" that is used in data analysis actually differs slightly from the χ that is used in the theory. A small correction[5] must be applied to make them consistent.

There are good reasons behind these discrepancies. Theory defines $\chi = (\mu(E) - \mu_0(E))/\mu_0(E)$, where $\mu_0(E)$ is the absorption coefficient of the central atom in its environment within the material, but without the scattering from neighboring atoms. This is sometimes called an "embedded" atom; it is noticeably different from absorption in the free atom. The absorption coefficient $\mu_0(E)$ is calculated by the theory but it is not accessible experimentally. The "χ" that is normally calculated in the data reduction instead is $(\mu_{\rm exp} - \mu_{\rm b})/\Delta\mu_{\rm exp}$, where $\mu_{\rm exp}$ is the experimentally measured absorption or fluorescence, $\Delta\mu_{\rm exp}$ is an estimate of the edge step, and $\mu_{\rm b}$ is a smooth background curve that is fit to the data to remove several extraneous contributions to the signal, which are described below.

Normally experimental $\mu_{\rm exp}$ data are not unique – they are only determined *modulo* an arbitrary multiplicative scale factor plus an arbitrary slowly varying function of energy. The multiplicative scale factor is due to factors such as sample concentration and thickness, detector solid angle, filter thickness, slit position (in fluorescence), amplifier gains, fill gas absorption, scan time per point, ... Although it is possible to carefully determine all of these factors and compensate for them, it is easier to lump them together into a constant factor that is scaled out by normalizing the data to unit edge step.

In addition to the scale factor, there are extraneous factors that vary slowly with energy. An intrinsic contribution to the measured absorption coefficient is the absorption from lower-energy edges by the atoms of interest themselves. Another contribution to the "background" is the (presumably smoothly varying) absorption from other elements in the sample.

The detector signals include slowly varying (in energy) multiplicative factors due to absorption from the ion chamber and sample cell windows,

[5] This is sometimes called a "McMaster" correction because it uses absorption cross section tables like those of McMaster *et al.* [25].

the sample substrate, and the air path between the ion chambers. There are also slowly varying multiplicative gain factors owing to the energy dependence of the ionization chamber sensitivities, which approximately are $E(1 - \exp(-\mu_{\rm fg} l)) \approx c/E^2$ where c is some constant.

In transmission measurements all of these *multiplicative* factors are transformed into *additive* background terms by taking the logarithm; these are later subtracted out in the "background" removal process. In fluorescence and electron yield modes, however, no logarithm is taken, and the effects of these slowly varying multiplicative corrections must be compensated for, or they will cause systematic errors in the Debye–Waller factors. The reason why all such slowly varying multiplicative corrections appear as corrections to σ^2 is explained next.

5.2.3 Systematic errors and Debye–Waller factors

Any spurious sufficiently slowly varying factor $f(E)$ multiplying μ will introduce an error in σ^2 and possibly the higher even order cumulants. This is a consequence of the fact that the data are normalized, and $E - E_0$ is quadratic in k.

Consider such a function $f(E)$. The data are normalized, so that the value immediately above the edge E^+ is set to 1.0. We assume the energy variation of $f(E)$ to be slow enough that we can make the approximation $E^+ \approx E_0$. Expanding $\ln f(E)$ in a Taylor series to first order in $E - E_0$ we find that the effective correction factor is $f(E)/f(E_0) \approx \exp(f'(E_0)/f(E_0)(E - E_0)) = \exp(f'(E_0)/f(E_0)\hbar^2 k^2/2m) = \exp(\alpha k^2)$, after expanding to first order in $E - E_0$, where $\alpha = f'(E_0)/f(E_0)\hbar^2/2m$. Therefore the correction appears as a small (usually negative) increment to σ^2. Expanding to higher order introduces small increments to the cumulants of fourth and higher order.

In practice, often $f(E)$ is approximately a power law: $f(E) \propto E^\beta$. In that approximation $f'(E_0)/f(E_0) = d\ln f/dE/E_0 = \beta/E_0$ and the apparent $\Delta\sigma^2 = (1/2)\hbar^2\beta/(2mE_0) = 3.81\beta/(2E_0)$. For the iron K-edge at 7112 eV and $\beta = -2$ (as for the fill gas absorption mentioned above) this gives a spurious contribution $\Delta\sigma^2 \approx -.001\text{Å}^2$, which has the effect of increasing the amplitude slightly with increasing k. Sample self-absorption effects in fluorescence mode give errors of the same sign and similar magnitude.

5.2.4 Energy calibration errors

Sometimes the energy calibration of a monochromator may be off by a substantial amount, say 20 eV. This can be detected by measuring the absorption edge of an element such as Cu, with a known edge energy, and the best procedure is to get it right during the experiment. If the calibration is off one can still fix the problem after the data are acquired. However, if one just adds a constant to the energy arrays to compensate for the error, there will be a progressively increasing error over a 1 KeV data range, because the relation between energy and angle is nonlinear.

We use Bragg's law expressed as $E \sin\theta = g$, with $g_{\mathrm{hkl}} = hc/2d_{\mathrm{hkl}} \approx$ 1977.1 eV for Si(111).[6] The error in energy scale can be computed as

$$g \csc\left(\arcsin\left(\frac{g}{E}\right) + \arcsin\left(\frac{g}{E_0 + \Delta E_0}\right) - \arcsin\left(\frac{g}{E_0}\right)\right) - (E + \Delta E_0),$$

where E is the energy (eV); E_0 is the known correct reference energy; ΔE_0 is the difference between the apparent edge position and E_0.

This is a larger effect for low energies, close to the g_{hkl}, which is the lower energy limit for that hkl. For example, with Si(111) crystals, over the range from 2500 to 3500 eV, a 30 eV monochromator calibration error gives a 50 eV error at 1 KeV above the edge, even after shifting the energy scale to make the discrepancy zero at the edge. The energy error increases approximately linearly with $E - E_0$, which is equivalent to rescaling the k axis by a constant factor $1+\epsilon$, where ϵ can be a few percent. This introduces systematic errors in the distance determination of a few percent, which is a serious error for EXAFS.

It is clear that this effect should not be ignored. To correctly compensate for energy calibration errors, instead of just shifting the energy arrays, it's better to convert the energy data back to monochromator angle (using $E \sin\theta = g$), calculate the angle offset error, subtract it out, and then convert the arrays back to energy space. Alternatively the angle offset $\Delta\theta$ parameter can be adjusted to give the correct energy calibration in the following expression:

$$E_{\mathrm{corrected}} = \frac{Eg}{\sqrt{E^2 - g^2}\sin(\Delta\theta) + g\cos(\Delta\theta)}.$$

5.2.5 Dead-time corrections

Dead-time effects and corrections are discussed in Chapter 3. When using pulse-counting detectors at a synchrotron source it usually will be necessary to perform dead-time corrections. Limiting the count rate to only a small fraction of its maximum in the hope of attaining linearity makes inefficient

[6] $g_{\mathrm{hkl}} \approx (1977.1)\sqrt{(h^2 + k^2 + l^2)/3}$ eV for silicon crystal reflection hkl.

use of expensive equipment and beam time; it also may lull one into a false sense that the signals are linear when they are not. If multielement detectors are used, dead-time corrections should be done separately on a channel by channel basis. Correction for the nonlinear dependence on flux substantially reduces noise and artifacts in the data, many of which are due to fluctuations in the beam not dividing out properly [27, 40].

The simplest approach to dead-time corrections (if it's not done in hardware) is to experimentally measure the calibration curve and fit it with a flexible but smooth curve (e.g. cubic spline) to eliminate noise jitter. The curve may or may not fit one of the standard models (e.g paralyzable), so if those are used, the fits must be checked to verify that they adequately represent the calibration curve.

5.2.6 Deglitching

A number of experimental problems such as Bragg spots, sample inhomogeneities, monochromator instability, source instability, and a number of other problems can cause localized disturbances or artifacts called "glitches" in spectra. Most can be eliminated by optimizing the experiment. If the beam (or a hot-spot within it) moves over the sample, inhomogeneities in the sample may introduce artifacts in the data that are localized in energy (glitches) or steps in the data. If the glitch is only a single energy point it can usually be removed without concern, but if it is spread out over several points one must be concerned about altering the data in a deleterious way. In that case it is best to remeasure the data, or rely on other scans. The aberrant data point can either be deleted or explicitly replaced with an interpolating function between nearby points. Since the data usually are interpolated to a different grid in subsequent processing, one should not be deluded into thinking that it is necessarily better to delete the point; in that case one is simply blind to how the program does the interpolation.

It is possible to automatically deglitch spectra if they suffer from single point glitches [103]. An approximate noise spectrum can be generated by subtracting a smoothed version of the μ spectrum from the spectrum itself. The first $\mu_{n+1}-\mu_n$ and second $\mu_{n+2}-2\mu_n+\mu_{n-2}$ finite differences of the noise spectrum and μ can be calculated. Aberrant points can be identified if the size of the first or second finite differences are larger than some multiple of the rms noise spectrum finite difference. Single point glitches can be replaced by the average of their neighbors. It is best to exclude the edge itself from this process, and also the near pre-edge if there are sharp transitions there, but the algorithm is suitable for most post-edge (EXAFS) and pre-edge use.

Similarly it is possible [103] to semi-automatically screen spectra in the following way. The spectra should be identical except for noise and an additive background, and a multiplicative scale factor. One spectrum can be "matched" to another by floating, in a linear fitting process, the first spectrum times a scale factor, plus an adjustable background function (e.g. a low-order polynomial in $E - E_0$). An energy shift can be also be adjusted in the fit, but renders the fit nonlinear in the parameters. After finding the best fit, the mean square residual gives a measure of the essential minimum difference between the spectra. Next average all of the spectra together, and match each spectrum in turn with the mean spectrum. Each comparison has a misfit error associated with it, the mean square residual. The distribution of error helps identify the defective spectra, the outliers. Once the spectra on the tails of the distribution are identified, exclude them from the set, calculate a new mean spectrum, and repeat the process. If the data are not too noisy it may be helpful to modify the algorithm to compare the first or second derivative of the spectra to enhance sensitivity to jumps and spikes.

In the absence of automated methods, it is a good idea to visually check each spectrum separately for flaws (jumps, discontinuities, glitches). If multielement detectors are used, it is best to check each channel separately, instead of summing first. Often a problem or Bragg glitch will occur in one channel but not others.

If you have any doubts about a spectrum, it is best to not use it. You can spend an eternity trying to analyze defective data, and even longer trying to justify having done so.

5.2.7 Signal averaging

When using multielement detectors one needs to add up the various detector channels into a total spectrum. After checking the channels for artifacts, and removing them if necessary, they can be summed. If the signal to noise ratio varies dramatically between them, the data should be weighted so the noisy ones have less weight. This can be accomplished by choosing weights that are inversely proportional to the variance of the data.

Sometimes data are obtained on a fine grid in energy or monochromator angle and one may wish to re-map the data onto a different grid. Direct interpolation from one grid to the other may not preserve the statistical weight of the point. It is straightforward, however, to map from one grid to another. Define the desired energy grid points, divide the initial energy space into regions around those grid points, and then average the data in those regions to produce the new values.

5.2.8 Estimating E_0

The energy threshold E_0 is a parameter that represents the $k = 0$ continuum level at which electrons have just enough energy to propagate through the material. The existence of a sharp value for E_0 is an idealization related to making the muffin-tin approximation. In the simple case we have $\hbar^2 k^2/2m = E - E_0$.[7]

When analyzing the data, we first need to approximate the location of this point with respect to the measured spectrum. Physically it should be close to the Fermi level, i.e. the energy below which the states are full of electrons. For molecular systems, it is between the highest occupied molecular orbital (HOMO) and the lowest unoccupied molecular orbital (LUMO). If there are pre-edge transitions, as in many transition metal complexes, E_0 would be around the lower part of the edge. Often the first inflection point on the edge is used; another approximation to E_0 is the point at which the absorption is half of the edge step. All of these metrics are sufficient because we just need an initial guess; a few electron volts error doesn't matter, because the energy zero is fine-tuned in later analysis.

Some analysis programs will automatically estimate E_0 for you. The Mathematica code example below simply looks for the maximum derivative in the data, after smoothing in energy to remove the high-frequency structure. It is not difficult to cook up more robust (and more complex) methods, but almost any algorithm is sure to fail on some data set, and the output of automated programs should always be visually checked.

Changes to E_0 have their main effect at low k. The spacing between the EXAFS wave crests increases with energy, making any E_0 shifts that are small compared to the spacing at that value of k unimportant. The energy spacing between successive oscillations of the path of effective length (i.e. including the slope of phase shifts) $2R$ is

$$\frac{\pi}{\tau R^2}(2kR + \pi) = \frac{\pi}{\tau R^2}\left(2R\sqrt{\tau(E - E_0)} + \pi\right);$$

($\tau = 2m/\hbar^2 = 2$ in atomic units, and $\tau \approx 0.2625$ in eV-Å units). This says that the spacing between EXAFS oscillations of a signal with $R = 4$ at $k = 1$ is about 8 eV, while at $k = 10$ it is around 60 eV. This tells us that at low k a 3 eV shift is not negligible. The effect on phase will be discussed further below.

[7] Using a complex scattering potential and self-energy introduces some subtleties; for example, FEFF chooses $k = 0$ at the Fermi energy E_f. The documentation for the theoretical programs that are used should be consulted for specific definitions.

5.2.9 Normalization to unit edge step

In the normalization process we wish to scale the data so that the edge jump is set to one. In effect this scales the data to a per-atom basis, and it factors out irrelevant (for data analysis) quantities, such as the sample thickness and concentration.

Normalization can be accomplished by performing a low-order polynomial (in $E - E_0$) fit to data in the pre-edge region, doing a similar post-edge fit, and calculating the difference between them at the edge energy. It is essential that the functions that are used extrapolate well, and that the fitting ranges that are chosen exclude the XAFS right at the edge which is highly dependent on local structure, which is exactly what we don't want the normalization to depend on.

It might seem reasonable to use a smooth post-edge fit (or similar) to the data as an estimate for $\mu_0(k)$, and use that to normalize the data, to make the experimental and theoretical definitions of $\chi(k)$ more consistent. This should not be done however – the experimental spectrum normally contains contributions from a variety of sources. Dividing by them will distort the EXAFS amplitudes in uncontrolled ways. It is much better to normalize to unity and then account for the energy dependence of $\mu_0(E)$ afterward.

5.2.10 Interpolation to k-space

To preserve the statistical weight of the original data points it is best to re-bin the original data in energy space into corresponding small k-space intervals, and then average the data in each bin. In k-space ($k = \sqrt{\tau(E - E_0)}$, $\tau \approx 0.2625$ in eV-Å units) this can be done by defining a uniform k-grid $k_n = n\delta k$ and then averaging the points in energy space that lie between $(1/\tau)(k_n - \delta k/2)^2$ and $(1/\tau)(k_n + \delta k/2)^2$. As described above, a k-space grid of 0.05Å^{-1} is sufficiently oversampled for general purposes. Once this has been accomplished, the data can be interpolated onto the desired k-space grid, if they are not already there. The downside of interpolating is that it statistically correlates nearby points.

5.2.11 Background subtraction

The pre-edge and post-edge fits are essentially to determine the normalization constant. The real background subtraction occurs after that point. An operational separation between "background" and "signal" is that the EXAFS oscillations are rapid enough to be separated in frequency from the slowly varying background. Care must be taken to ensure that the frequency

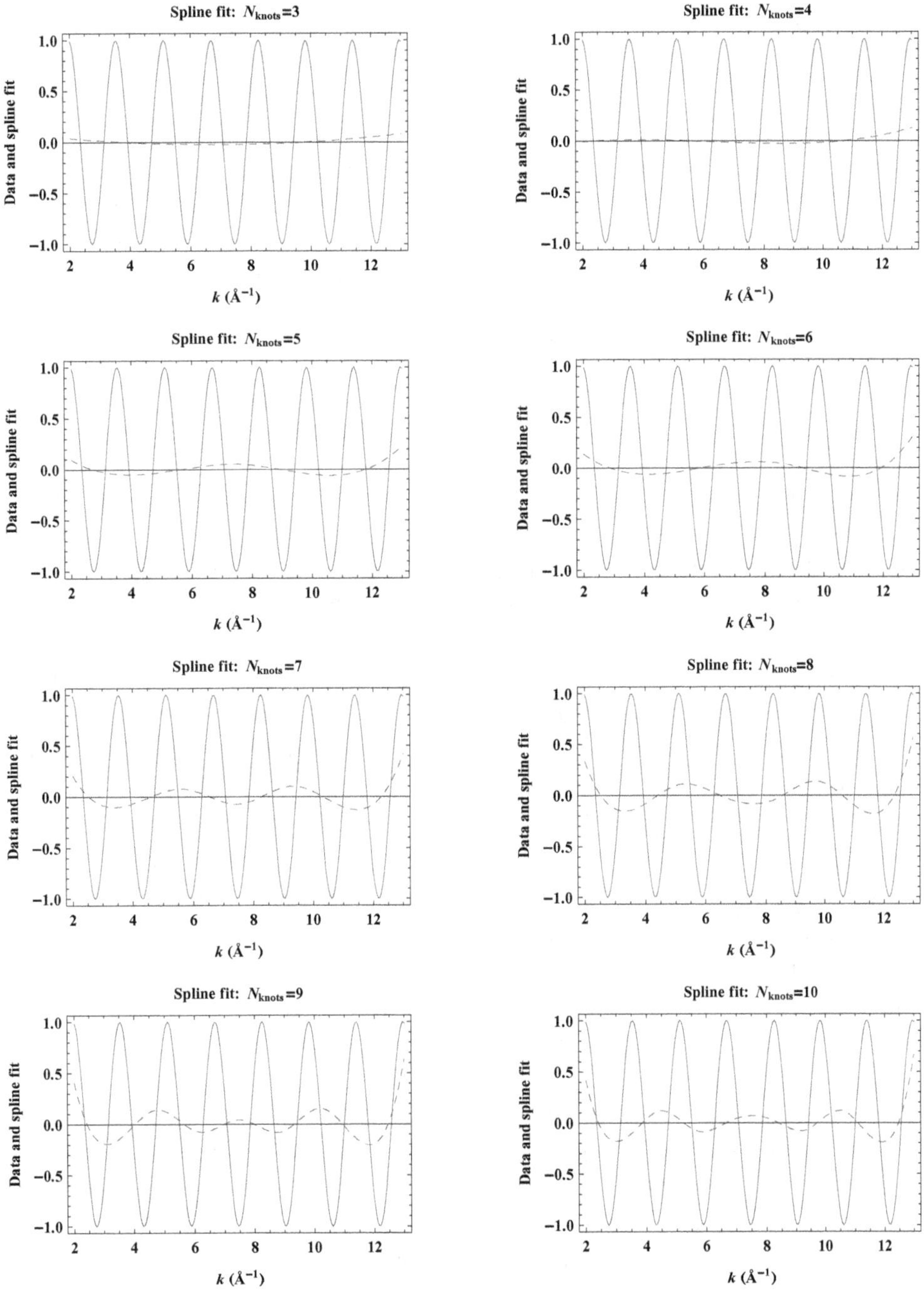

Fig. 5.8 This numerical experiment illustrates how flexible a cubic spline function is for background subtraction: how many knots are permissible before the background spline follows the first shell data? In this example there are 7 oscillations within the data range, but even with 7 spline regions (8 knots) it does not follow the data at the base frequency. Behavior at the endpoints is still problematic unless a window function or constraints are used however.

components of whatever background function is chosen are well localized below about $R = 1$. If one were to fit the data with low-order polynomials in $E - E_0$, and subtract the best fit from the data, the background fit would tend to follow the high-k data (which vary slowly in energy space) more than the low-k data, which would alter the EXAFS amplitude functions. A simple way to prevent this problem is to convert to k-space, and do the background fit there.

The residuals in the fits to the data generally should be weighted as a power of k (say k^3) to get a good background at high k. It is better to weight the fit residuals than the data.

Multielectron excitations [38, 39] can generate small steps in the absorption coefficient in the EXAFS range. Since they are small they can be subtracted out, if they are known to be present. GNXAS has the capability to do this. Often such excitations are present but show up mostly as a troublesome background. Sharp steps in the absorption pose a problem because their frequency content is widely spread in r-space. If there is a low-r background that refuses to go away, multielectron excitations are a possibility to look into.

One approach to calculating a background function is to use smooth curves such as cubic splines to fit the data above the edge (in k-space). These are piecewise cubic polynomial functions with the following properties: the values, first derivatives, and (optionally) second derivatives are continuous at the points (called "knots") where the polynomials joint up. Each cubic polynomial can mimic one cycle of a sine wave, but the constraint conditions between them greatly reduces the flexibility of the spline. It is crucial to avoid accidentally subtracting out part of the signal during background subtraction. If too many knots are used, this can easily happen. This is illustrated in Figure 5.8. This numerical experiment suggests that, if the function value and the first and second derivatives of the spline are continuous at the knots, the number of spline regions can slightly exceed the number of oscillations before they follow the data. If there are N periods of first shell oscillation in the data, it is suitable to use N knots.

Smoothing splines define a spline region between every adjacent data point (every point is a knot), which gives them the ability to follow almost anything, an alarming prospect when doing background subtraction. They are subjected to an additional constraint: the mean square curvature of the function has a specific user-supplied value, which is a smoothing parameter. If this mean square curvature is constrained to zero, it's a straight line. If the smoothing parameter is chosen properly one can get excellent background fits. However, if this parameter is chosen improperly, one can subtly alter their data without knowing it. For this reason such functions must be used with care. Cook and Sayers proposed a successful method to reduce the low-frequency background using cubic splines, while simultaneously ensuring that the first shell data were not modified by it.

Autobk (part of IFEFFIT, which is used by Athena, SixPack, and others) takes a related but different approach. It does not actually fit the data, rather it uses a spline function chosen so that it cannot follow the data too closely, and it varies the spline parameters to minimize only the low-frequency contributions that are below a cutoff distance (e.g. 1Å). It also estimates the edge energy and normalization. By default it attempts to remove background down to very low k, right into the edge region.

One potentially useful option that Autobk supports is that it allows the user to minimize the *difference* at low r between experimental data and theoretical data, rather than just zeroing out the low r signal, which is not strictly correct, because the data actually must have a small contribution there.

It is reasonable to ask why background subtraction needs to be done at all, if the data are going to be Fourier transformed (see below) anyway. The reason is this: Fourier transforming over a finite k-range does not perfectly localize signals in r-space. Even a signal that is a perfect sine wave (i.e. sharply defined frequency) but that exists over a limited k-range will under Fourier transformation give an r-space peak of finite width. If the low-frequency components were not subtracted out approximately before Fourier transformation, this "spectral leakage" would contaminate the data.

The last method described here is implemented in the Mathematica function ReduceData in the code listing below. An example of the steps in the procedure as applied to MnO is shown in Figures 5.9–5.13. After normalization and conversion to k-space, a low-order polynomial in k is fit to the data above the edge (with weighted residuals). This zeroth approximation is subtracted out of the data, which are Fourier transformed. The low-r data (below the cutoff, e.g. 1Å), which are the residual background, are then inverse transformed back to k-space and subtracted from χ, producing the next approximate χ. The process is repeated, as it must, because the Fourier projection onto the low-r subspace although good, is not perfect. Inspection of the intermediate Fourier transforms shows negligible collateral damage to the data from truncation effects. This method zeroes out the low-r, which, as described above, is not strictly correct, but it could be trivially modified to zero the difference between the data and theory.

5.2.12 Data reduction example

Here I give an example of a data reduction process, and a fairly concise Mathematica function that implements it. There evidently are many ways to improve the code, but here the intent is to keep it simple. The code does the following: reads the data; estimates E_0 from the location of the maximum of the smoothed first derivative; normalizes the data from the difference between post-edge and pre-edge fits; interpolates to k-space; subtracts an approximate post-edge background; and refines the background by iteratively subtracting the slowly varying components of frequency less than the r-space cutoff "rcut." It also optionally plots and prints the results. The variables are **inputfile**, **De** (optional small energy shift), **kmn** (desired minimum k e.g. 2), **kmx** (desired maximum k e.g. 12), **wgt** (Fourier k-weighting exponent e.g. 3), **rcut** (r-space cutoff e.g. 1.0 (Å)), **maxiter** (maximum iterations e.g. 10), **outflg** (logical flag turns on/off output e.g. True).

Example *Mathematica* 7 code to reduce the $\mu(E)$ to $\chi(k)$

```
ReduceData[inputfile_,De_,kmn_,kmx_,wgt_,rcut_,maxiter_,outflg_]:=
(If[!FileExistsQ[inputfile],(Print["File not found in: ",Directory[]];Return[])];
Print[fname=inputfile];
(*define simple window functions*)
fermi[x_,x1_,w_]:=1/(Exp[-(x-x1)/w]+1);
box[x_,x1_,x2_,w_]:=fermi[x,x1,w]-fermi[x,x2,w];
kmin=kmn;kmax=kmx;dk=.05;nptsca=2048;rmax=8;
dr=Pi/(nptsca*dk);rawdat=Import[inputfile,"Table"];
If[Length[rawdat]<100,(Print["Data too short: ",Dimensions[rawdat]];Return[])];
header=Take[rawdat,3];rawdat=Drop[rawdat,3];If[outflg,Print[TableForm@header]];
{etmp,xmutmp}=Transpose[rawdat];rawdat=Transpose[{etmp+De,xmutmp}];
dataemin=rawdat[[1,1]];dataemax=rawdat[[-1,1]];
(*optional cutoffs can be included here if needed*)
cutoffemin=dataemin;cutoffemax=dataemax;
inputdat=Select[rawdat,(#[[1]]>cutoffemin&&#[[1]]<cutoffemax&)];
(*estimate E0 from max derivative of smoothed data*);
xmufun=Interpolation[inputdat];{emin,emax}=xmufun[[1,1]];
de=.1;efgr=Range[emin,emax,de];xmufgr=Map[xmufun,efgr];
der=Drop[(RotateLeft[xmufgr,1]-xmufgr)/de,-1];
(*broaden derivative to make less sensitive to noise etc*)
kern=RotateLeft[Table[Exp[-.001(m-100.)^2],{m,1,200}],100];
kern=kern/Total@kern;der=ListConvolve[kern,der];
e0=efgr[[Position[der,Max[der]][[1,1]]]];
(*find normalization constant*);
preedge=Fit[Table[{e,xmufun[e]},{e,Max[e0-100,emin],e0-30,1}],
{1,e-e0,(e-e0)^2},e];
postedge=Fit[Table[{e,xmufun[e]},{e,e0+100,Min[e0+500,emax],1}],
{1,e-e0,(e-e0)^2,(e-e0)^3},e];
below=preedge/.{e->e0};above=postedge/.{e->e0};normconst=above-below;
If[outflg,Print[Grid[{{"E0","below","above","normconst"},
{e0,below,above,normconst}}]]];
(*interpolate to uniform grid in k-space anticipating fourier filtering*)
kofe[e_,e0_]:=Re[Sqrt[(.2625)(e-e0)]];
eofk[k_,e0_]:=e0+k^2/(.2625);kmax=Min[kmax,kofe[emax,e0]];
data=Table[{k,xmufun[eofk[k,e0]]/normconst},{k,kmin,kmax,dk}];
```

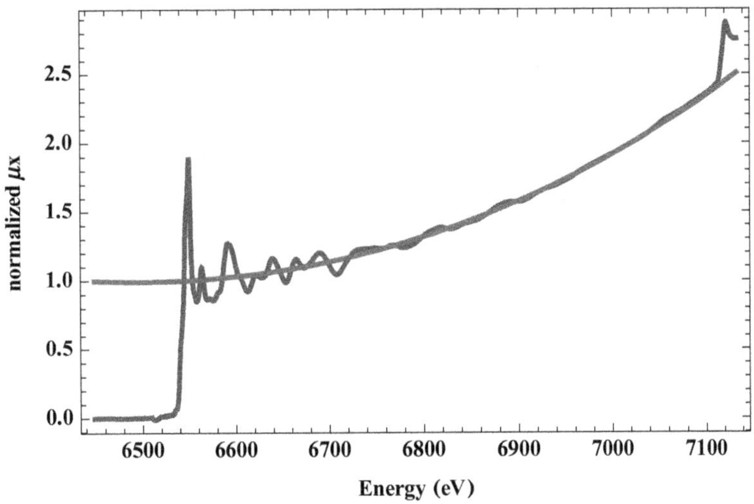

Fig. 5.9 Normalization of MnO data. The absorption feature above 7100 eV is a thin $CoSO_4 \cdot 7H_2O$ film used as an internal energy-shift reference.

```
If[outflg,Print[Plot[{(xmufun[e]-preedge)/normconst,(postedge-preedge)/normconst},
{e,emin,emax},PlotRange->All,PlotStyle->Thick,LabelStyle->Bold,Frame->True,
FrameLabel->{"Energy (eV)","normalized \[Mu]x"}]]];
{xdat,ydat}=Transpose[data];
(*do rough background subtraction*);
bkgerr=(Total[xdat^wgt(ydat-(a+b xdat+c xdat^2+d xdat^3))^2]//Simplify);
soln=NMinimize[bkgerr,{a,b,c,d}][[2]];bkg=(a+b*xdat+c*xdat^2+d*xdat^3)/.soln;
If[outflg,Print[ListPlot[{Transpose[{xdat,bkg}],data},Joined->True,PlotStyle->Thick,
Frame->True,LabelStyle->Bold,
FrameLabel->{"k(^-1)","Rough BKG fit to \[Mu]x"}]]];
(*cleanup residual background with iterative fourier filtering*);
kbar=(kmin+kmax)/2;kwindow=Table[1,{k,kmin,kmax,dk}];
kgrid=Table[k,{k,kmin,kmax,dk}];ydat=(ydat-bkg)*xdat^wgt;
data=Transpose[{xdat,ydat}];
oldydat=ydat;ncut=Floor[rcut/dr]+1;nrmax=Floor[rmax/dr]+1;
rgrid=Table[(n-1)*dr,{n,1,nptsca}];npts=Length[ydat];
bkgest=Table[0.,{npts}];Clear[bkgerror];
bkgerror[y_]:=Total[Take[Abs[Fourier[PadRight[y,nptsca]]],ncut]];
rwindow=RotateLeft[Table[box[n,1,2*ncut,.5],{n,1,nptsca}],ncut-1];
Do[newydat=ydat-bkgest;ttl="iter="<>ToString[iter];
ydat=newydat;err[iter]=bkgerror[ydat];
bkgest=Take[Re[InverseFourier[rwindow*Fourier[PadRight[kwindow*ydat,nptsca]]]],npts],
{iter,1,maxiter}];
If[outflg,Print[ListPlot[Array[err,maxiter],PlotStyle->PointSize[.02],
Frame->True,LabelStyle->Bold,FrameLabel->{"iter","Bkg err"}]]];
If[outflg,Print[ListPlot[Transpose[{xdat,ydat}],Joined->True,PlotRange->All,
Frame->True,LabelStyle->Bold,FrameLabel->{"k(^(-1))","chi"}]]];
If[outflg,Print[ListPlot[Transpose[{xdat,bkgest}],Joined->True,PlotRange->All,
Frame->True,LabelStyle->Bold,FrameLabel->{"k(^(-1))","bkgerr"}]]];
Return[{e0,Transpose[{xdat,ydat}]}])

FT[chi_]:=Fourier[PadRight[PadLeft[chi,Floor[kmax/dk]+1],nptsca]]
```

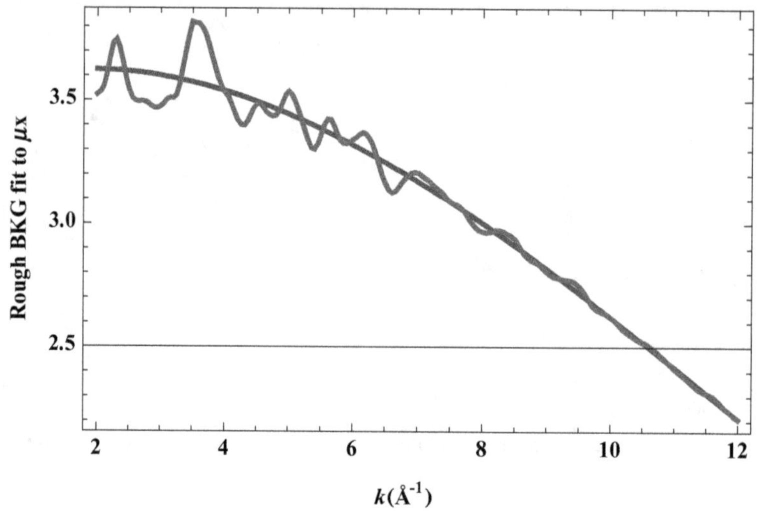

Fig. 5.10 Rough background subtraction of MnO data.

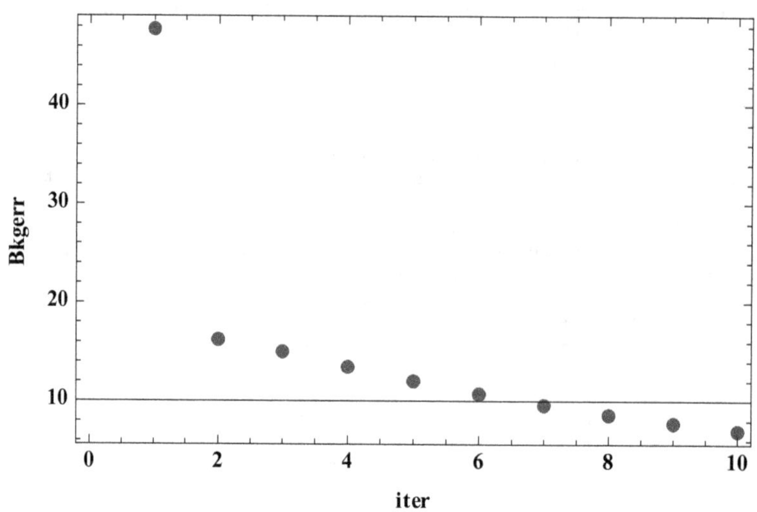

Fig. 5.11 Convergence of residual background removal below 1Å.

For comparison data reduction is also carried out by Athena. The graphical interface and results are shown in Figures 5.14–5.16. The results are substantially the same, except for a sign convention in the imaginary part of the Fourier transform. The Mathematica program above uses the

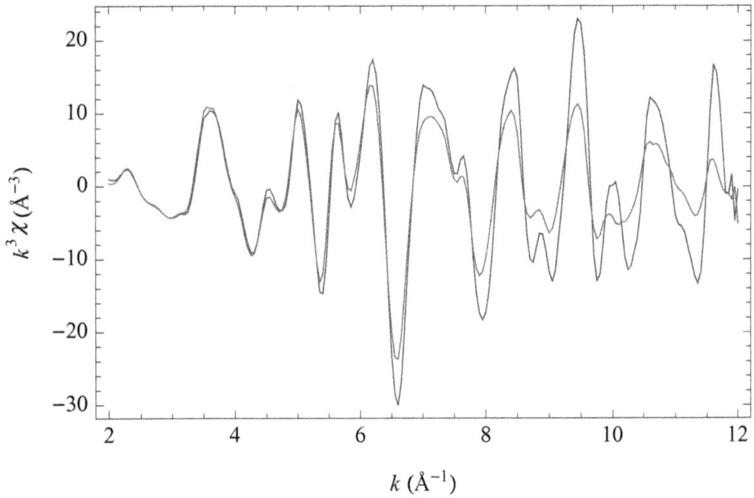

Fig. 5.12 $k^3\chi$ for MnO at sample temperature 80 K and 300 K. The EXAFS at the higher temperature damps out more quickly because of the larger thermal Debye–Waller factors.

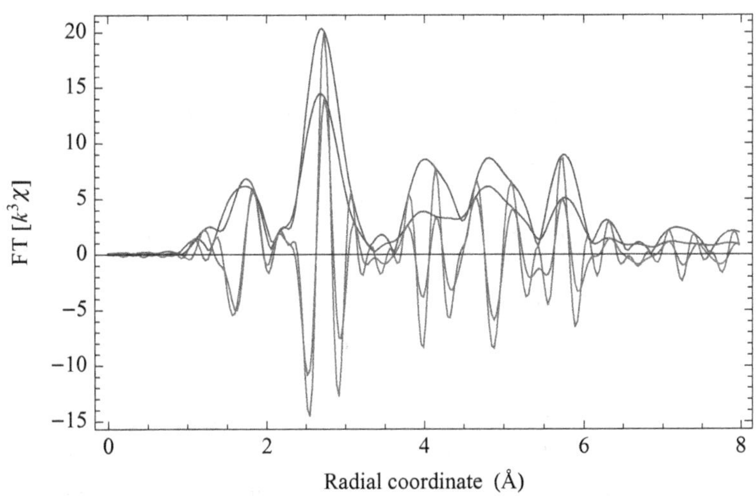

Fig. 5.13 Fourier transforms of $k^3\chi$ (Å^{-3}) for MnO at sample temperature 80 K and 300 K.

convention that if one transforms a sine function $\sin(kR)$, the imaginary part and the modulus peak at the same R.[8]

[8] To switch to Athena's sign convention, one could just take the complex conjugate of the complex array in r-space after transforming, or change all "Fourier" function calls to "InverseFourier," and vice versa.

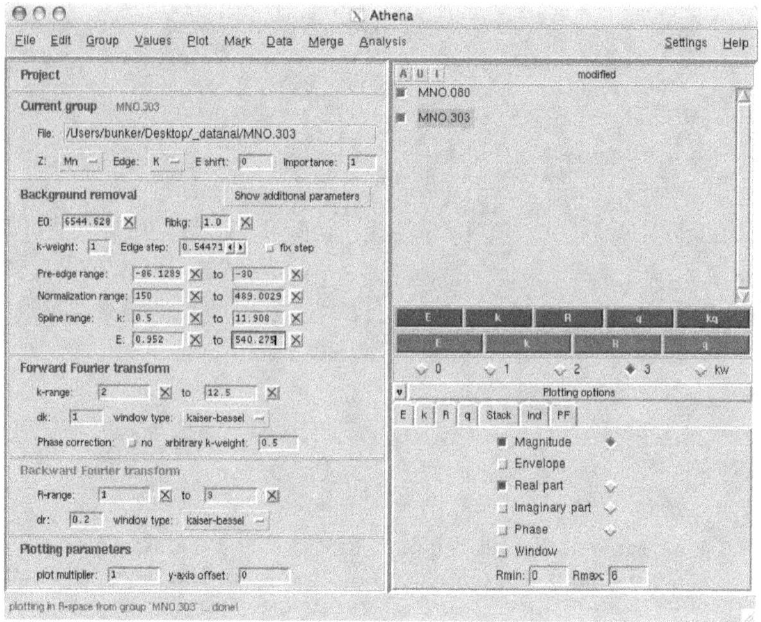

Fig. 5.14 Athena graphical interface.

5.2.13 Difficult normalization

Weng *et al.* [110] have reported an improved reliability of normalization by mapping the experimental data onto standard tables of cross-sections. This is done by scaling the experimental data by a multiplicative constant, and adding a variable background curve of low-order polynomials in $E - E_0$. A further term is optionally included to model contributions associated with the tails of the fluorescence line, which is greatly broadened by the Ge detector resolution function.

This procedure does indeed work well, and a simplified form is presented below. Results for MnO are shown in Figures 5.17–5.18. Instead of using tabulated functions, three user-supplied numbers are used as input: the calculated cross sections right below and right above the edge, and a power law exponent that is close to 3. The absorption of most materials decreases roughly as $1/E^3$. For MnO the exponent is close to 2.8 both above and below the edge. These three numbers are sufficient to generate a reference cross section "frame" to which the experimental spectrum can be matched.

Mathematica 7 code to map experimental μ onto reference cross-section

```
(*modified from Weng, Waldo, Penner-Hahn*)
(*note: energies must be absolute, not relative to edge*)
xsbelow=47.;(*input: value right below edge, cm^2/g*)
```

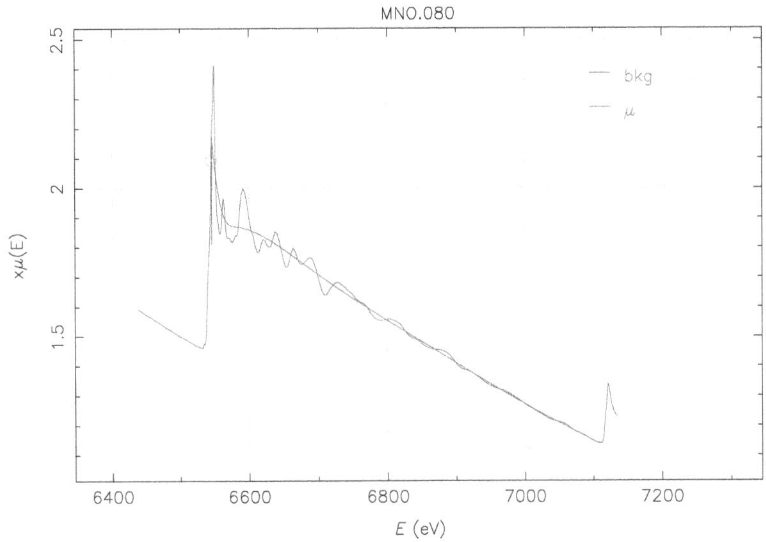

Fig. 5.15 Athena/Autobk background subtraction for MnO using default parameters.

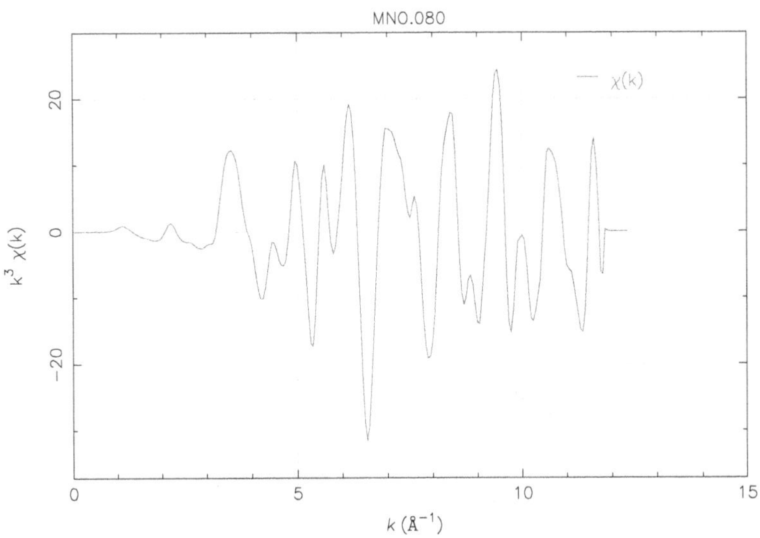

Fig. 5.16 Athena/Autobk $k^3\chi$ for MnO using default parameters.

```
xsabove=367.;(*input: value right above edge, cm^2/g*)
xp=2.8;(*input: decay exponent; or just 3*)
ecut=7080;(*cutoff energy optional; to disable, set to Infinity, or 10^5*)
{egrid,xmu}=Transpose[Select[inputdat,#[[1]]<ecut&]];
xs[e_]:=If[e<e0,xsbelow*(e0/e)^2.8,xsabove*(e0/e)^2.8]
xmuref=xs/@egrid;e=.;scaledxmufun:=
(a*xmu+b + c (egrid-e0)+d (egrid-e0)^2 +e (egrid-e0)^3+f (egrid-e0)^4);
```

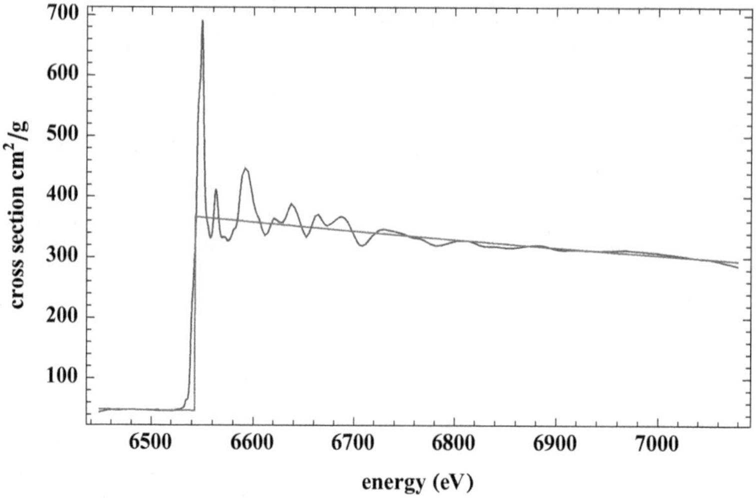

Fig. 5.17 MnO spectrum mapped onto "cartoon" cross section.

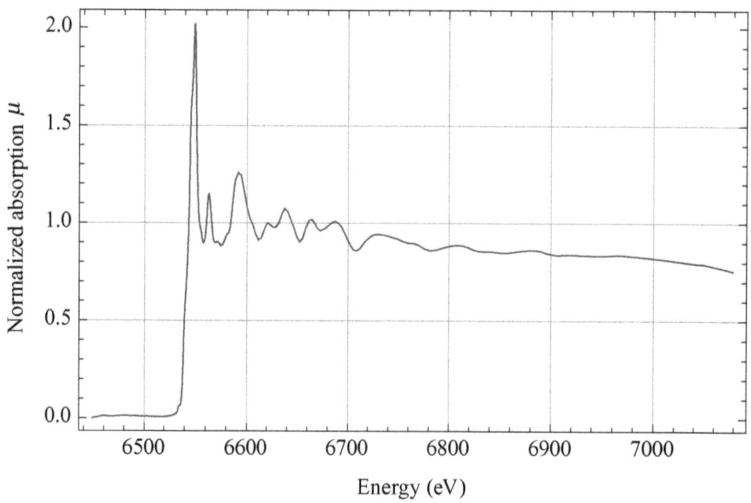

Fig. 5.18 MnO spectrum mapped onto "cartoon" cross section and normalized, ready for background subtraction.

```
(*Exclude immediate near-edge*)
fermi[x_,x1_,w_]:=1/(Exp[-(x-x1)/w]+1);
box[x_,x1_,x2_,w_]:=fermi[x,x1,w]-fermi[x,x2,w];
ewgt[e_]:=(1-box[e,e0-10,e0+10,1])(1-box[e,ecut,ecut+1000,1]);
(*give some weight to pre-edge residual by weighting with 1/sqrt(cross section)*)
reswgt=Map[ewgt,egrid](1/Sqrt[xmuref]);
residual=reswgt*(Map[xs,egrid]-scaledxmufun);errfun=residual.residual;
soln=NMinimize[errfun,{a,b,c,d,e,f}];
scaledxmu=scaledxmufun/.soln[[2]];
normdata = (scaledxmu - Min[scaledxmu])/(xs[e0 + .1] - xs[e0 - .1]);
```

5.2.14 Refining E_0

There are many situations in which one needs to adjust the relative E_0 of data sets in k-space with respect to each other, with respect to theory, or even different scattering paths with respect to each other. This is easily done, but there are a couple of subtleties. Shifting E_0 changes the k scale. Usually the data are weighted by a power of k, multiplied by window functions, etc. Such weighting functions should be divided out before changing the E_0. The k array then is transformed by $k' = \sqrt{k^2 - \tau\delta E_0}$, after which it is no longer a uniform grid, so the (k', χ) arrays then need to be interpolated back to the desired uniform grid. Once that is accomplished, the weighting function can be reapplied.

If E_0 is changed, it effectively adds a term α/k to the phase, where $\alpha = \tau R_{\text{eff}}\delta E_0$ and R_{eff} is half of the slope of the phase shift when plotted vs. k. Adding this term by shifting E_0 alters the slope of the phase, which appears as a small shift in distance. This introduces a fitting correlation between δE_0 and distances.

Since the phase shift decreases with energy above the edge, a tell-tale sign of a relative E_0 shift between two $\chi(k)$ functions is a phase discrepancy that decreases as k increases. In contrast, distance shifts cause phase differences that increase with k.

5.2.15 Beats: when is a "shell" a shell?

In EXAFS it is common to refer to groups of atoms at a similar distance as a coordination "shell." Because of geometrical constraints (finite size) and chemical bonding it is normal for the nearest neighbors to group at similar distances, making up the first shell of atoms. It is also typical for the radial distribution of atoms to have a minimum following the first shell maximum, and then rise to another maximum (second shell) at a larger distance. In amorphous materials this radial correlation decays away rapidly with distance, but in crystals it persists to large distances, even when thermal motion is taken into account. EXAFS is insensitive to such long-range order, however, except indirectly through its modest sensitivity to vibrational structure.

There is some ambiguity about the definition of a shell – the term is used loosely, and inconsistently, depending on the context. We see peaks in Fourier transforms, and often refer to the first significant peak at low r as the "first shell," and the next Fourier peak as the "second shell." The peaks may or may not correspond well to the actual radial distribution. Very often some of the atoms do not separate nicely into groups by distance, and in such cases the appearance of the FT can be sensitively dependent on the transform range, weighting, and other details.

When the atoms in a shell are not all of the same type, they must be treated differently in the fitting process. I prefer to call these "sub-shells," but others may call them different shells. The terminology is not standard and can be confusing because of inconsistent usage.

Another common point of confusion is "disorder." Because the focus in EXAFS is on the central atom and the radial structure around it, a person may call a spread in radial distances "disorder," when in fact the structure may be very well-ordered in three dimensions.

More fundamentally, one should ask how close atoms have to be together to be lumped into the same shell. This depends on the distance resolution of the experiment, which depends on the k range. Two shells can only really be resolved if their χ signals can get π radians out of phase somewhere in the data range. The phases of the two contributions could also have different constant terms in their phase shifts, if they are from different atoms. To nail down that ambiguity, there must be a *change* of at least π radians over the data range, which implies the minimum resolvable ΔR is given by $\Delta R = \pi/(2\Delta k)$, where Δk is the width of the k-space window. If one knows that the shells consist of the same atomic type it is sufficient that $\Delta R = \pi/(2k_{\text{max}})$. If the scatterers are of the same type, the position of the beat provides the difference in distance between the two subshells as $\Delta R = \pi/(2k_{\text{beat}})$. The depth of the minimum tells you the ratio of the amplitudes. If the amplitude drops to zero at the beat, the two contributions have equal amplitudes there, and the phase jumps by $\pm\pi$.

We can easily analyze beats quantitatively. At a beat the two χ contributions are completely out of phase and interfere destructively. The amplitude at that point is the difference between the two amplitudes. It is easy to show that the combined amplitude A and phase ϕ are given by

$$
\begin{aligned}
A &= \left(A_1^2 + A_2^2 + 2A_1A_2\cos\Delta\phi\right)^{1/2} \\
\phi &= \arctan\left(\frac{A_1\sin\phi_1 + A_2\sin\phi_2}{A_1\cos\phi_1 + A_2\cos\phi_2}\right) \\
&= \bar{\phi} + \arctan\left(\left(\frac{A_1 - A_2}{A_1 + A_2}\right)\tan\left(\frac{\Delta\phi}{2}\right)\right),
\end{aligned}
$$

where $\bar{\phi} = (\phi_1 + \phi_2)/2$ and $\Delta\phi = \phi_2 - \phi_1$.

The phase $\phi(k)$ exhibits a positive or negative jump in the phase at a beat. The size and sharpness of the jump is related to the ratio of the amplitudes.

The sign of the jump depends on whether the larger amplitude shell is at a larger distance (positive phase jump), or the large amplitude is at a smaller distance (negative phase jump). For example, a distribution with four atoms at 2 Å and two atoms at 2.3 Å will exhibit a negative phase jump. This can be pictured using a phasor diagram.

The parameter σ is a measure of the spread in distance in a shell. If $k\sigma << 1$ we can always approximate the Debye–Waller factor as Gaussian $\exp(-2k^2\sigma^2)$, but if it is not, other terms enter the amplitude and phase (cumulant expansion). These can provide additional information about the distribution of atoms, such as the skewness. For example, consider the distribution of atoms just described, 4@2.0 and 2@2.3. This distribution is skewed outward, and has positive third cumulant. At low k the phase (minus the atomic phase shifts) is linear, but at higher k the cubic term in the phase causes the phase to curve downward, anticipating the negative phase jump which is coming at the beat. Contributions from the higher-order cumulants grow as k increases. At the beat cumulants of all orders will contribute.

This tells us that a shell is a shell if it can be described over the data range by just a few cumulants. If many cumulants are needed, as is the case at a beat, it will be necessary to conceptually decompose the shell into subshells, each of which is described by its own small set of cumulants.

5.2.16 Fourier methods

Appendix A gives an elementary mathematical introduction to Fourier series, transformation, filtering, and the convolution theorem. A simple method for deconvoluting window distortions is also presented.

Before diving into Fourier methods, it should be mentioned that wavelets are another class of representations of EXAFS data that describe it using a two-dimensional transform [105]. The wavelet transforms show some promise of providing a more meaningful visual representation of the data in r-space. This approach is still fairly new however and does not yet seem to be widely used.

Fourier Transforms (FTs) represent a signal in terms of its frequency content. Any signal can be expressed as a sum of sine and cosine waves of different frequencies and amplitudes. The FT equivalently describes a signal as a sum of terms of the form $\exp(\pm ikr)$, with complex coefficients.

EXAFS data basically consist of sums of damped sine waves, and the rapidly oscillating contributions correspond to longer path lengths. An FT of the data therefore gives something analogous to a map of the distribution of distances (or half-pathlengths when multiple scattering is considered).

Interpretation of FTs can be subtle however – the FTs should not be considered to be (or described as) "radial distribution functions." That notion should be thrown overboard at the outset. The reason is that the FT contains extraneous information such as scattering phases and amplitudes, as well as other terms such as cumulants. More sophisticated methods such as regularization (see below) can generate estimates of radial distribution functions, however.

Much of the appeal of EXAFS stems from the intuitive r-space representation, in which peaks correspond (more or less) to groups of atoms at different distances. In most crystalline and molecular structures interatomic distances near a particular atom fall loosely into coordination shells, which are groups of atoms at similar distances. Because of the finite size of atoms, it is not surprising that if there is a group of atoms that is chemically bonded to the central atom, there might be fewer atoms slightly further away, and then another wave of atoms at somewhat greater distance. Even the random packing of spheres gives such behavior. At large distances the correlation should decrease, even in a crystal, because of thermal motion. At large distances the radial distribution should approach a constant value.

It is equally valid to represent the data in either k-space or r-space; one should use the best representation for the task at hand. Some people dislike Fourier methods because they can be tricky, and it's not hard to make mistakes in transformation and filtering. It is also possible to do EXAFS analysis without using Fourier methods at all. However they are a very useful part of the data analysis toolkit.

EXAFS data $\chi(k)$ consist essentially of sums of damped sine waves. The damping is due to factors in the EXAFS amplitude: the scattering amplitude, which has structure characteristic of the scattering atoms, but decreases roughly as $1/k^2$ at high k; an additional $1/k$ factor; and the Debye–Waller factor $\exp(-2k^2\sigma^2)$. If $\chi(k)$ is FTed without any pre-weighting, one gets peaks in r-space, but broad distorted ones. The r-space peak width represents the range of r-space contributions that interfere destructively with each other to produce the k-space amplitude decay.

If $\chi(k)$ is weighted by a suitable power of k, typically k^3, the EXAFS amplitudes flatten out into something resembling sine waves of constant amplitude. When these are FTed, they produce peaks with a width in r-space that are determined by the transform range in k-space, $k_{\max} - k_{\min}$.[9] If peaks are closer together than that width, they interfere with each other, and

[9] This is evident from the convolution theorem (see Appendix 1) since the peaks are essentially the FT of the k-space window.

cannot be separated clearly. If one attempts to do so, misleading conclusions will be drawn from the data.

Usually data are represented as peaks in the modulus of the FT. The FT is actually a complex function (it has real and imaginary parts), and half of the information is contained in the phase. The modulus is most often shown; it is more useful to present both the modulus and the imaginary part, as in Figure 5.13. The relative positions at which the imaginary part and the modulus have their maxima is affected by the constant part of the scattering phase shift, which can be helpful for distinguishing backscatterers.

The derivative of the phase shift is helpful for identifying interference effects, as shown in Figure 5.19. Note the correlations between the spikes in the phase derivative in r-space and interferences between the peaks in the modulus. These indicate that the peaks are not completely separated, and the tails are interfering with each other. If such interferences are ignored, serious errors can occur when performing single shell fits over a restricted range in r-space, or inverse transforming limited r-space regions back into k-space.[10]

Roughly speaking, the heights of the transform peaks represent the average k^n weighted EXAFS amplitude of the shell's signal over the transform range in k-space. The EXAFS amplitude is affected by the nature of the scattering atoms, their distributions in space, and the k-weighting and windows that are used.

The position of the peak in r-space represents the average slope of the phase over the transform range in k-space. The interatomic distance term $2kr$ makes the dominant contribution to the phase slope, and therefore the peak position is largely due to that. The central atom phase shift has a negative slope, however, which causes an average apparent distance shift of -0.4 or -0.5 Å to the peaks in the transform. The magnitude of the central atom phase shift varies considerably for different elements, but the average phase slope is similar, so the apparent distance shift is fairly consistent for different central atoms. This is accurately taken into account in later data analysis. The scattering phases, and other terms such as the third cumulant C_3 and phase jumps due to interferences, also can shift the peak positions.

As mentioned above, the widths of the FT peaks are determined mostly by the range over which the data are transformed, i.e. the k-window width. The peak widths are not directly related to the width of the actual spatial distribution of atoms. That information however can be obtained from the Debye–Waller factors and cumulant expansion, or by direct determination

[10] Inverse transform k-space is called q-space in Athena.

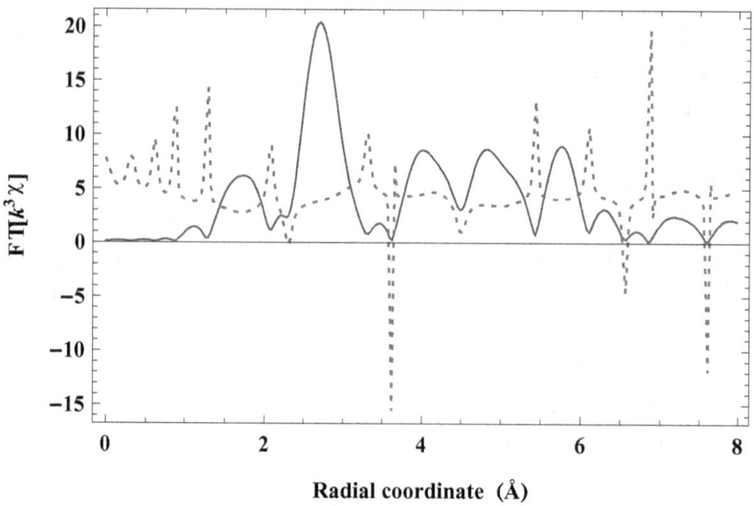

Fig. 5.19 Modulus and the derivative of r-space phase after elimination of 2π discontinuities. The derivative is multiplied by (0.25) for graphical clarity. Notice that the jumps in the phase (± spikes in the phase derivative) occur where there are residual interferences between adjacent peaks.

of the radial distribution function using regularization methods or other means.

5.2.17 Optical transform

The scattering amplitudes of heavy atoms have a complex structure that complicates the appearance of Fourier transforms by introducing multiple peaks, even if the atoms are at a single distance. Typically scattering amplitudes show sharp minima that are due to electron interference effects (Ramsauer–Townsend resonances) when scattering from the atom. Figures 4.7 and 4.8 show examples of such amplitude structure. In the first three rows of the periodic table these minima occur at low-k, so they are outside or near the lower limit of the transform range. Heavier atoms have them in locations that may complicate interpretation of the FT.

From a signal processing point of view, the minima are equivalent to those produced by interferences between two or more nearby frequencies, like the side-bands in radio communications. The FT of the amplitude gives the appearance of multiple close distances. The convolution theorem (Appendix 1) tells us the FT of a product of functions is the product of the FTs of the functions, so even an ideal sine wave transformed over a large window will look like the FT of the heavy atom's scattering amplitude, and give multiple peaks.

To simplify interpretation of FTs, if a shell of atoms consists of a single type of atom, it can be helpful before transforming to divide out the scattering amplitude from the data and subtract the scattering phase (which also shows structure related to that in the amplitude).

In practice one divides the data by the complex scattering amplitude. If the initial data are real-valued (consist of both positive and negative R components), the positive R term will be corrected as intended, and the negative R one will be doubly affected, but when inverse transforming it is discarded. Pre-compensation in this manner is sometimes called an "optical transform" for historical reasons.

5.2.18 Ratio method

The simplest form of EXAFS analysis is the Ratio Method [4, 61, 53]. It is primarily useful for single scattering analysis of shells that contain only a single type of atom. The ratio method has mostly been superceded by more elaborate and flexible fitting methods using theoretical data, but it still has its uses. For example, it is useful for making precise relative comparisons between experimental data sets, e.g. those of a sample measured at different temperatures. The temperature dependence of the Debye–Waller factors provides information about the stiffness of the bonds and the mass of the atoms.

One useful aspect of the ratio method is that it is inherently a relative measurement, a comparison between two similar spectra. A multitude of errors, both known and unknown, are eliminated in such comparisons. For example, errors due to amplitude suppression near the edge due to the tails of the monochromator resolution function will tend to cancel out. Similarly, Fourier filtering distortions in the back-transformed data cancel out surprisingly well if the two spectra being compared are filtered with exactly the same k-windows, and r-windows of the same width.[11] The overall scale (constant factor) and the phase of the oscillations are irrelevant to the window distortions, because the FT is a linear operation, and multiplying by a complex constant doesn't change the filtering distortions. If the shells

[11] If the entire transform peak and all sidelobes were inverse transformed, with no other peaks interfering with it, the undistorted spectrum would be recovered. In practice we are always separating shells from each other, and we do not have the luxury of large inverse transform windows. The price of cleanly obtaining single shell data is the introduction of some distortion in the amplitudes. Fortunately these distortions can mostly be made to cancel out or can be deconvoluted using the method shown in Appendix 1.

of the spectra have very different σ^2, however, the transform peak widths may different enough to introduce systematic errors. These can be reduced by pre-weighting one of them with an artificial $\Delta\sigma^2$ to make the spectral amplitude function more similar. This value can then be subtracted out afterward.

The ratio method in its simplest form is based on the basic EXAFS equation. Consider an "unknown" sample and a standard sample whose structure is known. If they have the same type of atom in a shell, taking the ratio of their Fourier filtered EXAFS amplitudes will cause the scattering amplitude to cancel out. Here we assume that the average distances are close enough that the spherical wave effects and the mean free path term also cancel out. In that case the logarithm of the ratio of two EXAFS amplitudes $\ln(A_2(k)/A_1(k)) = \ln(N_2/N_1) - 2\ln(R_2/R_1) - 2k^2(\sigma_2^2 - \sigma_1^2) + \cdots$ where higher-order terms involve the cumulant expansion (Appendix 2), which can be neglected if the disorder is small enough. In that case, the log-amplitude ratio gives a straight line when plotted vs. k^2. The y-intercept gives the log of the coordination number ratio and the slope gives the difference in variance of the distance to atoms in the shell. Similarly the distance can be determined from the phase difference $\phi_2(k) - \phi_1(k) = 2k(R_2 - R_1)$, where the central atom and scattering phase shifts are presumed to cancel out between standard and unknown. When plotted vs. k this is a straight line of slope $2(R_2 - R_1)$ and zero intercept. Since the distances of the standard are presumed known, this allows you to determine the distance in the unknown, and then the distance ratio in the log-amplitude ratio equation, and then coordination number of the unknown sample. If σ_1^2 is known, σ_2^2 also can then be determined. If the temperature dependence is being studied in a material, even the $\Delta\sigma^2$ is sufficient to obtain useful information about bond stiffness, Einstein and Debye temperatures, and related quantities.

In practice there is always a small (a few eV) relative E_0 shift between the standard and the unknown. Such shifts introduce a term proportional to $1/k$ into the phase, which shifts the slope and intercept of the phase linear fit. The relative E_0 error can be adjusted by shifting the E_0 of just one of the spectra. To determine the correct amount, just ensure that the y-intercept of the phase difference plot goes zero, as it should be.

It is essential to use error bars when performing the linear fits to the log-amplitude ratio and phase difference. The filtered data only can be considered linear within the error band. The fit should give less weight to high-k data because errors in the data tend to increase with k. Such error bars can be generated by comparing the amplitudes of different scans on the same sample. The data within about half an oscillation of the window edge should be excluded from the fit because of window distortions.

5.3 Data fitting

5.3.1 Introduction

There is a vast literature on probability and statistical data analysis. Much of it is philosophical, and some of it is contentious. Good books on this subject include [112, 113]. In statistics there are alternative "frequentist" and "Bayesian" points of view, philosophies, and indeed different definitions of probability. The frequentist probability is defined in terms of frequencies of (perhaps hypothetical) repeated measurements, and the Bayesian with degrees of plausibility: it is a quantitative theory of how new information should affect one's state of knowledge. The mathematical formalism of probability is generally accepted, but it admits different interpretations, and occasionally different applications.

The Bayesian approach has been gaining popularity in recent years. Its name derives from Bayes' theorem, which states that the conditional probability (in the notation of James [112])

$$P(\theta|\mathbf{X}) = \frac{P(\mathbf{X}|\theta)P(\theta)}{\int P(\mathbf{X}|\theta)P(\theta)d\theta},$$

where $\mathbf{X}$ is a vector of measurements, θ represents theoretical variables, and $P(a|b)$ is the conditional probability, i.e. the probability of a being true if b is known to be true. Viewing probability as an expression of our state of knowledge, rather than a frequency of events, this equation gives us a way to determine the probability that a given set of theoretical parameters has a range of values if we have some initial (prior) knowledge of the possibilities, and we have additional information from measurements.

EXAFS analysis generally uses standard statistical methods or variants thereof, but certain assumptions that are typically made in elementary statistics may not be valid, in particular the statistical independence of data points, and normal (Gaussian) distribution of experimental errors. The distribution of errors in the data is an experimental question; assumption of a normal distribution is unwarranted until it is measured. In most EXAFS data analysis the statistical independence of data points is compromised because fitting usually is done to processed data, and the processing (e.g. interpolation) statistically correlates the data points. It is possible to account for the correlations through the (nondiagonal) data covariance matrix, but this is not supported by existing programs, to my knowledge. Also interpolation to a fine grid may give the user (and programmer) the impression that the number of degrees of freedom in the data is greater than it really is. For this reason it is not unusual for EXAFS data analysis programs to modify in an ad hoc manner standard statistical methods in an attempt to address the problem and to prevent overfitting of the data.

A sensible way to avoid this complication is to fit the unprocessed (uncorrelated) data [100]. This holistic approach does not preclude using also the traditional reductionist approach to data analysis in order to gain insight into the data, but it does provide a viable and arguably better alternative. Instead of dismantling the data, by subtracting an estimate of the background etc., one leaves the data alone, and instead attempts to limit the dependence of the conclusions on the background.

Recent work by Krappe, Rossner [114, 115, 116, 117] *et al.* and Klementev [118, 119, 120] are good examples of data analysis from a Bayesian perspective. It is inescapable that in all data analysis the conclusions will depend on the correctness of the assumptions and prior knowledge that go into it. In particular rather broad assumptions are made in Bayesian estimation not only about the probability density functions for the experimental errors but also *the unknown parameters.*

Here we will attempt to convey basic ideas that are needed to do XAFS analysis as it is currently practiced; the methods are expected to evolve. In XAFS analysis we construct a theoretical model of the data containing unknown parameters, and we wish to determine their values. Parts of the theoretical model are well established, such as the basic theory of EXAFS. Other parts of the model have been well tested, such as the electron scattering contributions in EXAFS that are computed by FEFF, GNXAS, EXCURV, etc. Other parts of the model are less well determined; they must be defined by the data analyst. In particular it must be decided how to most usefully (and economically) describe the distribution of atoms and the disorder in the structure. Formulation of these aspects of the model is the most subtle and tricky part of data analysis.

5.3.2 Errors

We always fit to data with some presumably known or determinable experimental uncertainties from random noise and other influences. These can be divided into random noise (such as photon counting) and nonrandom errors. One essential job of the experimenter is to track down and eliminate the nonrandom (systematic) errors, but they cannot be completely eliminated.

The theoretical model contains undetermined parameters. The set of parameters can be regarded as coordinates in a space called parameter space. If we have four variables, say N, R, σ^2, and δE_0 that we wish to determine in fitting, we have a four-dimensional parameter space. In many cases we have a twenty- (or more) dimensional parameter space.

When fitting data we systematically vary the hypothetical values of the "unknown" (i.e. only partially known) parameters to give the best agreement with the experimental data. We can't (and shouldn't) achieve perfect agreement with the data, because of the noise and other uncertainties ("errors"), but we can determine how far we can vary the parameters around the best-fit values while still getting an acceptable fit to within the experimental (and theoretical) errors. This defines a volume in parameter space in which the true values presumably reside. Identifying and describing the volume (which may not be simply connected) is the job of data analysis.

First we need a way to quantify the comparison between the experiment and theoretical fit. There is an infinite number of ways to do this, with various advantages and disadvantages, but among the simplest is to calculate the squared deviation between the data and theory. The deviation ("residual") is squared so that both positive or negative deviations contribute positively to the error statistic.[12] If there are many data points the squared deviations are summed: $\epsilon^2 = \sum_i (y_i - T(x_i, \{\alpha\}))^2$, where y_i is the ith measured data point, T represents the theory function, which depends on the independent variable x_i and the set of parameters $\{\alpha\}$.

Since the noise and errors are not necessarily the same for all measurements, we should wish to give less weight to the measurements of greater uncertainty. This can be done in many ways, but if the errors are independent and normally distributed the optimal weights are the inverse variances σ_i^2 of the experimental measurements:

$$\epsilon^2 = \sum_i \frac{(y_i - T(x_i, \{\alpha\}))^2}{\sigma_i^2}. \tag{5.1}$$

If the distribution of measurement errors is a normal distribution with variance σ_i^2, then the error function consists of a sum of N random normal deviates of zero mean and unit variance, which follows a chi-square distribution. There is of course no guarantee that the distribution of experimental errors is a normal (Gaussian) distribution.

An alternative approach is to use a likelihood function. Given a set of parameters $\{\alpha\}$, there is a certain probability density function (pdf) that

[12] The absolute value is also useful, giving estimates that are less sensitive to outliers.

describes the probability a certain spectrum y will be obtained; call the probability density $f(y|\{\alpha\})$. If a set of independent hypothetical spectra y were measured, the joint probability would be the product of the pdfs for each measurement: $P(\{y\}|\{\alpha\}) = \Pi_n f(y_n|\{\alpha\})$. If we then replace the hypothetical spectra by the measured spectra $\underline{y}_n$ we have the likelihood function, which is a function only of the unknown parameters $\{\alpha\}$.

The logarithm of the likelihood function is maximized as a function of the parameters to determine the best fit. In the case of a normal distribution of errors, the pdfs are Gaussian, and maximizing the log-likelihood function is equivalent to minimizing ϵ^2.

5.3.3 Finding the best fit

A good reference on numerical methods is Numerical Recipes [111].

In the simplest fitting cases the function depends linearly on the parameters. For example, if one wishes to fit with a linear combination of functions (e.g. polynomials) the fit parameters are the unknown coefficients. In that case the error function (5.1) depends quadratically on the parameters. If viewed in parameter space, the constant surfaces of the function are hyper-ellipsoids. The orientation and lengths of the axes of the hyper-ellipsoid of suitable size gives the correlations and parameter errors. The minimum can be found by taking the derivatives with respect to each parameter and setting the result to zero. This gives a set of linear equations that can be solved for the parameter values at the minimum point.

Linear fitting problems have the great virtue that they have only one minimum. Nonlinear fitting problems generally will have multiple local minima, and there will be at least one global minimum that has the lowest value. In two parameters the local minima can be pictured as a surface, like a mountain range, and the minima are like mountain lakes. Then picture a mountain lake in twenty dimensions. We are interested in the best fit, which is the global minimum, but any region in the parameter space that is consistent with the data (and that also is physically plausible) should be considered.

The method described above for the linear fitting problem doesn't work for the nonlinear case unless you are close to a local minimum. One reasonable approach is to use a gradient search, which just iteratively calculates the negative local gradient of "downhill" and follows it to the bottom. For complicated surfaces this is often non-optimal, and a class of hybrid "conjugate gradient" methods such as Levenberg–Marquardt has been invented with good performance. In essence this is equivalent to gradient search far from the minimum; it smoothly converts to the linear search procedure described above when close to the minimum.

Such algorithms usually work very well for finding local minima, but they get trapped there; they can't find the global minimum. Performing parallel minimizations using Levenberg–Marquardt with many random starting points in the parameter space can be an effective way to find all the multiple minima. There are also a number of other algorithms that are designed to avoid getting trapped in local minima. Examples include the simplex method [122], Simulated Annealing [121], Reverse Monte Carlo [124], and the genetic algorithm Differential Evolution [123].

5.3.4 Parameter correlations

Some fitting parameters have similar effects to other fitting parameters. For example, changing the coordination number parameter multiplies the data by a constant scale factor, so the amplitude is changed uniformly as a function of k. Changing the σ^2 also affects the amplitude, but it has a greater effect at high k than at low k. The similar effect of these parameters implies that they are correlated. If you perturb one of them, the other one can adjust to partially make up for it. The correlation introduces larger uncertainties in the parameter errors. A similar correlation is observed between δE_0 and the distance R. Both affect the slope of the phase, but in different ways. The shorter the range of data in k space, the more highly they will be correlated.

Multiple parameters also show correlations. Consider, for example, a situation in which one is floating the positions of two subshells of the same type of atom. If the distances are always far enough apart (in the fitting process), there is no problem, but if the distances approach each other too closely, and σ^2 of one or both shells is also a floating parameter, there will emerge a strong correlation between $R_2 - R_1$ and σ^2. The fitting problem becomes ill-determined, and probably numerically unstable. This can be prevented by imposing constraints between the distance variables so that they can't approach each other too closely.

An extreme form of parameter correlation (which is trivial to avoid) is if the constant S_0^2 and coordination numbers are simultaneously floated. The two scale factors have exactly the same effect on the data, and the fitting problem will be ill-conditioned. One of them should be fixed.

5.3.5 Error analysis

The canonical approach to error estimation [112] is to determine the region around the best fit that contains the true value with a certain probability β. The projection of that volume onto an axis corresponding to a parameter gives the parameter errors (for the chosen value of β).

In the linear case a simple method is to calculate the matrix of second derivatives ("Hessian")

$$H_{ij} = \frac{\partial^2 \epsilon^2}{\partial \alpha_j \partial \alpha_k}$$

of the error function with respect to the parameters α. The first derivatives (gradient vector) are zero at the minimum. If the eigenvalues of the Hessian are all positive, it is positive-definite, and the point is a true minimum. Otherwise it is a saddle point (or a maximum). This shows how rapidly ϵ^2 increases as parameters deviate from the best-fit value. If ϵ^2 is properly scaled by the experimental measurement errors, the inverse matrix H^{-1} is the covariance matrix, and the diagonal elements are the squared parameter errors. The off-diagonal elements, when divided by the square root of the product of the corresponding diagonal elements, give the correlation matrix. This expresses the degree to which changes in one parameter can masquerade as changes in another parameter. In the linear case the parameter errors are symmetrical around the best-fit point.

In nonlinear fitting the situation is similar but the parabolic shape at the minimum may be only approximate, and the parameter errors may no longer be symmetrical about the best-fit value. It is possible to get reasonable error bars using the procedure described above for the linear case, but more accurate methods are available. A procedure derived in James [112] and implemented in the program Minuit from CERN (invoked with the minos command) involves choosing a parameter, floating all parameters except that one, and determining how far the chosen parameter may deviate from the best-fit value before it increases ϵ^2 by a fixed value that depends on the level of confidence required, and the normalization of ϵ^2. In this case the error bars may be asymmetrical about the best-fit value. The process is repeated for each parameter.

5.3.6 Independent data points

In these methods, over-fitting of the data will result in an increase in the error bars in parameters. Adding unnecessary parameters will introduce correlations that will cause the error bars to inflate, so that information about the original parameters is reduced. This raises the question: how many parameters can be sensibly fit? It can be argued that one should not put external limits on the number, but simply allow the confidence intervals on the parameter estimates to tell the story. Nevertheless it is instructive to estimate it and we will now do so.

EXAFS data exist over a finite range in both k-space and r-space, and therefore they have limited information content. We know from a Fourier series expansion that only a finite number of coefficients are needed to represent a signal of fixed bandwidth. A simple way to estimate the number is as follows. We know (from Appendix 1 on Fourier filtering) that all of the data in the FT exist on an array N points in size.[13] Suppose the k window width is Δk wide, and the r-space data are contained in total width ΔR. The grid spacings respectively are δk and δr. δr is related to the other parameters $\delta r = \pi/(N\delta k)$. The maximum r value in the r space is $(N/2)\delta r = \pi/(2\delta k)$, where the factor of two comes in because half the space is allocated to negative frequencies which offer no new information. The fraction of the total information contained within ΔR is then $2\delta k \Delta R$. The number of k-space data points in the window is (about) $\Delta k/\delta k$. But most of these data points actually carry no information, because the signal can be accurately represented within only a fraction of the total r-space. Therefore the actual information content in the data is number of k-space points $\Delta k/\delta k$ times the compression factor $2\delta k \Delta R$. This gives a total "independent point" count of $N_{\text{ind}} = 2\Delta k \Delta R/\pi$. Stern [125] has argued that a more precise value is $N_{\text{ind}} = 2\Delta k \Delta R/\pi + 2$; the extra +2 can be handy in a pinch.

In my view the expression above is an approximate but useful guideline to estimate how many parameters one can hope to fit. The estimate of the number of points in k-space is only roughly correct, because the signals are not sharply band-limited in r-space, and the original independent data points are correlated by interpolation to k-space (if that is done). The question of the information content of the data is best approached from the perspective of information theory.

5.3.7 Path by path fitting of EXAFS

A standard way to calculate EXAFS $\chi(k)$ is to calculate the spectrum using a multiple scattering expansion. Here we use FEFF8.4 as an example. Some of these paths are shown in Figures 4.3–4.5 in Chapter 4. This highly portable Fortran 77 program (at version 8.4 at the time of this writing) calculates all significant multiple scattering paths up to a finite path length. It also

[13] The final result will not depend on N so it should apply also in the continuum limit as $N \to \infty$.

can calculate XANES using full multiple scattering, and it can calculate self-consistent potentials, although it does take time to do so for a large cluster. FEFF8.4 also can estimate the effective σ^2 values for each path in various approximations. The inputs (read from a plain text file) required are essentially the atomic numbers and (x,y,z) coordinates of each atom in the cluster, and various control parameters. The program is quite flexible in what it calculates; much of the information is available in plain text files, by selecting the appropriate PRINT directive.

The multiple scattering paths each have a degeneracy N (the number of equivalent atoms in single scattering paths), an effective distance R (half path-length), and a mean square variation parameter σ^2, just like the canonical EXAFS equation. Once calculated, each of these paths can be easily modified to give a different N, R, σ^2, and even E_0 (if that is appropriate). The spectrum for a structure that is close to the reference structure can be computed without re-running FEFF, which makes it fast to do within a fitting loop. This is similar to the classical approach in which the spectrum of a standard compound was modified to fit a similar material of unknown structure. Artemis/IFEFFIT [96] work this way.

In this style of fitting, the user needs to devise mathematical expressions for these "numerical path parameters" as functions of a smaller set of fitting parameters that are adjusted by IFEFFIT to give the best fit to the data (usually over a selected region in r-space). For the data analyst the task is to mathematically relate the changes in geometry of all the paths to the limited set of fitting parameters that are chosen. This process can be simple or very complicated. Determining the disorder parameters in particular is a bit of an art. In complicated geometries, symbolic math programs like Mathematica can be very useful for automatically generating the math expressions for each path. A simple example is shown in Figure 5.20.

This approach is particularly good for analyzing simple structures in which only a few paths (mostly single scattering) are important, and also for refining complex structures that are pretty much known, except for a few details. An example would be studying distortions in a perovskite structure. The method is less robust if deviations from the reference structure are large. FEFF estimates which paths it thinks are important, a calculation that is highly sensitive to geometry. If the tweaked structure departs too much from the original one from which the paths were generated, the paths may change. It is sensible to periodically rerun FEFF in such cases, but then the path lists and path expressions must be reconstructed as well. If it is necessary to do it, this tedious process is best handled by suitable computer programs.

```
In[1]:= struct[a_, b_, c_] := Sphere /@
          {{0, 0, a}, {b, 0, 0}, {0, b, 0}, {-b, 0, 0}, {0, -b, 0}, {0, 0, a + c}};
        coords = Flatten[Apply[List, struct[a, b, c], 2], 1]; nrm[v_] := Sqrt[v.v];
        R[path_] := FortranForm[Total[nrm /@ Differences[coords[[path]]]] // Simplify]
        Manipulate[Graphics3D[struct[a, b, c]], {{a, 1}, 0, 4}, {{b, 2}, 0, 4}, {{c, 2}, 0, 4}]
```

Out[4]=

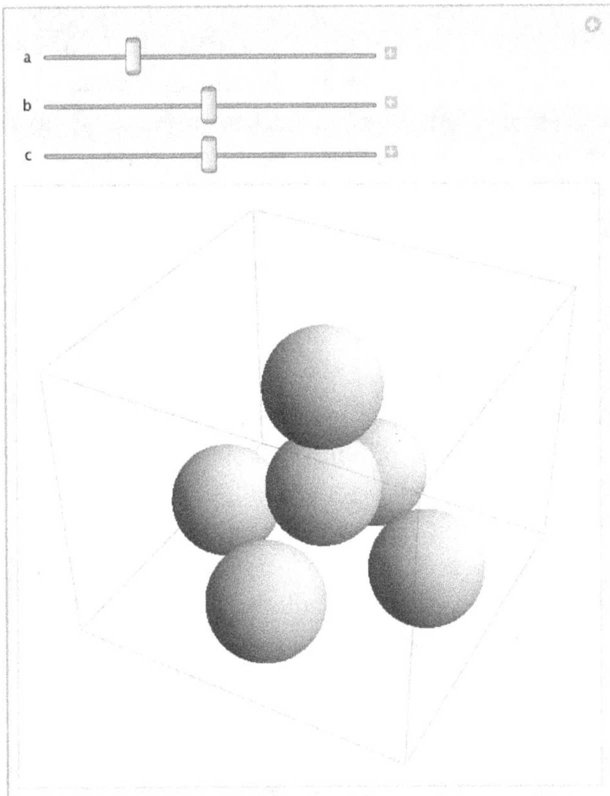


```
In[5]:= R[{1, 2, 1}]

Out[5]//FortranForm=
        2*Sqrt(a**2 + b**2)

In[6]:= R[{1, 2, 3, 2, 1}]

Out[6]//FortranForm=
        2*(Sqrt(2)*Sqrt(b**2) + Sqrt(a**2 + b**2))

In[7]:= R[{1, 2, 6, 3, 5, 1}]

Out[7]//FortranForm=
        2*(Sqrt(b**2) + Sqrt(a**2 + b**2) + Sqrt(b**2 + (a + c)**2))
```

Fig. 5.20 This Mathematica 7 code fragment illustrates the automatic generation of math expressions (equations) for the path lengths as a function of the geometrical parameters, which are needed as input to Artemis/IFEFFIT. Here the path is specified as a list of atom numbers that are involved in the scattering path. Atom 1 is the central atom. The output is the Fortran code. The code fragment also illustrates an interactively rotatable 3D model of the simple 3 parameter structure whose dimensions can be interactively modified using the sliders.

Another approach, which I routinely use for multiple scattering calculations, is brute force: for each hypothetical structure in a fitting process, run a full FEFF calculation.[14] FEFF happily computes new paths as needed for the hypothetical structure, and no user-driven assumptions about which paths are important are made. Each function evaluation in this case takes a few seconds, and a fitting run of several thousand fitting evaluations takes a few hours, but the process does not require human intervention as long as a robust minimizer such as Differential Evolution is used. I have written Mathematica programs that read in the experimental data; generate a generalized structure incorporating the structural degrees of freedom of interest; generate a FEFF input file and exports it; execute FEFF8; harvest the results (e.g. chi.dat, xmu.dat files); and calculate a fitting error by comparison to the experimental data. These operations are encapsulated as a function of the fitting parameters, which are minimized using the Differential Evolution algorithm including physical constraints. This process could also readily be implemented in Python or many other languages.

Dimakis and Bunker [88] have used an approach similar to this that additionally uses precomputed (using Density Functional Theory) multiple scattering σ^2 values for each scattering path within metal-amino acid coordination complexes, for the purpose of fitting metalloprotein sites ab initio. The point of this is to avoid exceeding the allowed number of fitting parameters (100 paths may mean 100 σ^2 parameters, which usually exceeds the information content in the data). Although reasonable ad hoc assumptions or constraints can be made to limit the degrees of freedom, it seems more defensible to calculate the DWFs, validate them, and then use them as supplementary information in the fitting process.

5.3.8 Fitting unknown structures

When studying structures about which little is known, it is usually best to use a divide-and-conquer strategy, by modeling the data shell by shell. The first shell is dominated by single scattering, even though the tails of multiple scattering contributions at larger r may leak in. These are smaller corrections and initially can be neglected. If the first shell shows a large amount of disorder, cumulants, regularization, or parametric fitting with robust optimizers can be employed.

Sometimes the noise level in the data can be substantial, and it may become difficult to tell what is real and what is not, e.g. distinguishing

[14] Recomputation of the atomic densities and perhaps even the partial wave phase shifts can be turned off after the first cycle to save a little time if desired.

between real peaks and truncation ripple in a FT. In such cases it is helpful to divide the set of scans in half, generate two independent partial sums of the scans, and also the total sum, and carry the three sums through the analysis. This provides a simple but effective estimate of the uncertainties due to noise at all steps in the analysis.

Fourier transforms can be helpful in identifying the types of atoms that are present indirectly through the k-dependence of their amplitudes and phases. Heavier (high Z) atoms tend to scatter more strongly at higher k, so if the k weighting exponent used in the FT is increased, their FT peaks will increase more than do the low Z scatterers. It is often forgotten that if only the modulus is inspected, half of the data is hidden because the FT is a complex quantity with a phase. The constant part of the scattering phase affects the imaginary (and real) part of the FT, but not the modulus. Plotting the modulus and imaginary part simultaneously can be used (roughly) to identify the nature of the scattering atoms in a shell. Interferences can complicate the interpretation, but it is still simple and useful. Detailed fitting provides the definitive answer.

If the sample is measured at multiple temperatures (say, 80K and 300K), the atoms that have strong bonds with the central atom will show less temperature dependence than weakly bound atoms. This temperature dependence of σ^2 can be useful to distinguish strongly bound in-plane atoms vs. weakly bound axial atoms in many systems. If it is supported by the software that is used, it is a good idea to fit several scans simultaneously.

I often find it useful to first do a rough analysis of the data to get an overview, and then refine the analysis in later steps. When modeling, it is best to start with the simplest hypothetical structure, and if the fit is inadequate for all sensible choices of the parameters, then introduce more complexity into the model. Robust minimizers like Differential Evolution [123] can be very helpful for determining whether or not the hypothetical model can possibly fit the data, because it is essential that the fitting algorithm not get trapped in a local minimum, and prematurely reject the simpler hypothesis.

For example, if one shell of oxygen atoms doesn't give a good fit, for any sensible coordination number, add another subshell of oxygen atoms, or another type of atom that might conceivably be present. If the apparent disorder is high enough, one can try fitting to a cumulant expansion; if the disorder is too high for a cumulant expansion to converge, try a more complex distribution of atoms such as a Gamma distribution, or use regularization.

Once the first shell is adequately fit, try the second, if it can be cleanly isolated in r-space. Often the higher shells cannot be well isolated in a Fourier transform, in which case the interfering shells must be fit as a unit. The process is similar to that in the first shell, but multiple scattering contributions may be needed also.

After approximate solutions are obtained, it is a good idea to fit the whole spectrum together as a unit, in order to account for residual overlap between the shells.

5.3.9 Regularization methods

In situations in which multiple scattering can be neglected, and there is only a single type of backscattering atom, the transformation from radial probability density $p(r)$ to $\chi(k)$ is a linear transformation:

$$\chi(k) = \int_{r=0}^{\infty} A(k,r)p(r)dr,$$

where the physics of EXAFS is embedded in the kernel $A(k,r)$. An example of such a Kernel is shown in Figure 5.21. Normally this will be discretized so that χ and p are vectors, and A is a matrix with $\chi_n = \sum_m A_{nm}p_m$. If more than one type of atom is present, with some repackaging of matrices[15], the equation retains the same form.

It might seem reasonable in that case to invert the matrix A (assuming that it has an inverse) and take the scalar product with $\chi(k)$ to produce $p(r)$. Even if you can do that, however, you may get rubbish. The basic reason is that there is a space of functions that, if you apply A to them, give zero, or something close to it. That space of functions is called the nullspace of the kernel. For example, if a kernel is smooth, multiplying by a very rapidly oscillating function will give something close to zero. Similarly, if the kernel does not represent a space of very slowly varying functions, it cannot represent them.

This shows that the kernel has nothing to say about the functions that lie in its nullspace. If one inverts the matrix and uses it in a simplistic way, any linear combination of functions in the nullspace can end up in the reconstructed $p(r)$, generally resulting in rubbish. It is a matrix analog of dividing by zero.

This sort of problem is exactly what the singular value decomposition (SVD) is suited to. It enables you to identify the nullspace of a matrix and effectively cast out functions belonging to it according to the size of the

[15] In the case of more than one backscatterer, one can simply introduce a distinct $p(r)$ for each type of atom, concatenate the p arrays to form a master p array, and construct a new A consisting of a block-diagonal combination of the A's for each atom type.

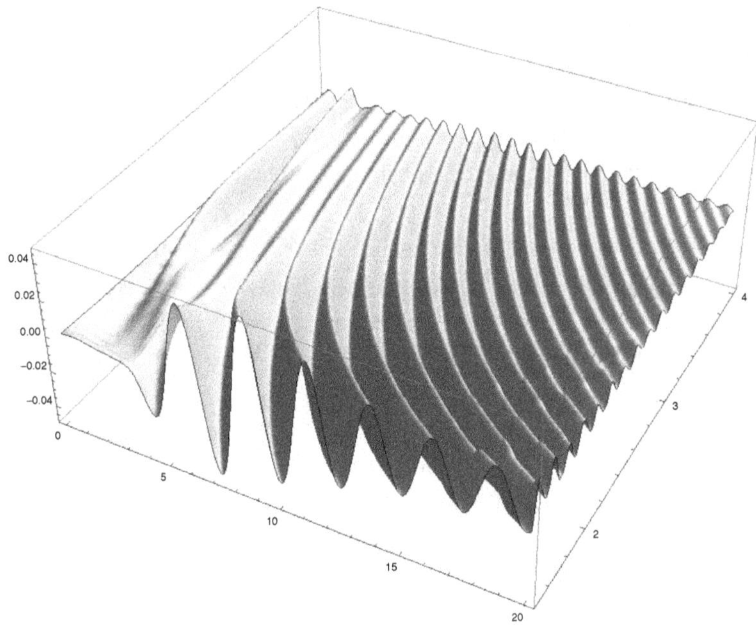

Fig. 5.21 k-weighted kernel for Fe-Se single scattering over the range $k = .05 - 20.0$ Å^{-1} and $R = 1.5 - 4.0$ Å. The kernel was automatically generated using a Mathematica program that created FEFF8.4 input files for selected geometry, executed FEFF, and assembled the results.

eigenvalues (singular values). This approach to inversion is referred to as taking the pseudoinverse of the matrix A. It is quite robust, and it does work well (up to a point) for reconstructing $p(r)$ [103]. For example, it delivers peaks in the RDF in the right places. However, since $\chi(k)$ may not extend to $k = 0$, and noise is present in the data, there is no guarantee that $p(r)$ will be positive definite, and in fact often is not.

There are many ways to tweak the inversion process so as to regularize the solution and render it better-behaved. The general idea is that small perturbations in the data should not give rise to large deviations in the data under inversion. Regularization methods also provide ways to incorporate physical constraints on the result, such as smoothness, and positivity. Normally there is at least one regularization parameter that is used to perturb the nearly singular kernel matrix in favorable (e.g. smooth) directions. Typically there is a tradeoff between smoothness of the solution and accuracy of the fit.

The general problem of solving such "ill-posed" problems was addressed by Tikhonov and others independently. There is a vast literature with specialized journals on the subject of inverse problems. Regularization methods have a broad range of application in the sciences; XAFS is just one of them. They are helpful for combining data from complementary techniques [131], and deconvolution [119].

One particularly useful reference with application to EXAFS is the book by Vasin and Ageev [131]. Babanov *et al.* [126, 127] first applied regularization methods to determination of radial distribution functions by EXAFS. Subsequently hybrid regularization/fitting techniques were demonstrated [128], principally to address the need for error estimates.

Although the formalism of regularization theory is beyond the present scope of this book, implementation of regularization methods can be relatively compact and straightforward. A simple illustration of this is shown in Figures 5.22 and 5.23. This implementation was adapted from Khelashvili [130]. A synthetic test $p(r)$ consisting of five Gaussian peaks of various widths, heights, and locations was first generated. The following operations were then carried out (and the figures generated) by the Mathematica code shown. The $\chi(k)$ corresponding to the known $p(r)$ was generated by taking the dot product with the kernel matrix; synthetic random noise was added to simulate experimental noise; and the RDF was reconstructed by iterative Tikhonov regularization. A projection operator (just a function, in this case) onto the set of positive real numbers was used to ensure non-negativity of $p(r)$. The α and β regularization parameters (which affect stability and smoothness) were set to nominal values. The results depend on the regularization parameters; several automatic methods of setting them have been developed, for example one described in ref [130].

Alternative methods such as Landweber–Friedman regularization have been used for determining RDFs from EXAFS [129, 130]. This approach has the virtue of automatic selection of regularization parameters, and, of particular importance, generation of error estimates. Regularization methods are under active development; in particular, newer methods for reconstructing RDFs show promise of speeding convergence and improving accuracy.

Mathematica 7 code fragment: iterative Tikhonov regularization.

```
(*inputs: chi "chiexp"; kgrid, rgrid of kernel, spacings dk, dr; kernel "A"*)
alpha=.001;beta=.001; (*regularization parameters*)
nkdim=Length[kgrid];nrdim=Length[rgrid];
tau=dr/dk;h=1/(dr dk);
P[x_]:=(x+Abs[x])/2 ;(*projects onto nonnegative functions*)
Clear[g];g[0]=Table[0,{nrdim}];
II=IdentityMatrix[nrdim];
BB=2II-RotateRight[II]-RotateLeft[II];(*2nd derivative matrix, for smoothness*)
AT=Transpose[A];ATA=AT.A;
```

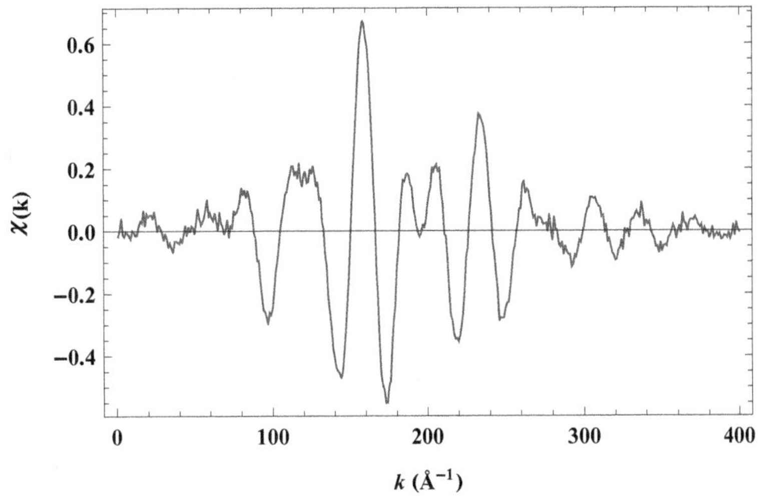

Fig. 5.22 $k^2\chi(k)$ corresponding to synthetic radial distribution of selenium atoms surrounding a central Fe (see Figure 5.23). Gaussian random noise is added to simulate experimental noise.

```
InvA[alpha_,beta_]:=PseudoInverse[ATA+alpha tau II+ beta h BB];
g[0]=InvA[alpha,beta].AT.chiexp;
g[n_]:=g[n]=P/@(InvA[alpha,beta].(AT.chiexp+alpha tau g[n-1] +
beta h BB.g[n-1]));
ListPlot[{gtrue,g[5]},PlotRange->All,Joined->{False,True},
PlotStyle->{Thick,Thick},LabelStyle->Bold]
```

A natural extension to basic regularization theory is to incorporate the effects of random noise on estimation; this "statistical regularization theory" was pioneered by Valentin Turchin. Statistical regularization methods incorporating ideas from Bayesian statistics have been introduced to XAFS from the nuclear physics community by Krappe and Rossner [114]. These and other inverse methods promise to play a growing role as part of the XAFS analysis toolkit.

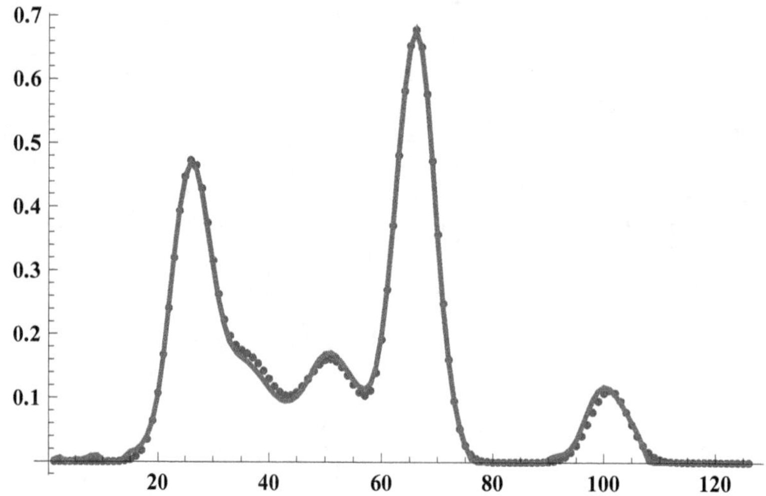

Fig. 5.23 "True" radial distribution function (dotted line) and the RDF reconstructed from the $\chi(k)$ using 5 cycles of iterative Tikhonov regularization, with regularization parameters $\alpha = \beta = 0.001$ (solid line).

6

Related techniques and conclusion

6.1 Related techniques

Many other techniques are closely related to XAFS. It is beyond the scope of this book to address them seriously – there are complete books on all of them – but here we briefly describe a few of them. It will be easy to find more information as desired from the name and acronym of the technique.

Most of the techniques of interest here involve sending photons with defined properties (E,$\vec{k}$, $\hat{\epsilon}$) into a sample, and then analyzing the properties of the photons that come out. These are essentially scattering processes. Depending on what is scanned and analyzed and the sort of phenomena they are used to study, the different protocols are given different names. A few of them are X-ray Magnetic Circular Dichroism (XMCD), Inelastic X-ray Scattering (IXS), Resonant Inelastic X-ray Scattering (RIXS), X-ray Raman Spectroscopy (XRS), Diffraction Anomalous Fine Structure (DAFS), X-ray Standing Waves (XSW), and X-ray Fluorescence Holography (XFH). Incidentally the program FEFF8.4 can calculate a number of these spectra in addition to XAFS.

Another class of techniques involves sending photons in, and detecting electrons that are emitted. The electrons have a limited path length in the materials, so any electrons that are detected emerge from near the surface. X-ray Photoemission Spectroscopy (XPS) and Angle Resolved PhotoEmission Fine Structure (ARPEFS) are examples. These techniques provide information about low-energy excitations, the structure of adsorbed species, and other surface effects.

Yet another class of techniques involves sending high-energy monochromatic electrons through a thin sample within a specially instrumented electron microscope, and measuring their energy as they come out. This is Electron Energy Loss Spectroscopy (EELS). Interestingly it can provide the equivalent information to XAFS, but with an electron microscope's spatial resolution. Multiple scattering of the electrons is a complication however.

Photons are characterized by their momentum $\hbar\vec{k}$, energy $\hbar\omega$, and angular momentum $\hbar$, which has projection ± 1 onto the direction of $\vec{k}$. This projection is the helicity, which determines whether it is left or right circularly polarized. The photon state can be expressed as a linear combination of helicities, which, depending on the coefficients, gives it an arbitrary plane, elliptical, or circular polarization.

The source (undulator, wiggler, bend magnet) produces X-rays with well-defined polarization characteristics. Planar undulators and wigglers produce beams that are highly polarized in a plane, usually the horizontal one. These do not produce circularly polarized X-rays – their paths curve back and forth and such effects cancel out. The trajectory through a bend magnet curves in only one direction. Exactly in the plane of the orbit the X-ray beam is highly plane-polarized, but out of the plane the beam picks up some degree of left and right circular polarization, depending on whether one is above or below the beam center. These circularly polarized beams are useful for X-ray Magnetic Circular Dichroism, XMCD, which is particularly useful for studying the magnetic structure of materials. Helical undulators and wigglers can be used to generate copious quantities of circularly polarized X-rays. X-ray optical devices (quarter-wave plates) also can be used to convert linearly polarized X-rays into circularly polarized ones, at least over a narrow range of energies.

Monochromators and focussing optics (Chapter 3) define the X-ray beam energy E and momentum vector $\hbar\vec{k}$. If we send in monochromatic photons, and measure photons at the same energy as a function of angle (a scattering/diffraction experiment) we probe the spatial periodicities in the sample. These experiments are (depending on characteristics of the sample) X-ray Crystallography or X-ray Diffraction (3D or 2D ordering), X-ray fiber diffraction (1D ordering), X-ray scattering (2D or 1D), X-ray solution scattering. The incident and scattered angles may be scanned and analyzed. The angle analysis can be done by scanning a slit, or using a position sensitive detector.

If we send in monochromatic photons and measure the intensities at different (mostly lower) energies, we probe the excitation spectrum of the sample, which provides information about the electronic quantum states. In Inelastic X-ray scattering and RIXS both the incident energy and analyzer energy are scanned, providing a two-dimensional map of the energy levels, of which XAFS essentially is a projection onto one axis. IXS can provide more complete understanding of the electronic structure in the XANES region. The downside is that the experiments are relatively photon-inefficient, being a second-order process in the matrix elements. The experiments are also considerably more difficult, and the equipment more specialized than XAFS. RIXS can be used to suppress the core-hole lifetime broadening, which can be very useful for understanding the near-edge spectra in some samples. Resonance effects near the edge greatly enhance the signal. Very high energy resolution IXS can be used to study extremely low energy excitations that correspond to vibrational structure in materials.

X-ray Emission Spectroscopy (XES) involves the high-resolution energy analysis of the fluorescence that is emitted by a sample, which provides information on chemical shifts of energy levels. X-ray Fluorescence Holography relies on analyzing the angular pattern of the fluorescence emitted by excited atoms at specific sites in a sample. The angular pattern is modulated by interference effects of the final state photon, and provides information about the spatial distribution of the excited atoms. Diffraction Anomalous Fine Structure is a hybrid XAFS/Diffraction technique. The X-ray elastic scattering amplitude is altered near X-ray absorption edges because of the potential for electronic excitation. The scattering amplitude acquires a complex contribution, the imaginary part of which is proportional to the X-ray absorption coefficient. The real and imaginary parts are related to each other through a Kramers–Kronig transform (mathematically, a Hilbert transform). Essentially one measures a diffraction pattern as a function of energy over and above the edge of an element of interest. The modifications to the X-ray scattering phase/amplitude as a function of energy provide site selectivity in crystals that XAFS alone cannot provide.

Altogether, these techniques, and others such as neutron scattering, are invaluable complements to XAFS. Much of the information presented in this book may be helpful in understanding those techniques as well.

6.2 Conclusion

In this book I have attempted to give a practical introduction to XAFS that is accessible to the nonspecialist, while pointing to more detailed treatments elsewhere. This is particularly true in the area of theory and areas that show promise of significant advances in the near future. I have attempted to pass along lessons learned in the past, some of which seem to have been forgotten, while being mindful of recent and potential future advances.

XAFS and related synchrotron radiation based techniques offer many opportunities for graduate students because of their range of applicability; the wide range of types of thinking and activities that they entail; and the opportunities for interdisciplinary and international collaborations. I hope this book provides a platform from which to dive into exciting new areas of research.

Appendix 1
Introduction to Fourier transforms in EXAFS

A1.1 Fourier integrals

This appendix provides an elementary introduction to Fourier methods in EXAFS.

Consider the elementary integral

$$\int_0^a e^{\gamma x}dx = \frac{e^{a\gamma}-1}{\gamma}.$$

Choosing $\gamma = 2\pi i n/a$ this becomes

$$\int_0^a e^{2\pi i n x/a}dx = \frac{ia\left(1-e^{2i\pi n}\right)}{2\pi n}.$$

If n is zero the integral evaluates to a which can be seen by taking the limit or using L'Hôpital's rule. If n is a nonzero integer the integral vanishes, because $\exp(i2\pi n) = \cos 2\pi n + i \sin 2\pi n = 1$, by Euler's identity $e^{i\theta} = \cos\theta + i\sin\theta$, which is valid for any θ.

Now suppose we wish to integrate the product of two complex exponentials

$$\int_0^a (e^{2\pi i n x/a})^* e^{2\pi i m x/a}dx,$$

where the $*$ indicates complex conjugation ($i \to -i$). Combining the exponentials we see this is of the same form as the previous integral, but with $n \to (m-n)$; therefore the integral is nonzero only when $n = m$. It evaluates to $a\delta_{nm}$ where the Kronecker δ_{jk} is defined as 1 if $j = k$, and is 0 otherwise. The two complex exponentials are said to be "orthogonal" to each other over the interval {0, a}, i.e. the integral of the product of exponentials of different frequency is zero; the integral of the product of an exponential with itself is nonzero.

A1.2 Orthogonality

The reason for calling such functions orthogonal may not be obvious to the reader. Imagine approximating a function $f(x)$ by N of its values on a uniform grid x_j: $f_j = f(x_j)$. If the function is well behaved, as $N \to \infty$ we can approximate it as closely as we like. Therefore the function $f(x)$ can be represented as an N-dimensional vector f_i, with N becoming indefinitely large. From this point of view the integral of the product of two functions $f(x)$ and $g(x)$ is proportional to the dot (scalar) product of their vector representations: $\sum_i f_i g_i$. If the functions are complex the dot product is defined as $\sum_i f_i^* g_i$. When the dot product of two vectors is zero, they are said to be orthogonal.

We now see the rationale behind calling two functions orthogonal (over some interval) if the integral of their product is zero (over that interval). Just as we can represent vectors in terms of a linear combination of basis vectors, we can represent functions as a linear combination of basis functions.

For the precise representation of arbitrary functions, it may take an infinite number of basis functions. Arbitrary functions can be expanded over intervals in terms of orthogonal polynomials such as Legendre polynomials. In a Fourier series the basis functions are sine and cosine functions with frequencies that are integer multiples of a base frequency. In Fourier analysis the basis functions are complex exponentials over an infinite range with a continuous frequency distribution.

Like ordinary vectors, functions can be thought to have a magnitude and a direction. The magnitude is the square root of the dot product of the function with itself, i.e. $\int f^* f\, dx = \int |f|^2\, dx$. The cosine of the "angle" between two functions is $(\int f^* g\, dx)/(\int |f|^2\, dx \int |g|^2\, dx)^{1/2}$. The functions that are intrinsically most distinct from each other are the ones that are mutually orthogonal.

A1.3 Fourier series

We have initially defined our function over the range {0, a}. Now consider the product of two trigonometric functions. By Euler's identity we have $\cos 2\pi n = (e^{i2\pi n} + e^{-i2\pi n})/2$ and $\sin 2\pi n = (e^{i2\pi n} - e^{-i2\pi n})/(2i)$. Writing out the products gives four terms that can be evaluated with the integral above, simplifying to:

$$\int_0^a \cos(2\pi mx/a)\cos(2\pi nx/a)\,dx = (a/2)\delta_{mn}$$
$$\int_0^a \sin(2\pi mx/a)\sin(2\pi nx/a)\,dx = (a/2)\delta_{mn}$$
$$\int_0^a \cos(2\pi mx/a)\sin(2\pi nx/a)\,dx = 0.$$

These show that sines and cosines that have "frequencies" that are integer multiples of $2\pi/a$ are orthogonal on the interval $\{0, \text{a}\}$.

We next attempt to expand an arbitrary function $f(x)$ in a Fourier Series:

$$f(x) = \bar{f} + \sum_{n=1}^{\infty} (a_n \cos(2\pi nx/a) + b_n \sin(2\pi nx/a)),$$

where $\bar{f} = (1/a)\int_0^a f(x)\,dx$, the average of f.

Multiplying both sides by $\cos(2\pi mx/a)$ and integrating, for $m > 0$ we have:

$$\int_0^a f(x)\cos(2\pi mx/a)\,dx =$$
$$\sum_{n=1}^{\infty}\left(a_n \int_0^a \cos(2\pi mx/a)\cos(2\pi nx/a)\,dx + b_n \int_0^a \cos(2\pi mx/a)\sin(2\pi nx/a)\,dx\right)$$
$$= (a/2)\sum_{n=1} a_n\delta_{mn} = (a/2)a_m;$$

This shows that integrating $f(x)$ against the basis function $\cos(2\pi mx/a)$ directly gives us the corresponding coefficient in the expansion for $f(x)$. Similarly, if we multiply by $\sin(2\pi mx/a)$ we obtain the coefficient b_m:

$$a_m = (2/a)\int_0^a f(x)\cos(2\pi mx/a)\,dx$$
$$b_m = (2/a)\int_0^a f(x)\sin(2\pi mx/a)\,dx.$$

Suppose a function is known to be band-limited, i.e. it has a highest and lowest frequency component. In that case, only a finite set of components are needed to reproduce the function; in other words, the signal has a limited information content. If the highest frequency is $n_{\max}$ (in units of $2\pi/a$), and the lowest is $n_{\min}$, a maximum of $2*(n_{\max} - n_{\min} + 1)$ Fourier coefficients are needed to fully describe the signal. If the band includes the zero frequency, the parameter $\bar{f}$ must also be included.

A1.4 Delta function

Next let us take the interval a to be indefinitely large. In that case the integer multiples of the base frequency $2\pi/a$ become quasicontinuous, and accordingly we will change to the continuous variable $2\pi n/a \rightarrow k$ over the range $\{-a/2, a/2\}$. The integral of the exponential can be directly evaluated as

$$\int_{-a/2}^{a/2} e^{ikx}\, dx = \frac{2\sin(ka/2)}{k} \equiv 2\pi\Delta_a(k)^1.$$

The function $\Delta_a(k)$ is plotted for different values of a in Figure A1.1. It can be seen that $\Delta_a(k)$ has a central maximum at $k = 0$ and oscillations of decreasing amplitude as one moves away from the center. The integral $\int_{-a/2}^{a/2} \Delta_a(k)\, dk = 1$. As the x-range a grows larger, the central maximum grows and becomes narrower, so that the area under the central maximum stays constant. If one were to calculate $\int_{-a/2}^{a/2} \Delta_a(k)F(k)\, dk$ for some function $F(k)$, only its value near $k = 0$ contributes. The oscillatory structure away from the central maximum averages to zero when multiplied by the well-behaved function $F(k)$. In the limit $a \rightarrow \infty$ the integral approaches $F(0)$.

This limiting behavior will be recognized as the defining characteristic of the Dirac delta function $\delta(x)$, in particular $\int_{-\infty}^{\infty} \delta(x)F(x)\, dx = F(0)$. Letting $x \rightarrow x - x'$ we have $\int \delta(x - x')F(x')\, dx' = F(x)$. Integrating a delta function centered at x_0 with a function $F(x)$ plucks out the value of F at x_0. It is an even function: $\delta(x) = \delta(-x)$.

For simplicity (despite its imprecision – the integral does not exist, and δ is not a proper function) we will use the form

$$\int_{-\infty}^{\infty} e^{ikx}\, dx = 2\pi\delta(x)$$

as a shorthand for the taking the limit of the integral as outlined above.

A1.5 Fourier transform

Consider the function $\tilde{f}(k) = \int_{-\infty}^{\infty} f(x)e^{ikx} dx$. The function $\tilde{f}(k)$ is the Fourier transform of the function $f(x)$. It is an alternate representation of

[1] The function $\sin(\pi y)/(\pi y)$ is often used in signal processing – it is called a *sinc* function. The same function appeared in Chapter 3 in the context of diffraction from crystals.

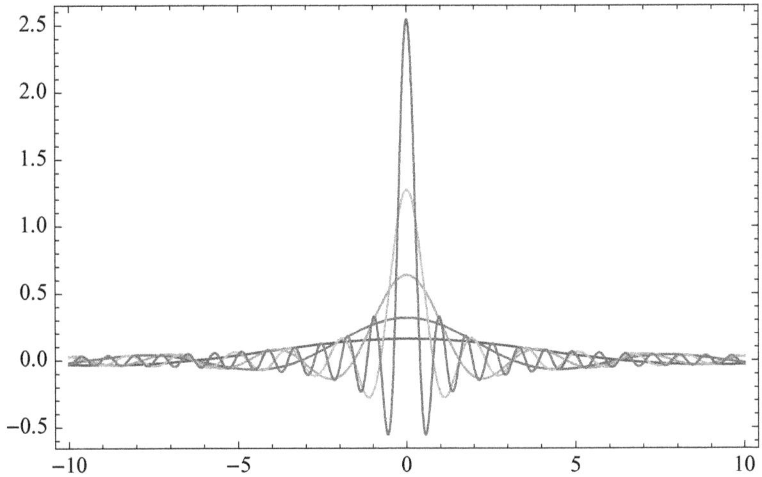

Fig. A1.1 The function $\Delta_a(k) = \frac{\sin(ka/2)}{(\pi k)}$ for $a = 1, 2, 4, 8, 16$. As the range a increases, the central peak centered at $k = 0$ becomes higher and narrower, while $\int_{\infty}^{\infty} \Delta_a(k)\, dk = 1$. As $a \to \infty$, $\Delta_a(k) \to \delta(k)$, the Dirac delta "function."

$f(x)$ that contains the same information. We can see this by carrying out the inverse transform

$$\begin{aligned}\int_{-\infty}^{\infty} \tilde{f}(k)e^{-ikx'}\, dk &= \int_{-\infty}^{\infty} \left(\int_{-\infty}^{\infty} f(x)e^{ikx}dx\right) e^{-ikx'}\, dk \\ &= \int_{-\infty}^{\infty} \left(\int_{-\infty}^{\infty} e^{-ikx'} e^{ikx}dk\right) f(x)dx \\ &= \int_{-\infty}^{\infty} 2\pi\delta(x' - x)f(x)dx = 2\pi f(x').\end{aligned}$$

We recover the original function $f(x)$ by back-transforming (using e^{-ikx} instead of e^{+ikx}); we also pick up a factor of 2π going back and forth. The forward and inverse Fourier Transforms sometimes are defined in a symmetrical manner with a factor $1/\sqrt{2\pi}$ multiplying the integral of both forward and inverse transforms.

A1.6 Convolution theorem

Convolution is a binary mathematical operation (written $*$) between two functions (say $f(x)$ and $g(x)$) that produces a new function $(f * g)(x)$ in the following way. First, at each point x', generate the function that is the value of g at that point, times a copy of the f function which is shifted so that it is located at x'. This function is $f(x-x')g(x')$. Next, add up (i.e. integrate) all the weighted, shifted replicas of $f(x)$ that are obtained for all values of the shift variable x'. These operations generate the convolution of the two functions $(f*g)(x) = (g*f)(x) = \int f(x-x')g(x')dx'$. Convolutions occur naturally in many contexts, such as broadening of spectra by resolution functions, and describing distributions in statistics.

The convolution theorem can be simply stated: the Fourier transform of the convolution of two functions is the product of their Fourier transforms: $FT[f*g] = FT[f]\cdot FT[g]$. This result shows that $f*g = g*f$, because multiplication is commutative. The corresponding result holds for inverse Fourier transforms. The convolution theorem was used in Chapter 2 to describe Bragg diffraction from crystals.

The theorem is straightforward to prove if one can assume (as we do) that the functions are well-behaved, so the order of integrations can be interchanged. We start with the convolution of f and g, and then FT it. An intermediate expression resolves to a delta function, which after integration, and taking the FT of both sides, simplifies to the result:

$$
\begin{aligned}
f(x) &= \frac{1}{2\pi}\int e^{-ikx}\tilde{f}(k)dk; \qquad g(x) = \frac{1}{2\pi}\int e^{-ik'x}\tilde{g}(k')dk'; \\
f*g &= \int f(x-x')g(x')dx' \\
&= \int \left[\frac{1}{2\pi}\int e^{-ik(x-x')}\tilde{f}(k)dk\right]\left[\frac{1}{2\pi}\int e^{-ik'x'}\tilde{g}(k')dk'\right]dx' \\
&= \frac{1}{(2\pi)^2}\int\int\int e^{-ikx}e^{i(k-k')x'}\tilde{f}(k)\tilde{g}(k')dkdk'dx' \\
&= \frac{1}{(2\pi)^2}\int\int e^{-ikx}\left(\int e^{i(k-k')x'}dx'\right)\tilde{f}(k)\tilde{g}(k')dkdk' \\
&= \frac{1}{(2\pi)^2}\int\int e^{-ikx}\left(2\pi\delta(k-k')\right)\tilde{f}(k)\tilde{g}(k')dkdk' \\
&= \frac{1}{2\pi}\int e^{-ikx}\tilde{f}(k)\tilde{g}(k)dk \\
&= FT^{-1}[\tilde{f}(k)\cdot\tilde{g}(k)] \\
FT[f*g] &= \tilde{f}(k)\cdot\tilde{g}(k).
\end{aligned}
$$

A1.7 Multidimensional Fourier Transforms

It is straightforward to generalize the FT to more than one dimension simply by Fourier transforming with respect to each independent variable. Suppose one has a function of two variables $f = f(x, y)$. The FT is

$$\tilde{f}(q_x, q_y) = \int\int f(x, y)(e^{iq_x x}\, dx)(e^{iq_y y}\, dy) = \int\int f(x, y)e^{i(q_x x + q_y y)}\, dx\, dy.$$

The quantity in the last exponential can be viewed as the scalar (dot) product of $\vec{q} \equiv (q_x, q_y)$ with $\vec{r} \equiv (x, y)$. Evidently the same argument applies to any number of variables and dimensions, and we can write

$$FT[f] = \int f e^{i\vec{q}\cdot\vec{r}}\, d\tau,$$

where the integral is over the appropriate area or volume over which the function is defined.

A1.8 Wave packets

A sine wave in space has the form $\sin(2\pi x/\lambda)$ where λ is the wavelength, or spatial period of the wave. With $k = 2\pi/\lambda$ the wave has the form $\sin(kx) = (e^{ikx} - e^{-ikx})/(2i)$. It has a constant amplitude and precisely defined spatial frequency k over the range $\{-\infty \leq x \leq \infty\}$. The sharp spatial frequency distribution is represented by a pair of delta functions in k space corresponding to the e^{ikx} and e^{-ikx} parts of the sine. If the wave is represented as a single complex exponential in x-space, its k-space representation is proportional to a single delta function.

If one wishes to confine the wave to a finite region it is necessary to add other waves to it with different spatial frequencies k, so that destructive interference makes the amplitude small outside the region of interest. For example, if one adds up sine waves with a gaussian distribution $\exp(-(k - k_0)^2/(2\sigma_k^2))$ of spatial frequencies k about a mean value k_0, the resulting "wave packet" is proportional to a gaussian[2] distribution in x: $\exp(-(x - x_0)^2/(2\sigma_x^2))$, where $\sigma_x = 1/\sigma_k$. This superposition of waves can be seen to be the convolution of the gaussian in k with $\delta(k - k_0)$. By the convolution theorem the result therefore is the product of the FT of the gaussian with the FT of the delta function, resulting in a gaussian amplitude wave packet in x. This is illustrated in Figure A1.2.

The inverse relation between the width in k-space and the width in x-space is a universal property of waves that has connections with Heisenberg's uncertainty principle in quantum mechanics.

[2] Gaussians are unusual in that their FTs are also gaussian. The hyperbolic cosecant (sech) is another such function.

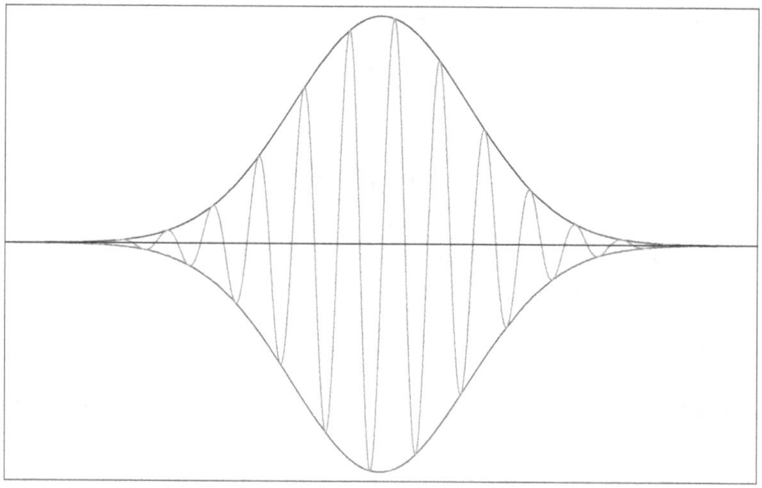

Fig. A1.2 Gaussian wave packet. The wave and ± its gaussian amplitude function are shown.

A1.9 Discrete FT and Fast FT (FFT)

When one carries out an FT of data on a regular grid it is called a discrete FT (DFT). The DFT is usually defined as $\tilde{f}_m = 1/\sqrt{N}\sum_{n=1}^{N} e^{i2\pi nm/N} f_n$ where N is the dimension of the array. The k and r grids are given by $k_n = (n-1)\delta k$, $r_m = (m-1)\delta r$, and $\delta r = \pi/(N\delta k)$.

The inverse DFT is the same, except for the usual change $i \to -i$ in the exponential. Instead of dividing by $1/\sqrt{2\pi}$ on each (forward and inverse) transformation, one divides by $\sqrt{N}$. Another difference with the continuous FT is that the domains of the functions are limited by the array dimension N, so there are maximum spatial frequencies and distances that can be represented. Frequencies higher than the maximum fold down "alias" to the lower frequency domain.

Another difference with continuous FTs is the periodicity of the DFT. It is evident from the form of the exponential $e^{i2\pi nm/N}$ that it is periodic; if $n \to n+N$ or $m \to m+N$ the results are unchanged. This can be pictured by imagining that the array wraps around into the shape of a ring and connects to itself so the $n=1$ and $n=N$ positions are adjacent. The zero spatial frequency component shows up as the first element of the array, and negative frequencies fall near the end of the array. The maximum positive frequency is found around $N/2$.

If one transforms a real function, the real part of the complex transform is an even function, i.e. it is symmetrical about the origin. The imaginary

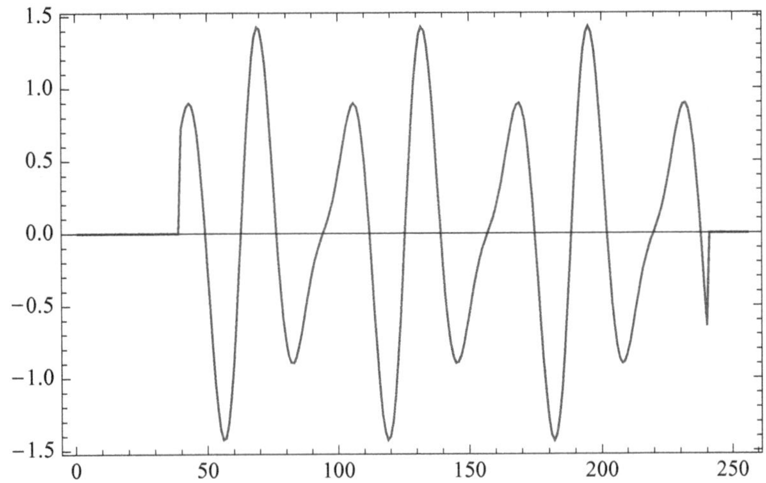

Fig. A1.3 Sample test function in k-space.

part of the complex transform is an odd function, i.e. it is antisymmetrical about the origin.

Fast Fourier Transforms (FFTs)[111] are a class of clever algorithms for rapidly computing DFTs. Their speed makes possible a number of techniques in signal processing and data compression that would not be practical otherwise.

The data are represented as complex numbers – even if $\chi(k)$ is real, the FT to r-space gives a complex result, so the data have both real and imaginary parts. The modulus of the data is frequently the only thing that is shown, but the phase of the data is equally important.

The data typically are mapped onto a uniform grid in k-space; as mentioned previously, $\delta k = 0.05$ Å^{-1} is a good choice that adequately samples the data in k-space. The r-space grid is then determined by $\delta r = \pi/(N\delta k)$. If one were to choose (say) $N = 256$ this would give a $k_{\text{max}} = 256 * .05 = 12.8$ Å^{-1}, which may or may not be adequate. However, the r-space grid δr would only be 0.2454 Å which is far too coarse to resolve much of interest. By artificially extending the upper k-limit by choosing $N = 2048$ the sampling in r can be made more reasonable, in this case 0.0307 Å. The effects of zero-padding are shown in Figures A1.5–1.8.

It might seem that this is just a cheap trick, but it is not difficult to show that this process just amounts to sampling the exact analytical Fourier transform on a finer mesh. This is illustrated in Figure A1.9, where the modulus of the exact analytical fourier transform of a test function is compared with

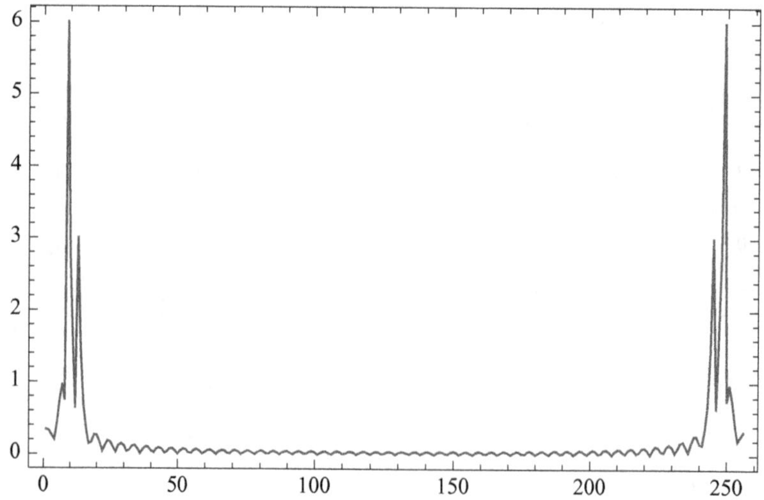

Fig. A1.4 Modulus FT of sample test function (256 point array). Note the negative frequency components at the end of the array.

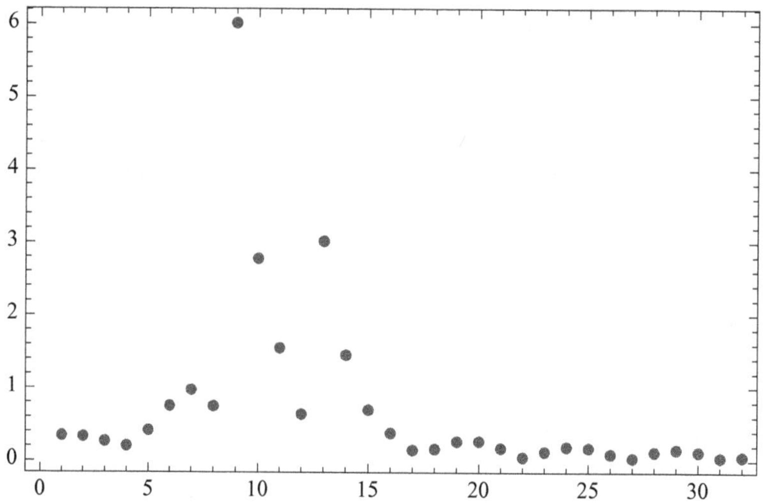

Fig. A1.5 Modulus of FT of sample test function (256 point array). Note the insufficient density of r-space points because of minimal zero-padding.

the DFT of the function zero padded out to 2048 points ($k_{\text{max}} = 102.4$). The test function is $\sin(4k) + 0.5\sin(6k)$ (equivalent to distances of 2.0 and 3.0 Å) that are truncated by a rectangular window between $k = 2$ and $k = 12$. The ringing observed is a consequence of the sharp truncation where the amplitude is large.

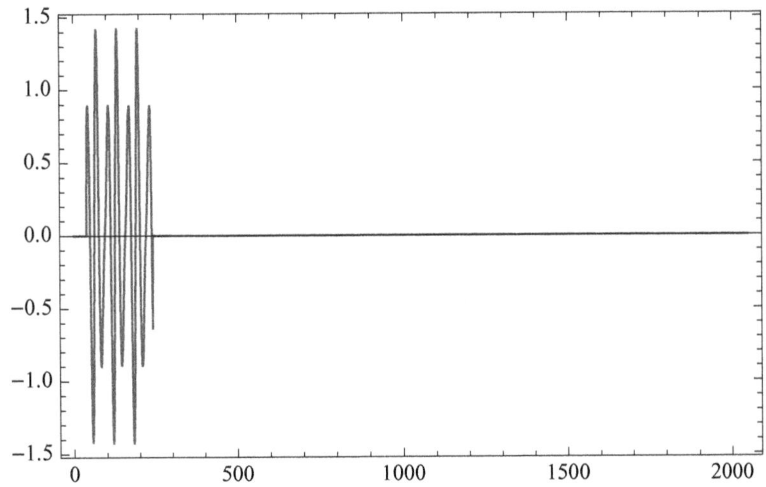

Fig. A1.6 Sample test function in k-space.

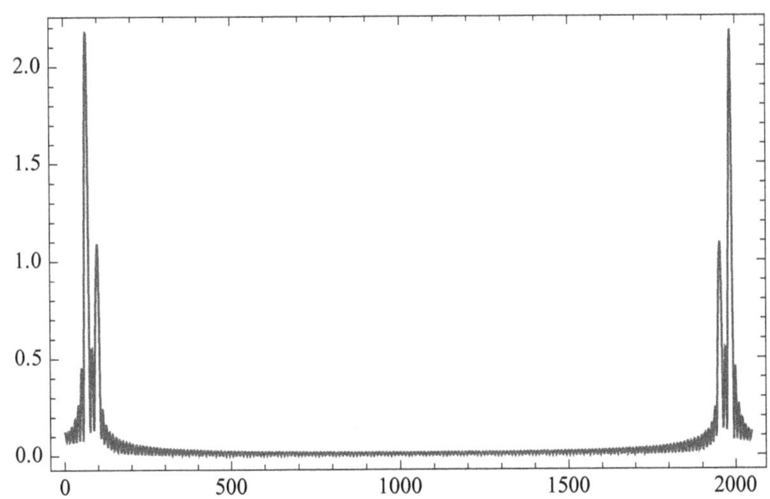

Fig. A1.7 Modulus FT of sample test function (2048 point array). Note the negative frequency components at the end of the array.

A1.10 Truncation errors

The "ringing" (sidelobes) observed in the FTs are a consequence of the sharp truncation that they suffer at the ends of the window. The object that is transformed is the product of the weighted $k^n\chi(k)$ and the k-window $W(k)$. The convolution theorem tells us that the FT of the product is the

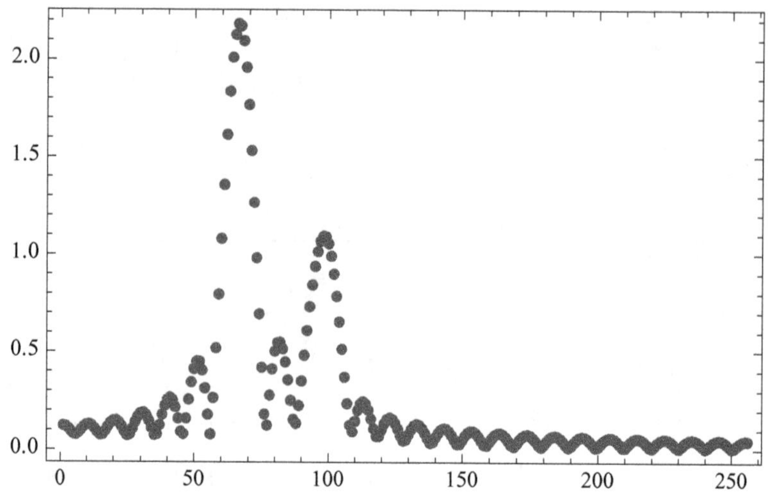

Fig. A1.8 Modulus of FT of sample test function (2048 point array). Note the adequate density of r-space points because of zero-padding.

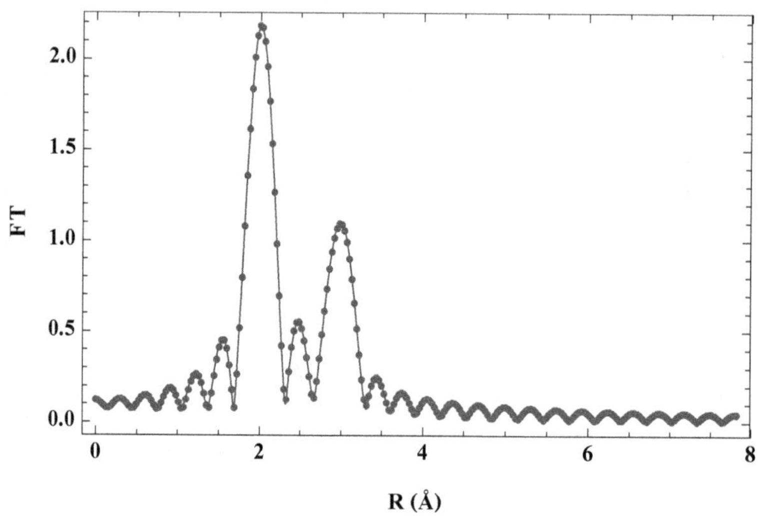

Fig. A1.9 Modulus of exact analytical Fourier transform of test function (solid line) compared to the modulus of 2048 point zero-padded discrete transform (points). Zero padding of the k-space array generates denser sampling of the function in r-space. The ordinate is rescaled to account for differences in normalization.

convolution of the FT of the data with FT of the window $W(k)$:

$$FT[k^n \chi(k)W(k)] = FT[k^n \chi(k)] * FT[W(k)].$$

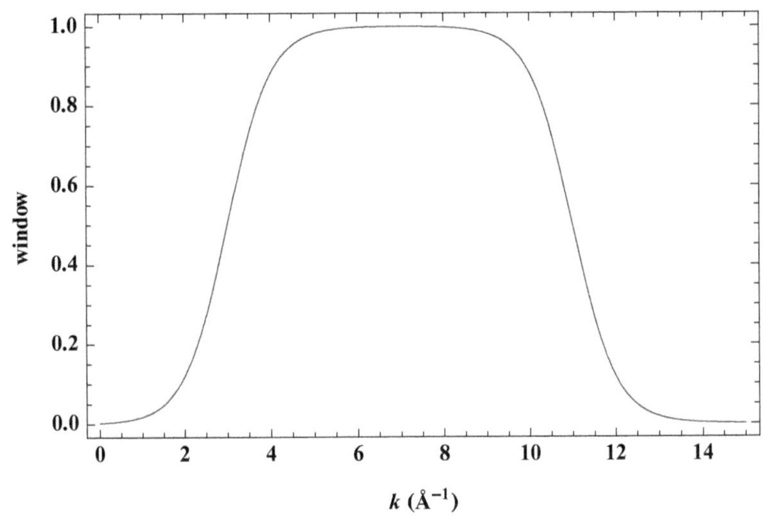

Fig. A1.10 Tapered k-window of form $(1/2)(\tanh((k - k_1 - k_w)/k_w) - \tanh((k - k_2 + k_w)/k_w))$ with $k_1 = 2$, $k_2 = 12$, $k_w = 1$.

The FT of a rectangular window is easily calculated to be

$$\tilde{W}(x) = \Delta k e^{i\bar{k}x} \frac{\sin(\Delta k x/2)}{(\Delta k x/2)},$$

where $\bar{k} = (k_1 + k_2)/2$ and $\Delta k = k_2 - k_1$. This function has a main peak centered at $x = 0$ whose width is inversely proportional to Δk, with sidelobes that decrease inversely as the distance from the peak. Since the FT of a pure sine wave is a pair of delta functions centered at some nonzero $\pm x_0$, the convolution of the FTs is just that of the window shifted so that is centered at $\pm x_0$. For more complex data it's more complicated, but it is apparent that the result still acquires sidelobes from $\tilde{W}(x)$.

This truncation ripple can be reduced if the data are weighted so that the amplitude rises and falls smoothly from/to zero at the window edges. Sometimes transform ranges can be chosen so the data naturally have that property, but more often it is accomplished by multiplying the data by a window function of the right properties. An example is shown in Figures A1.10 and A1.11.

There is a plethora of tapered FT windows that can be used, such as Hanning windows, etc. In practice there do not seem to be huge differences between them. The key issue is how much tapering they produce and how much that broadens the peaks in r-space. It is very important to be consistent

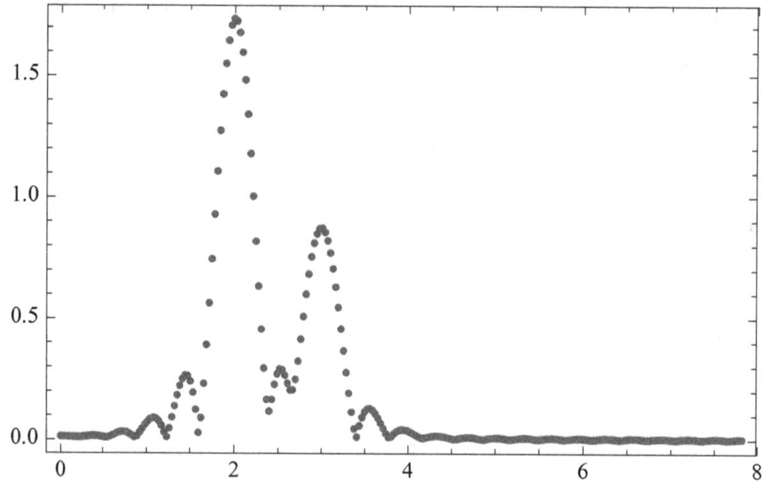

Fig. A1.11 Modulus of FT using tapered window. Notice the suppression of side-lobes, but at the cost of greater broadening of the peaks themselves.

in one's choice of windows when comparing data sets however if one wishes to minimize systematic errors.

A1.11 Fourier filtering

So far the discussion has been fairly general, not specifically tied to the EXAFS problem. At this point we will describe things in terms of r, the radial coordinate in EXAFS, rather than x, some arbitrary position coordinate. The only real difference is the factor of 2 stemming from the $2kR$ term in the EXAFS phase, but the following will be directed toward the EXAFS problem.

Once the data have been FTed to r-space, we can separate out the different contributions according to their distances. By zeroing out the contributions outside the r-range of interest, and doing an inverse transform FT^{-1}, we can select only the shell of interest. This is Fourier filtering.

If one wishes the filtered χ to be a real number, both positive and negative frequency components must be included symmetrically in the backtransform. However, it is often more useful to backtransform only one of them, which allows (after scaling by $2i$) us to determine the complex $\chi_c(k) = A(k)e^{i\phi(k)}$, the imaginary part of which is the ordinary $\chi(k) = A(k)\sin\phi(k)$. The amplitude is the modulus of the complex χ_c, and the phase is its argument (the arctangent of the ratio of the imaginary and real parts). Since the phase is only determined modulo 2π, if calculated in this way, it will usually be desirable to remove the discontinuities by suitable integer shifts of 2π so as to make the phase function continuous. This can done in a line or two of code.

A1.12 Uniqueness of amplitude and phase

It is reasonable to ask, from a signal processing/information theory point of view, how one can determine two separate functions $A(k)$ and $\phi(k)$ from the single function $\chi(k)$. What are the amplitude and phase of a potato? For a general function there is an infinite number of combinations of A and ϕ that give the same $A \sin \phi$.

But EXAFS data oscillate around zero. The functions A and ϕ will be well-determined if the positive and negative frequency components can be separated cleanly from each other. It must be remembered that they are actually closer to each other than it appears, because the array used in the DFT effectively wraps around on itself. Clean isolation will occur if the separation between the positive and negative frequency components in the data is much greater than the widths of the peaks. The separation increases with R of the first shell (which is the rapidity of oscillation in k-space), and the width is inversely proportional to the width of the k-space window. Putting these together, the amplitude and phase are well determined if there are many oscillations in the k-space window.

A1.13 Fourier filtering distortions

In the following we will assume the positive and negative frequency components do not interfere, so that we can determine the complex $\chi_c(k)$ unambiguously. We will drop the subscript so χ is the complex function.

We know that truncation effects cause distortions in the data. In the filtered data, these usually manifest themselves as reductions in amplitude near the k-window edges. This makes sense, because (by the convolution theorem), the filtered data are the windowed k-space data convoluted with the (inverse) FT of the r-space window. The narrower the r-space window, the wider is its FT in k-space, and the larger the region near the k-window boundary that is affected.

Fortunately in XAFS the data are analyzed using relative comparisons to experimental or theoretically synthesized data. The Fourier filtering distortions can be represented as a function multiplying χ that depends only rather weakly on χ, but depends strongly on the k- and r-windows.

The truncation effects will be small if $A(k)$ is small at the window cutoffs.

It does not matter if $A(k)\sin\phi(k)$ is zero there – if you change the phase of the data by $\pi/2$ (i.e. from a node to an antinode) the window distortions are the same. This is a consequence of the linearity of the Fourier transform:

$$FT[c_1\chi_1 + c_2\chi_2] = c_1 FT[\chi_1] + c_2 FT[\chi_2],$$

where c_1 and c_2 are arbitrary complex constants. Similarly, the window distortions will be the same if the data are multiplied by a scale factor, or even if they are shifted in distance, as long as the r-window is also shifted appropriately.

By using exactly the same k-window, and using the same width r-window, the distortions will largely cancel for two different data sets. If the data sets have very different amplitude decays it may be wise to pre-weight them so they are closer to each other, and then divide out the weighting factor after filtering.

A1.14 Method to remove Fourier filtering distortions

It is possible to carry out this process to higher order [52]. We represent the filtering process as a linear operator $\hat{P}$. It projects the data onto a subspace of functions of certain frequencies, but it is not a perfect projection operator.

We have the filtered χ in hand, but it has suffered distortions, which we describe as a multiplicative correction that is weakly dependent on χ. If we had an approximation (guess) χ_g to the complex single shell χ, we could determine the distortion function from $\hat{P}\chi_g/\chi_g$. This could then be divided out to produce a better estimate to the single shell χ. The process can then be repeated.

$$\chi_{n+1} = \hat{P}[\chi]\frac{\chi_n}{\hat{P}[\chi_n]},$$

where χ_n is the nth approximation to the undistorted single shell χ. We start the process with $\chi_0 = \hat{P}[\chi]$.

Interestingly, this procedure [52] actually seems to work. Code to implement this simple algorithm is given below, and demonstrated by filtering with grotesque windows, and then repairing the damage.

Mathematica 7 program to correct Fourier filtering distortions

```
(*test function*)
f[k_]:=(6 Sin[2 k 2.25-\[Pi]]+ 6Sin[2 k 2.7]) Exp[-.01 (k-3)^2]

(*the following stuff is just bookkeeping*)
kmin=2;kmax=12;dk=.05;(*data k-range*)
rmin=2.0;rmax=2.8;(*data r-range*)
rminplt=0;rmaxplt=6;(*plotting r-range*)
```

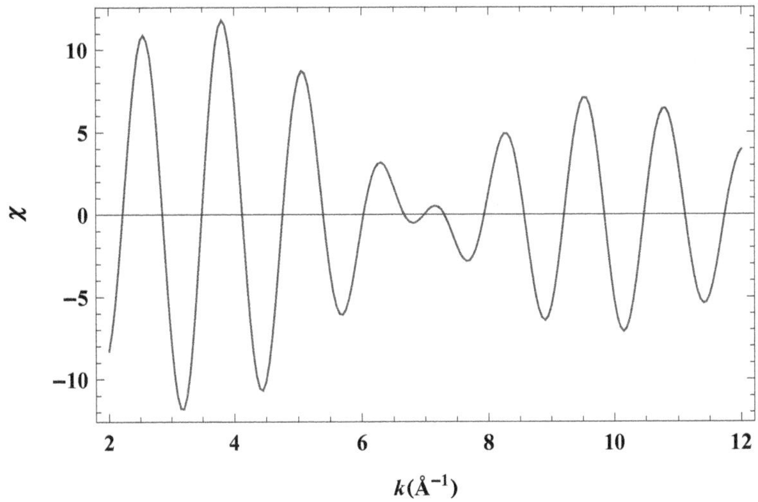

Fig. A1.12 The function χ before Fourier filtering. The truncation errors are severe because the amplitude is large at the window edges, and the truncation is sharp.

```
dr= \[Pi]/(nptsca*dk);nptsca=2048; (*npts in padded complex array*)
nkmin=Ceiling[kmin/dk]+1;nkmax=Floor[kmax/dk]+1;
nrmin=Ceiling[rmin/dr]+1;nrmax=Floor[rmax/dr]+1;
kgrid=Range[kmin,kmax,dk];(*FFT needs regular grids*)
rgrid=Range[rmin,rmax,dr];
kgridca=Range[0,dk*(nptsca-1),dk];(*grids on ca complex array*)
rgridca=Range[0,dr*(nptsca-1),dr];
kbox[k_]:=UnitStep[k-kmin]-UnitStep[k-kmax];(*rectangular window*)
rbox[r_]:=UnitStep[r-rmin]-UnitStep[r-rmax];
kwin=kbox/@kgridca;rwin=rbox/@rgridca;
FT[kdata_]:=Fourier[kwin*kdata];
IFT[rdata_]:=InverseFourier[rwin*rdata];
P[kdata_]:=IFT[FT[kdata]];(*filtering operator*)
kplt[kdata_]:=ListPlot[Take[Transpose[{kgridca,kdata}],
{nkmin,nkmax}],PlotRange->All,Joined->True,PlotStyle->Thickness[.003]];
rplt[rdata_]:=ListPlot[Take[Transpose[{rgridca,rdata}],
{Round[rminplt/dr]+1,Round[rmaxplt/dr]+1}],
PlotRange->All,Joined->True,PlotStyle->Thickness[.003]];
(*now map the data onto the zero padded complex array*)
chica=PadRight[PadLeft[f/@kgrid,Length[kgrid]+Ceiling[kmin/dk]],nptsca];
Clear[chi];

(*algorithm from g bunker thesis 1984*)
chi[0] = 2 I P[chica];(*chi[0] is filtered "experimental" data*)
chi[n_] := chi[n] = chi[0]*chi[n - 1]/P[chi[n - 1]];

Show[{kplt@chica, kplt@Im@chi[0], kplt@Abs@chi[0]}]
Show[{kplt@chica, kplt@Im@chi[50], kplt@Abs@chi[50]}]
```

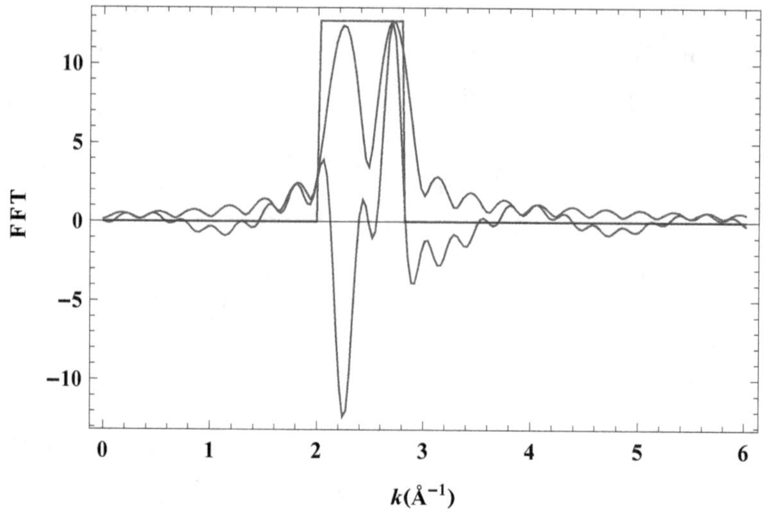

Fig. A1.13 Fourier transform of χ, with a grotesque r-space window for the inverse transform.

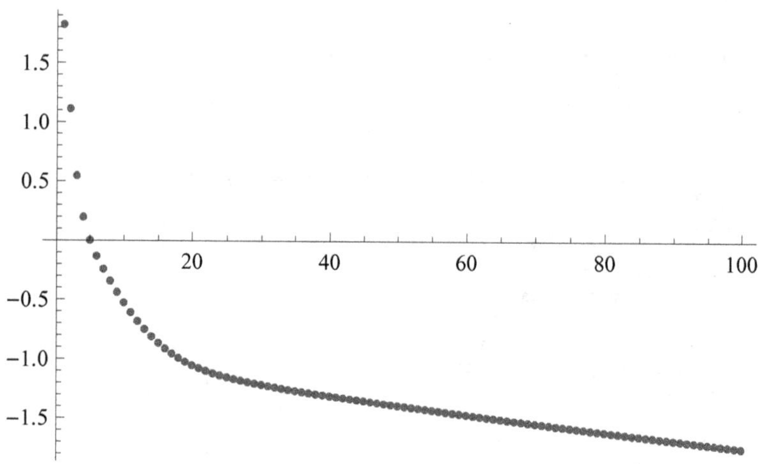

Fig. A1.14 Log of error plot to monitor convergence. Stopping right after the initial dropoff is probably sufficient.

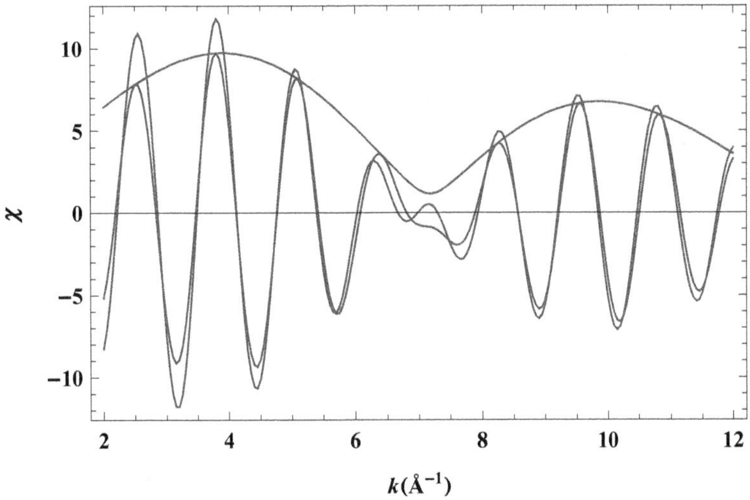

Fig. A1.15 Filtered χ vs. true χ. Amplitude of filtered data is distorted by grotesque windows.

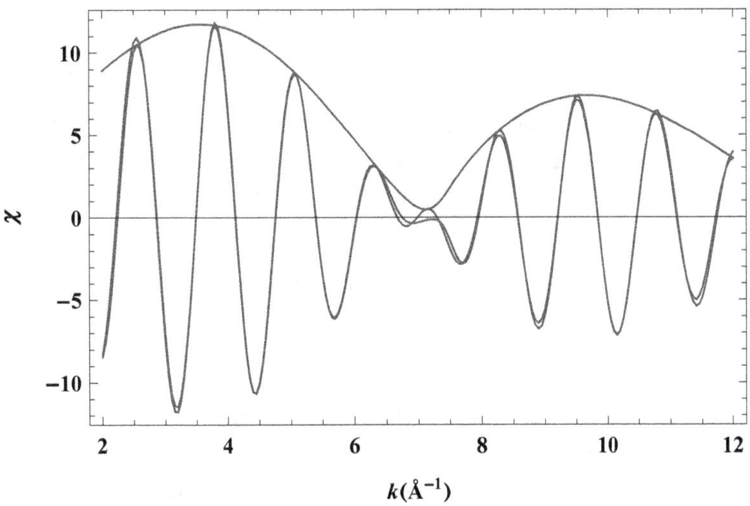

Fig. A1.16 Reconstructed χ vs. true χ. Iterative procedure mostly repairs the damage.

Appendix 2
Cumulants in EXAFS

The cumulant expansion[1] provides a useful way to describe moderately disordered distributions of atoms [89, 90]. If more than a few terms in the cumulant expansion are needed it is better to use alternative methods such as regularization. This appendix contains some modest extensions to early work [53, 52].

We will initially suppose the weak r dependence of $f_{\mathrm{eff}}(k;r)$ is accurately cancelled out by comparison with a suitable theoretical or experimental standard, so we can replace it by $f_{\mathrm{eff}}(k;\bar{r})$, which for simplicity we will call $f(k)$. Then, in the single scattering approximation, the EXAFS amplitude $A(k)$ and phase $\phi(k)$ (where $\chi(k) = \mathrm{Im}\,(\mathrm{A(k)e^{i\phi(k)}})$) for a distribution of identical atoms are given by

$$A(k)e^{i\phi(k)} = NB(k)\int_0^\infty \frac{p(r)}{r^2}e^{-2\gamma(k)r}e^{2ikr+i\delta(k)}\,dr,$$

where $Np(r)\,dr$ gives the number of atoms between r and $r+dr$, $B(k) = S_0^2|f(k)|/k$, $\delta(k)$ is the scattering phase $2\delta_c + \arg f(k) - l\pi$, which includes the central atom phase and the scattering phase, l is the final state angular momentum, and $\gamma = 1/\lambda_e$ is the inverse electron mean free path. Then

$$\ln\left(\frac{A(k)}{NB(k)}\right) + i(\phi(k)-\delta(k)) = \ln\left(\int_0^\infty \frac{p(r)}{r^2}e^{-2\gamma(k)r}e^{2ikr}\,dr\right). \quad \text{(A2.1)}$$

We define the "effective distribution"

$$P(r;k) = \frac{p(r)}{r^2}e^{-2\gamma(k)r}, \quad \text{(A2.2)}$$

which is a weighted version of the "real" distribution $p(r)$. Both the $1/r^2$

[1] J. J. Rehr first introduced cumulants to EXAFS in a paper that was not published. The present author [53] derived a recursion relation connecting them with power moments, some of their properties, and explored their utility for describing moderately disordered systems.

and exponential factors tend to shift the distribution to lower r, which is a larger effect the wider the distribution $p(r)$.[2] It is called the "effective" distribution because the distance and other parameters that are determined by EXAFS analysis describe it – not the real distribution. Once known, however, corrections can be applied to determine the moments of the real distribution.

Since γ (the inverse mean free path) is a function of k this complicates things. Appendix B of reference [53] presented a general method to handle k-dependent mean free path; below we will take a slightly different approach.

A2.1 Constant γ

First we will consider the simple case in which γ is taken as constant, so (for example) we can approximate it by its value $\gamma(\bar{k})$ in the middle of the k-range of interest. In the following section we consider the general case of k-dependent mean free path. Since the logarithm on the right side of Equation (A2.1) is well behaved at $k = 0$ we can expand it in a Taylor series in k:

$$\sum_{n=0}^{\infty} C_n \frac{(2ik)^n}{n!} = \ln\left(\int_0^{\infty} \frac{p(r)}{r^2} e^{-2\gamma r} e^{2ikr}\, dr\right). \tag{A2.3}$$

The expansion coefficients C_n are called cumulants; they are very useful for describing the shapes of distributions. The first cumulant gives the centroid of the distribution; C_2 gives the mean square width; C_3 is related to the skewness; C_4 measures how much weight is in the tails, compared to a Gaussian distribution. All cumulants higher than C_2 are zero for a Gaussian distribution.

Suppose one has a distribution that is the convolution of two other distributions. The convolution theorem (see appendix on Fourier methods) says the Fourier transform of the convolution of two distributions is just the product of the FTs of the distributions. Taking the logarithm shows us that the cumulants of the convoluted distribution is just the sum of the cumulants of the separate distributions. The familiar practice of adding the σ^2 "in quadrature" when broadening functions with Gaussians is a special case. The second cumulants of distributions of any shape just add together, as do all the other cumulants.

We can determine the nth cumulant by taking the nth derivative of Equation (4.6) and setting $k = 0$. The left side gives C_n, and the right side, after

[2] The distribution $p(r)$ itself is a projected version of the three dimensional distribution function. Because of that transformation, a distribution that is symmetrical (e.g. Gaussian) in $3D$ may be asymmetrical when projected to $1D$.

some calculation, gives an expression in terms of power moments of the effective distribution: $P_n = \int_0^\infty P(r) r^n \, dr$. These moments are unnormalized; we define the normalized moments: $p_n = P_n/P_0$. The cumulants can be expressed in terms of these power moments. The recursion relation [53] relating the moments and cumulants is

$$C_0 = \ln P_0; \quad C_{n+1} = \frac{dC_n}{dq}; \quad P_{n+1} = \frac{dP_n}{dq},$$

where the q is a formal parameter of no consequence. For example

$$C_1 = \frac{dC_0}{dq} = \frac{d \ln P_0}{dq} = \frac{1}{P_0}\frac{dP_0}{dq} = \frac{P_1}{P_0} = p_1.$$

Expressions relating the cumulants to the moments of any order can be generated in this way. The results of these calculations are shown in Table A2.1.

Cumulants have the very useful property that they are independent of origin for $n > 1$. The expressions simplify considerably if we take the origin to be the centroid (first moment) of the distribution, so that $p_1 = 0$. These expressions are given in Table A2.2; space allows us to present them up to order 8, which is usually more than enough. It can be seen that the second and third cumulants are just the central power moments $\langle (r - \bar{r})^2 \rangle$ and $\langle (r - \bar{r})^3 \rangle$, but the expressions for the fourth and all higher orders are more complex.

With the approximation that the mean free path is constant, the amplitude $A(k)$ depends only on even order cumulants, and the phase $\phi(k)$ depends only on the odd cumulants.

$$\ln\left(\frac{A(k)}{NB(k)}\right) \approx C_0 - 2k^2 C_2 + \frac{2k^4 C_4}{3} - \frac{4k^6 C_6}{45} + \frac{2k^8 C_8}{315} \cdots$$

$$\phi(k) - \delta(k) \approx 2kC_1 - \frac{4k^3 C_3}{3} + \frac{4k^5 C_5}{15} - \frac{8k^7 C_7}{315} \cdots \qquad \text{(A2.4)}$$

Here, C_2 is the mean square variation in distance about the mean – it is the variance, $C_2 = \sigma^2$. The magnitudes of C_n for $n \geq 2$ are of order σ^n or less. Thus the expressions above are essentially expansions in the parameter $k\sigma$. At small enough k, only a few terms are required. Retaining terms with cumulants up to C_2 gives the Debye–Waller term in the standard EXAFS equation. If the effective distribution were a Gaussian, all cumulants higher than 2 would be zero. Therefore the higher-order cumulants measure the deviation of the distribution from a Gaussian shape.

Table A2.1 *Cumulants and power moments (arbitrary origin).*

n	C_n
1	p_1
2	$p_2 - p_1^2$
3	$2p_1^3 - 3p_2p_1 + p_3$
4	$-6p_1^4 + 12p_2p_1^2 - 4p_3p_1 - 3p_2^2 + p_4$
5	$24p_1^5 - 60p_2p_1^3 + 20p_3p_1^2 + 30p_2^2p_1 - 5p_4p_1 - 10p_2p_3 + p_5$

Table A2.2 *Cumulants and centered power moments (origin at $\langle r \rangle$)*

n	C_n
1	0
2	p_2
3	p_3
4	$p_4 - 3p_2^2$
5	$p_5 - 10p_2p_3$
6	$30p_2^3 - 15p_4p_2 - 10p_3^2 + p_6$
7	$210p_3p_2^2 - 21p_5p_2 - 35p_3p_4 + p_7$
8	$-630p_2^4 + 420p_4p_2^2 + 560p_3^2p_2 - 28p_6p_2 - 35p_4^2 - 56p_3p_5 + p_8$

The zeroth cumulant C_0 is the log of the integral of the effective distribution $\ln\left(\int_0^\infty p(r)dr/r^2 \exp\left(-2\gamma r\right)\right)$; $\exp C_0 \approx \exp(-2\bar{r}/\lambda)/\bar{r}^2$. The first cumulant C_1 is simply the first power moment; the second cumulant C_2 is the mean square variation; the third cumulant C_3 measures the skewness; the fourth cumulant C_4 measures how much of the distribution lies out in the tails relative to a Gaussian (kurtosis). The cumulant C_2 is intrinsically positive, but the higher order cumulants can be positive or negative.

A2.2 k-dependent γ

Typically the mean free path drops to its minimum value of $\approx 3 - 5$ Å at $k \approx 2 - 5$ Å^{-1}, and then rises monotonically at higher k. Therefore the inverse mean free path has a maximum of ≈ 0.3 Å^{-1}, falling away monotonically to near zero, so that $\gamma/k << 1$ over the EXAFS data range.

Consider Equation (A2.1) with the energy-dependent γ. By combining the exponentials and defining $k' = k + i\gamma(k)$ it takes the form

$$\ln\left(\frac{A(k)}{NB(k)}\right) + i(\phi(k) - \delta(k)) = \ln\left(\int_0^\infty \frac{p(r)}{r^2} e^{2ik'r}\, dr\right). \qquad \text{(A2.5)}$$

Therefore it has a cumulant expansion in the variable k', with the effective distribution $p(r)/r^2$:

$$\sum_{n=0}^{\infty} C_n \frac{(2ik')^n}{n!} = \ln\left(\int_0^{\infty} \frac{p(r)}{r^2} e^{2ikr}\, dr\right). \tag{A2.6}$$

Expanding out the powers of $k' = k + i\gamma(k)$ in the cumulant expansion, and keeping terms up to order $(\gamma/k)^3$ (i.e. neglecting terms less than $\approx$ 0.1%), and collecting powers of k gives:

$$\ln\left(\frac{A(k)}{NB(k)}\right) \approx$$
$$\left(C_0 - 2\gamma C_1 + 2\gamma^2 C_2 - \frac{4\gamma^3 C_3}{3}\right) + k^2\left(-2C_2 + 4\gamma C_3 - 4\gamma^2 C_4 + \frac{8\gamma^3 C_5}{3}\right)$$
$$+ k^4\left(\frac{2C_4}{3} - \frac{4\gamma C_5}{3} + \frac{4\gamma^2 C_6}{3} - \frac{8\gamma^3 C_7}{9}\right) + k^6\left(-\frac{4C_6}{45} + \frac{8\gamma C_7}{45} - \frac{8\gamma^2 C_8}{45}\right)$$
$$+ \frac{2k^8 C_8}{315}; \tag{A2.7}$$

$$\phi(k) - \delta(k) \approx$$
$$k\left(2C_1 - 4\gamma C_2 + 4\gamma^2 C_3 - \frac{8\gamma^3 C_4}{3}\right)$$
$$+ k^3\left(-\frac{4C_3}{3} + \frac{8\gamma C_4}{3} - \frac{8\gamma^2 C_5}{3} + \frac{16\gamma^3 C_6}{9}\right)$$
$$+ k^5\left(\frac{4C_5}{15} - \frac{8\gamma C_6}{15} + \frac{8\gamma^2 C_7}{15} - \frac{16\gamma^3 C_8}{45}\right) + k^7\left(-\frac{8C_7}{315} + \frac{16\gamma C_8}{315}\right)$$

where these cumulants refer to the effective distribution $p(r)/r^2$ (i.e. with no exponential term). A general relation connecting the cumulants of $p(r)$ and $p(r)/r^2$ was previously derived by the author [52] and is given in the next section. This formally solves the problem of determining the cumulants including the k-dependent mean free path.

These equations would be of little use if we did not know $\gamma(k)$, but fortunately modern theoretical codes such as FEFF calculate it (in files feffxxxx.dat, in the variable lambda). Therefore these equations could be used to fit the phase and amplitude and determine the cumulants as a fitting problem that is linear in the parameters C_n. The even and odd cumulants do not cleanly separate into the amplitude and phase as they do with constant mean free path. Normally cumulants up to order 4 and terms up to γ^2

should suffice. If more than a handful of cumulants are required the speed of convergence of the cumulant expansion is questionable, and other methods such as regularization may be preferred.

A2.3 Relating cumulants of the real and effective distributions

The following is based on early work by the author [52]. First, observe that

$$\frac{d^2}{d\xi^2}\int \frac{p(r)}{r^2}e^{\xi r}\,dr = \int p(r)e^{\xi r}\,dr,$$

and, by definition of the cumulant expansion that this implies

$$\frac{d^2}{d\xi^2}\exp\left[S^{(e)}\right] = \exp\left[S^{(r)}\right]$$

where

$$S^{(e)} = \sum_{n=0}^{\infty}\frac{\xi^n}{n!}C_n^{(e)}$$

and similarly for $S(r)$, $C_n(r)$.

The superscripts (e) and (r) denote respectively the effective and real distributions for $\gamma = 0$. The final ingredient we need is the identity

$$C_m = \frac{d^2 S}{d\xi^2}|_{\xi=0}.$$

Taking successive derivatives of $\exp S$ we obtain (prime denotes differentiation):

$$\begin{aligned}
(\exp S)' &= (\exp S)S' \\
(\exp S)'' &= (\exp S)(S'' + S'^2) \\
(\exp S)''' &= (\exp S)(S''' + 3S'' + S'^3) \\
(\exp S)''' &= (\exp S)(S'''' + 4S'''S' + 3S''^2 + 6S''^2 + S'^4).
\end{aligned}$$

As $\xi \to 0$ each nth derivative turns into an nth order cumulant. The hierarchy is the same for both real and effective distributions, but they are shifted relative to each by two rows because of the two derivatives between them. Using the fact that the real distribution is normalized to 1, we have the equations:

$$
\begin{aligned}
1 &= C_1^2 P_0 + C_2 P_0 \\
C_1^{(r)} &= C_1^3 P_0 + 3C_2 C_1 P_0 + C_3 P_0 \\
\left(C_1^{(r)}\right)^2 + C_2^{(r)} &= C_1^4 P_0 + 6C_2 C_1^2 P_0 + 4C_3 C_1 P_0 + 3C_2^2 P_0 + C_4 P_0,
\end{aligned}
$$

where the quantities on the right refer to the effective distribution. Finally, these *exact* equations can be solved for the normalization constant and cumulants of the real distribution in terms of those of the effective distribution, yielding:

$$
\begin{aligned}
P_0 &= \frac{1}{C_1^2 + C_2} \\
C_1^{(r)} &= \frac{C_1^3 + 3C_2 C_1 + C_3}{C_1^2 + C_2} \\
C_2^{(r)} &= \frac{C_2 C_1^4 + 2C_3 C_1^3 + C_4 C_1^2 - 2C_2 C_3 C_1 + 3C_2^3 - C_3^2 + C_2 C_4}{(C_1^2 + C_2)^2}.
\end{aligned}
\tag{A2.8}
$$

These are identical to those presented previously [52]. The dominant term in each of these expressions is the one with the highest power of C_1 in the numerator; the next important is the one with the next higher power of C_1, etc.

If the reader wishes to generate the expressions to higher order, the following code can be used, setting nmax to whatever is needed.

Mathematica 7 code to relate cumulants of real and effective distributions

```
nmax=3;tableau=(Table[D[Exp@Se[q],{q,n+2}]==D[Exp@Sr[q],{q,n}],{n,0,nmax}]/.{q->0});
eqns=(tableau/.{Exp[Se[0]]->P0,Exp[Sr[0]]->1,Derivative[n_][Se][0]->Ce[n],
Derivative[n_][Sr][0]->Cr[n]});
vars=Append[{P0},Table[Cr[n],{n,1,nmax}]]//Flatten;soln=Solve[eqns,vars][[1]];
vars/.soln/.{Ce[n_]->Subscript[C,n]}//Together//TableForm
```

Appendix 3
Optimizing X-ray filters

Despite advances in energy dispersive detector technology, X-ray filters remain a useful tool for optimizing fluorescence XAFS experiments. They are essential when using fluorescence ionization chambers if diffractive analyzers are available.

The filter quality is an important but frequently ignored parameter that can determine the success or failure of a fluorescence XAFS experiment. In this appendix we define filter quality, describe its effects, and derive equations and rules of thumb for choosing the optimal filter thickness during an experiment. We will see that the optimal thickness depends on the background to signal ratio, the quality Q of available filters, and the effectiveness of the slits.

This appendix is concerned with minimizing high-energy (usually elastically scattered) background. Filters are used to preferentially absorb the X-rays that are scattered from dilute samples. Statistical fluctuations in the number of scattered photons are a principal source of noise in fluorescence EXAFS experiments, and it is therefore desirable to minimize the scattered background as much as possible, without attenuating the signal significantly. As described above, normally the main constituent of a filter is an element which has an absorption edge that falls between the strongest fluorescence lines and the absorption edge of the element of interest (see Figure 3.12). In many cases a suitable filter can be constructed using the element of atomic number (Z) one less than that in the sample (a "$Z - 1$ filter"). For example, one uses a Mn filter ($Z = 25$) for Fe ($Z = 26$) K-edge studies. A well-made filter strongly absorbs the elastically scattered background at the excitation energy, but it only slightly attenuates the desired signal at the (lower) fluorescence energy.

It must be noted that filters and slits are useful for discriminating against photons that are higher than the absorption edge of the filter – they offer little or no benefit if there is a source of background whose energy is close to, but below the filter edge. However, if there is a strong background that is well below the energy of interest, it can at times be useful to exploit the $1/E^3$ behavior of absorption coefficients by using a judiciously chosen number of aluminum foils to attenuate the low energy background. This can be easily done experimentally by optimizing the effective counts as a function of number of aluminum foils.

Unfortunately, for every photon that is absorbed by the filter, some fraction (the fluorescence yield, which is about 30% for Fe and elements of similar Z) is re-emitted by the filter as its own fluorescence. The fluorescence yield increases with the Z of the absorber, and generally is greater for K-edges than L-edges. The fluorescence radiation is re-emitted in all directions, so automatically half of the refluorescence goes away from the detector and is not collected. To a first approximation, elements are transparent to their own fluorescence (since the fluorescence is always below the absorption edge), and therefore the refluorescence is largely unattenuated as it escapes from the filter. Thus, for iron, about 15% of the scattered background that is absorbed by the filter will still be detected. Therefore an upper limit for the reduction in background is about $1/0.15 \approx 7$ or so. This is fine, but it is not enough for dilute specimens, for which the background to signal ratio may be 50 or 100.

Stern and Heald [16] proposed a clever method of preferentially blocking most of this refluorescence by using soller-type slits that focus on the sample. This trick works because the refluorescence is isotropically emitted from the filter (which is placed close to the detector), while the signal radiates from the sample, which is at a small but nonzero distance from the detector. It is not difficult to reduce the re-fluorescence by a factor of 10 or more using such slits, which results in a maximum background reduction of a factor of about 70. This is appropriate for most enzyme samples in the millimolar concentration range.

Filters unfortunately not only attenuate the background, they also attenuate the signal to some extent. For most pure elements of interest, the ratio of the absorption coefficient $\mu(E)$ above the K-edge to that below the edge is in the range 6–8. This imposes fundamental limits on how effectively a filter can work. We can define a figure of merit for a filter, the Quality, $Q = \mu(E_a)x/\mu(E_b)x$, where $\mu(E)$ is the absorption coefficient of the filter at energy E, x is its thickness, E_a is near the midpoint of the energy range of the scan (e.g. 7.5 KeV for Fe), and E_b is the average energy of the fluorescence lines (6.4KeV for the average Fe K_α fluorescence).[1]

For a filter constructed of a homogeneous foil of a pure element, Q is typically in the range 4–6, as mentioned above; there is a fundamental upper limit for a particular element. Most filters in practice are nonuniform, and also contain extraneous material (e.g. plastic, or Duco cement) that absorbs essentially the same fraction of signal and background. Thus, many filters in common use have quality factors of only $Q = 3$ or so, and it is well worth the trouble to do better than that. Mediocre filters can easily degrade the signal to noise ratio by a factor of 2 relative to good filters.

It is important to note that the quality of a filter is not proportional to its thickness, because the factor x divides out in the ratio $Q = \mu(E_a)x/\mu(E_b)x$. Two filters of the same Q stacked together are equivalent to one (thick) filter of the same quality Q.

It is straightforward to write an equation for the number of effective counts N_{eff}, in the case in which we have N_b background counts and N_s signal counts (i.e. sample fluorescence) collected by the detector:

$$N_{\text{eff}} = \frac{N_s^2}{N_s + N_b}. \tag{A3.1}$$

If we place a filter between the sample and the detector, the background is reduced by the factor $\exp(-\mu_a x)$, and the signal is reduced by the factor $\exp(-\mu_b x)$, where $\mu_a \equiv \mu(Ea)$ and $\mu_b \equiv \mu(Eb)$. The refluorescence is proportional to the number of photons absorbed, and a fraction η finds its way into the detector and are counted, where $\eta \approx 0.15$ if no slits are used, and $\eta \approx 0.02$ if slits are used ($\eta \approx 0$ for energy dispersive detectors if they are set to resolve only the fluorescence signal). Combining these observations, we can write the number of effective counts as

$$\frac{N_{\text{eff}}}{N_s} = (1-\eta)\frac{e^{-\mu_b x} + \gamma}{1 + A\left(\frac{e^{-\mu_a x}+\gamma}{e^{-\mu_b x}+\gamma}\right)}, \tag{A3.2}$$

where $A \equiv N_b/N_s$ and $\gamma \equiv \eta/(1-\eta)$. It is convenient to reexpress μ_a and μ_b in terms of the filter thickness $\Delta\mu x \equiv \mu_a - \mu_b$ and quality $Q \equiv \mu_a/\mu_b$, or $\mu_a = \Delta\mu x(Q/(Q-1))$, $\mu_b = \Delta\mu x(1/(Q-1))$. The number of effective counts depends on the background to signal ratio (before the filter is put in), the

[1] It is usually more convenient in practice to measure the absorption right above and right below the filter's absorption edge, and scale the result, accounting for the approximate $1/E^3$ energy dependence of the filter absorption coefficient. For example, suppose we are measuring Fe fluorescence, and the ratio of the measured absorption coefficients immediately above and below the Mn edge at 6.54 KeV is 6. Appropriate choices for E_a and E_b are 7.50 KeV and 6.40. Then $Q = 6(6.54/7.50)^3 \times (6.40/6.54)^3 = 6(6.40/7.50)^3 = 6(.62)$, or $Q \approx 3.7$.

thickness $\Delta\mu x$ and quality μ_a/μ_b of the filter, and the effectiveness of the slits η. We plot this function for various typical values of these parameters in the figures below. These curves can be used during your experiment to choose the optimum filter thickness, provided you have previously measured the thickness and quality of your filters.

We can find the optimum value by taking the derivative of Equation (A3.2) with respect to $\Delta\mu x$, setting it to zero, and solving for $\Delta\mu x$. However, unless somewhat dubious further approximations are made, this gives a transcendental equation that has no analytical solutions. Numerical solutions however can be readily obtained, and are presented in the appendix as tables and graphs. The figures have the additional advantage of showing how soft the optimum filter thickness is.

Simple analytical solutions are readily obtained when η approaches zero. This is appropriate when a solid state detector is used to reject filter refluorescence,[2] or if the slits are able to block nearly all of the filter refluorescence. In that case ($\eta = 0$) the optimum filter thickness is given by the simple expression:

$$\Delta\mu x(\text{best}) = \ln\left[\frac{N_\text{b}}{N_\text{s}}(Q-2)\right]. \tag{A3.3}$$

This equation indicates that if the filter quality is 2.0 or less, one is better off without it: it actually degrades the signal to noise ratio. Another consequence of this equation is that, for a filter quality of 3.0, the best signal to noise ratio is obtained when the signal and background are about equal. If the filter quality is 4, the optimum S/N ratio occurs when the background is half the size of the signal. Although the equation is approximate, it is useful during the experiment for estimating how thick a filter to add or remove in order to improve the signal to noise ratio. For example, suppose that you are already using some filters, and the background has been reduced to 3 times the signal, and the filters you have left are of quality $Q = 4$. The equation indicates that you should add a filter of $\Delta\mu x \approx 1.8$ to obtain optimal S/N.

The exact expression (Equation (A3.2)) is plotted vs. filter thickness in the accompanying figures. Using these curves is probably the best way to estimate the proper filter thickness under given experimental conditions. Note how strongly the filter quality affects the optimum achievable S/N ratio. One can readily double the effective number of counts by using a high quality ($Q = 5$) filter rather than a mediocre one. The thickness of the filter must also be correct, however, and it is very important that slits be used if

[2] It is often beneficial to use filters and slits in conjunction with solid state detectors, so that scattered background radiation does not saturate the detector. In this case the strategy is to use only just as thick a filter as needed to prevent saturation.

the filter is close to the detector. Also it is striking how low the efficiency of use of signal photons actually is: clearly there is a need for energy dispersive detectors of large solid angle and higher maximum count rate. Having made the point that filter quality is a very important experimental parameter, we now describe the most important factors that affect filter quality. Considering only X-ray characteristics, the ideal filter would be a uniform, self-supporting foil of the appropriate element. In most cases this is not practical, however. Usually the appropriate foil thickness is only a few microns, and such a foil usually is easily damaged under experimental conditions. Excellent filters can often be constructed by thermally evaporating metals on a 0.5–1.0 mil Kapton plastic film (1mil = 0.001 inch). The quality Q is higher for thicker filters because the absorption of the substrate is present regardless of the coating thickness. In general, one wants to minimize the absorption of signal by the support and other material. Uniformity is very important, because thin regions will allow leakage of background photons through the filter and degrade its effectiveness, particularly for thick filters.

One can alternatively use fine particles of the appropriate element applied to adhesive tape (Scotch Magic Transparent tape works well), or uniformly suspended in a slurry of Duco cement and thinner that is painted (or sprayed) on a thin support. One can use the cement itself as a support, but care should be taken that it not be too thick and absorb too much of the fluorescence signal.

If fine particles are used, it is very important that they be fine enough: the particle size should be not be much larger than one absorption length (μ^{-1}). For example, if one were to use MnO to construct a filter for use at the Fe K edge, the particle size should be not much larger than 5 microns. A −400 mesh sieve permits particles smaller than 38 microns to pass through. Although these particles make up a very fine dust, they are still about ten times larger than needed. In short, one must first calculate what one absorption length is, and then figure out some way to obtain particles that size. Just grinding up the material until the particles seem small usually results in a low-quality filter (Q = 2–3). Large particles do not effectively filter out the background, and moreover, they unnecessarily block out the signal. Finer particles can be separated by sedimentation, for example by mixing particles with methanol or acetone, stirring the mixture up, letting the larger particles fall to the bottom for a minute or so, and decanting and drying the ultrafine particles that remain suspended. Paradoxically it is sometimes better to use a compound or complex salt (which has more junk in it), because particles of the appropriate size are more easily prepared. In this case attenuation by the other elements in the filter should of course be minimized. For example, a few oxygen atoms may be ok, but a few sulfurs are not. The final step in filter preparation is to measure: (A) the filter thickness $\Delta\mu x$, and (B) the filter quality $\mu x_{\text{above}}/\mu x_{\text{below}}$. These numbers, and the equations above, and the accompanying figures then allow one to systematically optimize the filter thickness during the experiment.

To determine filter quality Q, it is necessary to make absolute measurements of the filter absorption coefficient, rather than the relative measurements that are normally carried out in XAFS experiments. It is straightforward and fast to do so by the following procedure. Move the monochromator to the fluorescence energy and measure the ion chamber reading with the filter placed in front of the detector (I_b), and also without it (I_{b0}). Then move to an energy in the middle of the EXAFS scan range (say 400 eV above the edge) and again measure the fluxes with (I_a) and without (I_{a0}) the filter. After subtraction of dark currents (offsets) (and correction for air absorption if needed) the filter quality can be calculated as $Q = \ln(I_a/I_{a0})/\ln(I_b/I_{b0})$. The measurements should be done without large time delays between the readings so that the incident flux does not appreciably change. Alternatively the incident flux can be measured simultaneously and each flux above can be normalized by the incident fluxes. Several measurements on different parts of the filter should be taken to check uniformity. As in a transmission measurement, care should be taken to eliminate harmonic contamination of the beam, otherwise filter quality will be underestimated.

The following tables and graphs can be used for optimizing use of filters prior to and during an experiment. To use these curves, you need to know your background to signal ratio. The background is presumed to come almost entirely from elastically scattered X-rays. The signal can be estimated to be the signal right above the edge minus the signal right below ($S = A - B$); the background is essentially the signal right below the edge (B). Since a ratio is taken, the absolute scale doesn't matter, but amplifier offsets/dark currents should be subtracted out before taking the ratio. You also need to know your approximate filter quality ($Q = 4$ or 5 is probably a good guess if you don't know), and the refluorescence rejection factor η (2% is probably a good guess).

The curves are plotted vs. $\log_2(B/S)$; if the background to signal ratio is ≈ 32, it's $\approx \log_2(32) = 5$. The curves and tables tell us that for $\eta = 0$ the optimum filter thickness $\Delta\mu x \approx 4.2$, and for $\eta = 0.05$ the optimum $\Delta\mu x \approx 2.3$. If $\eta \approx 0.02$ then the optimum filter thickness should be a little over 3, so that's the one to use. The curves then tell us that the effective counts under those conditions are about 10% of the total counts. This can be

- **Basic Definitions**

```
Ns = Ns0 (Exp[-mubelow] + Eta (1 - Exp[-mubelow]));
Nb = Nb0 (Exp[-muabove] + Eta (1 - Exp[-muabove]));
mubelow = x / (Q - 1); muabove = x Q / (Q - 1);
Neffective = Ns^2 / (Ns + Nb);
Neff[x_, ns_, nb_, q_, eta_] := Evaluate[Neffective /. {Q → q, Eta → eta, Ns0 → ns, Nb0 → nb}];
```

- **The following code generates a series of tables and curves for various filter qualities (q between 3 and 7) slit selectivities (η from 0 to .15), and background to signal ratios, which vary in powers of two from ratios of 1/1 up to 256/1, or Log2[Nb/Ns]=0 to 8. The code also finds the optimal values of filter thickness vs Log2[Nb/Ns].**

Note that the fluorescence yield is included in the parameter η, so η=0.15 corresponds to a fluorescence yield 0.3, 2π solid angle, and no slits.

```
Reff[x_, A_, q_, eta_] := Evaluate[Neffective /. {Q → q, Eta → eta, Ns0 → 1., Nb0 → A}];
Reffcurves[Q_, eta_] := Plot[Table[Reff[x, 2^n, Q, eta], {n, 0, 8}], {x, 0, 8}, PlotRange → All,
   Frame → True, ImageSize → 4 * 72,
   FrameLabel → {"Δμx", "Neff/Ns", "Q=" <> ToString[Q] <> "  η=" <> ToString[eta]}];
best[A_, Q_, eta_] := FindMinimum[-Reff[x, A, Q, eta],
    {x, 3, 0, 10}, AccuracyGoal → 6][[2, 1, 2]];
exp2[x_] := {2^x[[1]], x[[2]]};
bestvalues[Q_, eta_] := Table[{n, best[2^n, Q, eta]}, {n, 0, 8}];
hline = StringJoin@Table["_", {80}];
Do[(Print[hline]; Print["Q=", q, "   η=", η];
  opt = bestvalues[q, η]; If[NumberQ[opt[[1, 2]]], opt, Drop[opt, 1]];
  Print["Nb/Ns \t Opt. Δμx:"]; Print[TableForm[Map[exp2, opt], TableSpacing → {1, 8}]];
  Print["Optimum Δμx of filter vs z=Log₂(Nb/Ns)=\n", bff = Fit[opt, {1, z, z^2, z^3}, z]];
  bfplt = Plot[bff, {z, 0, 8}];
  Print[GraphicsRow[{Reffcurves[q, η],
     Show[bfplt, ListPlot[opt, Joined → False, PlotStyle → PointSize[.02],
       Axes → False, PlotRange → All], Frame → True, ImageSize → 4 * 72,
      FrameLabel → {"Log2(Nb/Ns)", "optimum filter thickness Δμx",
        "Q=" <> ToString[q] <> "  η=" <> ToString[η]}]}]]),
 {q, 3, 7}, {η, 0, .15, .05}]
```

Fig. A3.1 Mathematica 7 code to generate tables and plots for filter optimization.

used to estimate the amount of time needed to signal average. The shape of the curves shows how rapidly performance deviates away from the optimum.

If a solid state detector can be used as the back-end detector, it effectively acts like a set of perfect slits. The solid angle may be reduced compared to a Lytle detector however. In that case the $\eta = 0$ curves are appropriate and much better efficiencies can be achieved.

Q=4 η=0.

Nb/Ns	**Opt. $\Delta\mu$x:**
1	0.693148
2	1.38629
4	2.07944
8	2.77259
16	3.46574
32	4.15888
64	4.85203
128	5.54518
256	6.23832

Optimum $\Delta\mu$x of filter vs z=Log$_2$(Nb/Ns) =
$0.693148 + 0.693147\, z + 6.89505 \times 10^{-8}\, z^2 - 4.63815 \times 10^{-9}\, z^3$

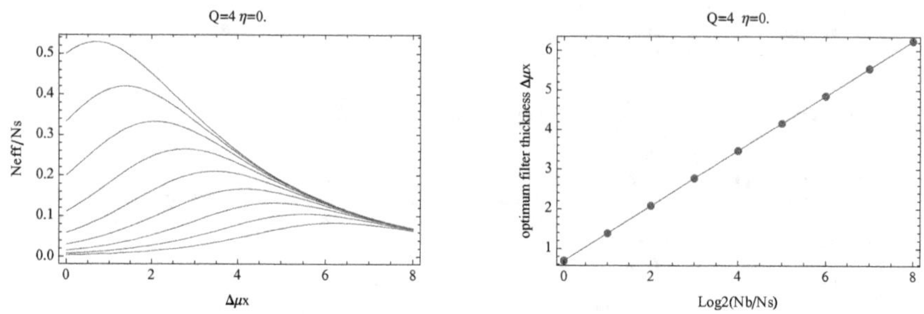

Fig. A3.2 Filter optimization $Q = 4$, $\eta = 0$.

Q=4 η=0.05

Nb/Ns	**Opt. $\Delta\mu$x:**
1	0.637155
2	1.20095
4	1.64785
8	1.95683
16	2.14534
32	2.25063
64	2.30643
128	2.33518
256	2.34977

Optimum $\Delta\mu$x of filter vs z=Log$_2$(Nb/Ns) =
$0.630136 + 0.671191\, z - 0.0895802\, z^2 + 0.00406667\, z^3$

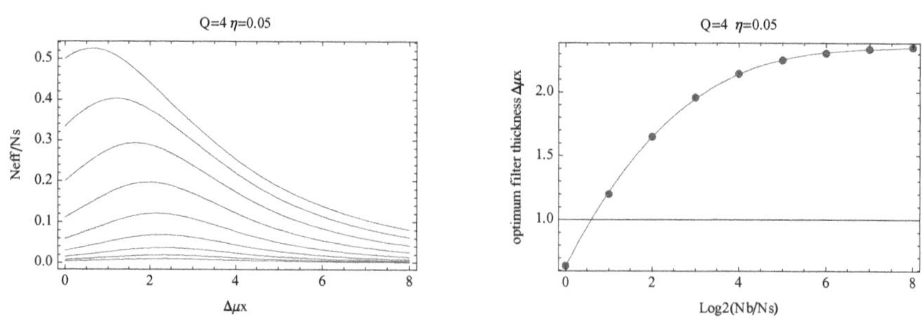

Fig. A3.3 Filter optimization $Q = 4$, $\eta = 0.05$.

```
Q=4   η=0.1

Nb/Ns        Opt. Δμx:

1             0.591932
2             1.07726
4             1.42623
8             1.64645
16            1.77232
32            1.83993
64            1.87502
128           1.8929
256           1.90192
```

Optimum $\Delta\mu x$ of filter vs $z=\mathrm{Log}_2(\mathrm{Nb/Ns})=$
$0.593401 + 0.564447\,z - 0.0840739\,z^2 + 0.0042549\,z^3$

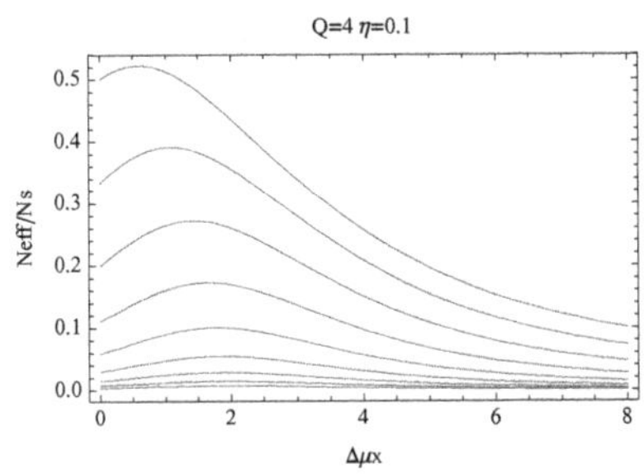

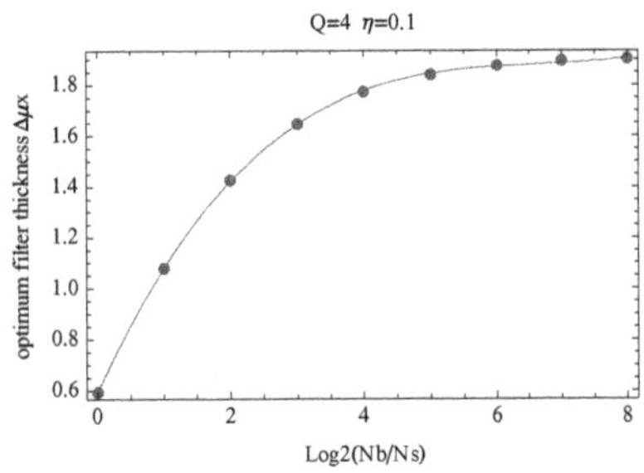

Fig. A3.4 Filter optimization $Q = 4$, $\eta = 0.10$.

```
Q=4   η=0.15

Nb/Ns        Opt. Δμx:

1             0.554243
2             0.985245
4             1.27836
8             1.45538
16            1.55376
32            1.60579
64            1.63257
128           1.64615
256           1.653
```

Optimum $\Delta\mu x$ of filter vs $z=\mathrm{Log}_2(\mathrm{Nb/Ns})=$
$0.559102 + 0.493472\,z - 0.0767958\,z^2 + 0.00403852\,z^3$

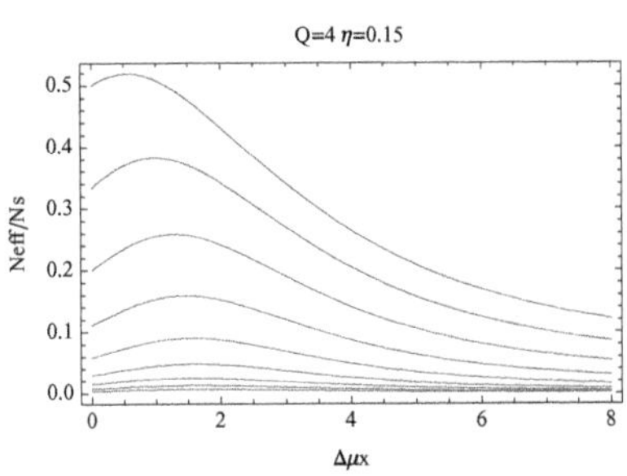

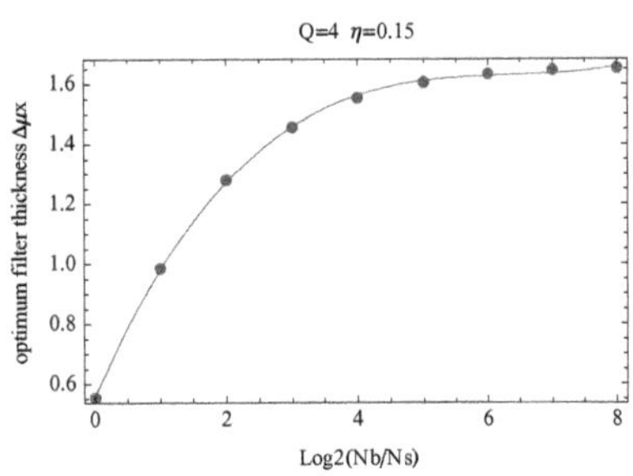

Fig. A3.5 Filter optimization $Q = 4$, $\eta = 0.15$.

Q=5 η=0.

Nb/Ns	Opt. Δμx:
1	1.09861
2	1.79176
4	2.48491
8	3.17805
16	3.8712
32	4.56435
64	5.2575
128	5.95064
256	6.64379

Optimum Δμx of filter vs z=Log₂(Nb/Ns) =
$1.09861 + 0.693147\,z + 3.22352\times10^{-8}\,z^2 - 1.29452\times10^{-9}\,z^3$

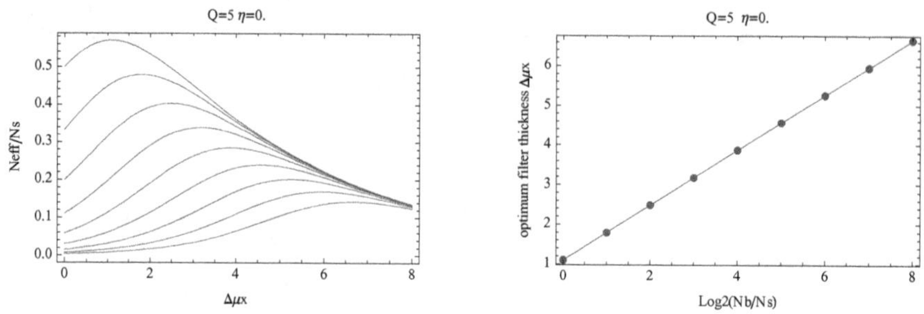

Fig. A3.6 Filter optimization $Q = 5$, $\eta = 0$.

Q=5 η=0.05

Nb/Ns	Opt. Δμx:
1	1.0204
2	1.58423
4	2.03972
8	2.36361
16	2.56683
32	2.68273
64	2.74496
128	2.77725
256	2.7937

Optimum Δμx of filter vs z=Log₂(Nb/Ns) =
$1.01182 + 0.671503\,z - 0.0860052\,z^2 + 0.00373307\,z^3$

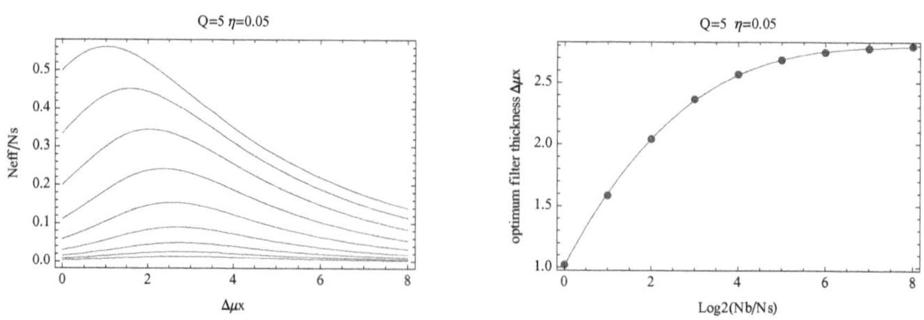

Fig. A3.7 Filter optimization $Q = 5$, $\eta = 0.05$.

```
Q=5   η=0.1

Nb/Ns       Opt. Δμx:

 1            0.956812
 2            1.44137
 4            1.79693
 8            2.02657
 16           2.16032
 32           2.23306
 64           2.27107
 128          2.29052
 256          2.30035
```

Optimum $\Delta\mu x$ of filter vs $z = \text{Log}_2(\text{Nb}/\text{Ns}) =$
$0.956788 + 0.565868\,z - 0.0822983\,z^2 + 0.00407678\,z^3$

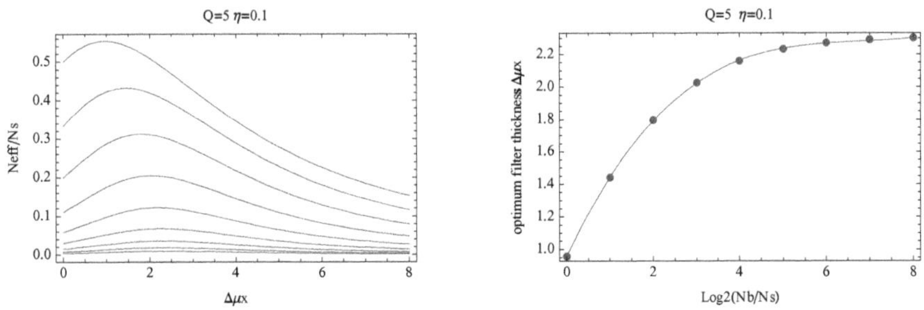

Fig. A3.8 Filter optimization $Q = 5$, $\eta = 0.10$.

```
Q=5   η=0.15

Nb/Ns       Opt. Δμx:

 1            0.903471
 2            1.33306
 4            1.6309
 8            1.81422
 16           1.91754
 32           1.97266
 64           2.00116
 128          2.01566
 256          2.02297
```

Optimum $\Delta\mu x$ of filter vs $z = \text{Log}_2(\text{Nb}/\text{Ns}) =$
$0.907114 + 0.494198\,z - 0.0755988\,z^2 + 0.0039187\,z^3$

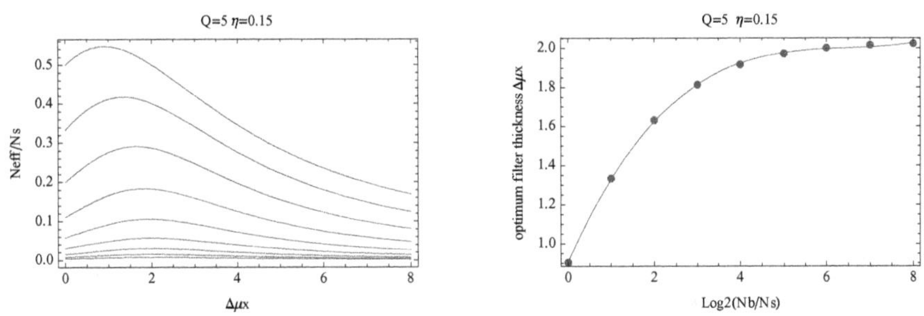

Fig. A3.9 Filter optimization $Q = 5$, $\eta = 0.15$.

Q=6 η=0.

Nb/Ns	Opt. $\Delta\mu$x:
1	1.38629
2	2.07944
4	2.77259
8	3.46574
16	4.15888
32	4.85203
64	5.54518
128	6.23832
256	6.93147

Optimum $\Delta\mu$x of filter vs z=Log$_2$(Nb/Ns) =
$1.38629 + 0.693147\,z - 1.96168\times10^{-8}\,z^2 + 1.43354\times10^{-9}\,z^3$

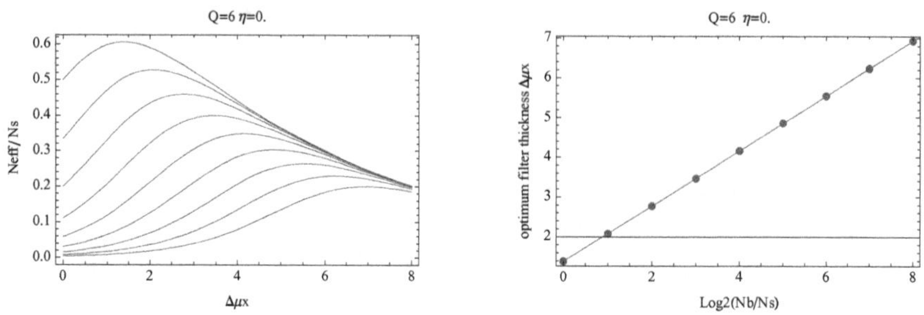

Fig. A3.10 Filter optimization $Q = 6$, $\eta = 0$.

Q=6 η=0.05

Nb/Ns	Opt. $\Delta\mu$x:
1	1.29752
2	1.86437
4	2.32878
8	2.66577
16	2.88162
32	3.00671
64	3.07456
128	3.10996
256	3.12806

Optimum $\Delta\mu$x of filter vs z=Log$_2$(Nb/Ns) =
$1.28777 + 0.674966\,z - 0.0835934\,z^2 + 0.00348895\,z^3$

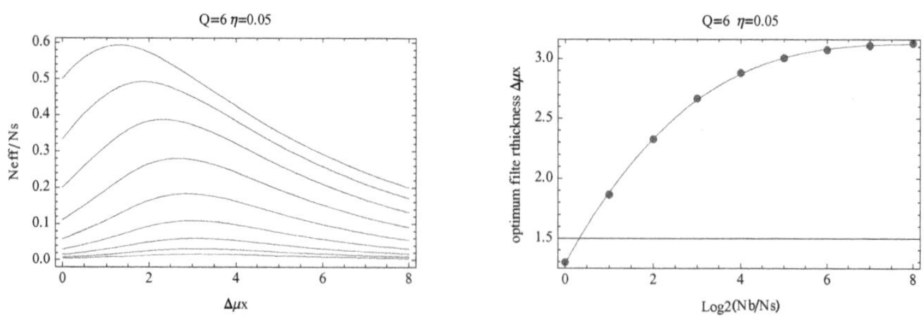

Fig. A3.11 Filter optimization $Q = 6$, $\eta = 0.05$.

```
Q=6   η=0.1

Nb/Ns         Opt. Δμx:

1              1.22456
2              1.71231
4              2.07558
8              2.31417
16             2.45507
32             2.53242
64             2.57307
128            2.59392
256            2.60448
```

Optimum $\Delta\mu x$ of filter vs $z = Log_2(Nb/Ns) =$
$1.22344 + 0.571214\,z - 0.0815375\,z^2 + 0.00396958\,z^3$

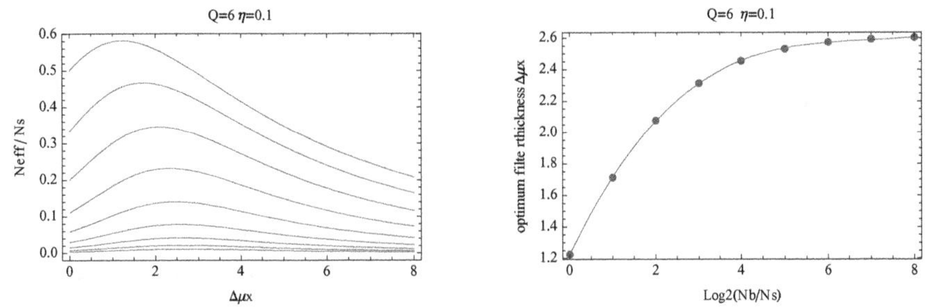

Fig. A3.12 Filter optimization $Q = 6$, $\eta = 0.10$.

```
Q=6   η=0.15

Nb/Ns         Opt. Δμx:

1              1.16284
2              1.59522
4              1.89915
8              2.0888
16             2.19678
32             2.25475
64             2.28484
128            2.30018
256            2.30792
```

Optimum $\Delta\mu x$ of filter vs $z = Log_2(Nb/Ns) =$
$1.16563 + 0.49906\,z - 0.0753545\,z^2 + 0.00386262\,z^3$

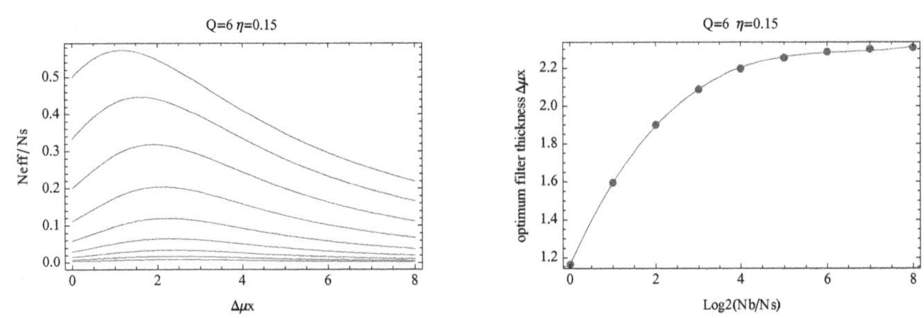

Fig. A3.13 Filter optimization $Q = 6$, $\eta = 0.15$.

Appendix 4
Reference spectra

There is perennial interest in reference XAFS spectra. This appendix contains a small sample of μ data taken by the author and collaborators. Pre-edge fits are removed to improve visibility, but that is all.

The first three spectra are from study of multiple scattering in gases [109]. A Si(111) monochromator was used on a bend magnet beamline (X9) at NSLS. This study permitted isolation of the Cl multiple scattering signal in molecular gases simply by taking linear combinations of raw spectra. Weak backscattering from hydrogen was also noted. The triangle path scattering is evident right above the "white line"; similar features are seen in ZnS_4 tetrahedral sites such as are present in ZnS and various enzymes, such as Aspartate Transcarbamoylase. The $GeCl_4$ spectrum contains single scattering plus multiple scattering among the Cl; GeH_3Cl contains mostly single scattering from the Cl and weak scattering from H, (the weak multiple scattering between the central atom and the Cl and multiple scattering among the hydrogens are neglected). GeH_4 principally has only weak backscattering from the hydrogens.

The other spectra were taken with a Si(220) monochromator and an 18 KG wiggler at SSRL. The next three spectra are $KMnO_4$ at 80K, 160K, 300K sample temperature. Note the temperature dependence of the high-frequency structure at the edge. This is due to single scattering contributions from non-nearest neighbor atoms, which are loosely bound, and therefore have large temperature dependent σ^2. In contrast, the nearest neighbor oxygens are very tightly bound, and show very little temperature dependence.

The subsequent spectra are K_2MnO_4, Mn_2O_3 (Bixbyite structure), Mn_3O_4 (Hausmannite), MnO_2 (Pyrolusite), and MnO (Manganosite) at 80K. The temperature dependencies of these materials are less striking than $KMnO_4$ so other temperatures are not shown. The pre-edge transition in K_2MnO_4 is much smaller than $KMnO_4$ because the bond distance is longer, as a consequence of the extra electron occupying an antibonding orbital. The state also becomes partially filled, which slightly decreases the intensity.

The next spectrum shown is a $Mn^{3+}(CN)^6$ hexacyanide coordination complex (potassium counterion). This exhibits large multiple scattering contributions because the second shell atoms are in line with the first shell atoms, and the distances between them are short. A clear beating-type behavior between the first and second shell signals is seen in the spectra.

Ferrocene (FeC_{10}) is an interesting structure in which the Fe is sandwiched between two pentagonal rings of carbon (cyclopentadiene). The ten nearest neighbor carbon atoms form the first shell, which is all that is visible at 300K. Notice the absence of higher shell signals.

The last spectrum highlighted here is a square planar tetraimidazole complex. This has four nearest neighbor nitrogen atoms in a plane, and two more Cl ligands at much greater distance, which are effectively missing. The sharp peak (often associated with a Cu $4p_z$ orbital) on the rising part of the edge is typical of such planar complexes. Similar transitions are seen in Heme complexes if there are no axial ligands. If the axial ligands are close, this peak is suppressed. The double-humped structure just above the edge is characteristic of imidazole rings and hemes, arising from interferences between first shell and third shell atoms.

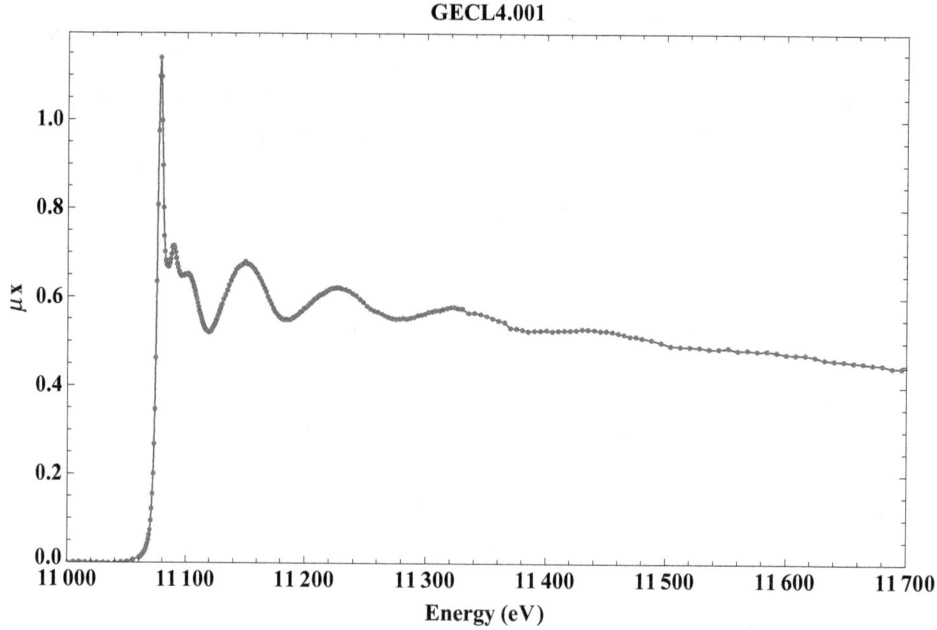

Fig. A4.1 $GeCl_4$ gas at room temperature.

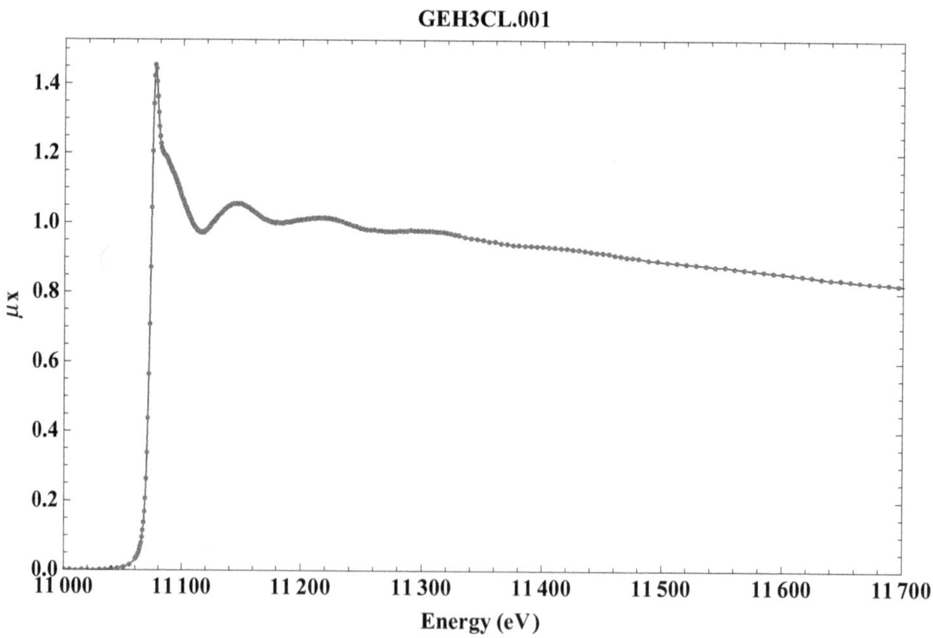

Fig. A4.2 GeH_3Cl gas at room temperature.

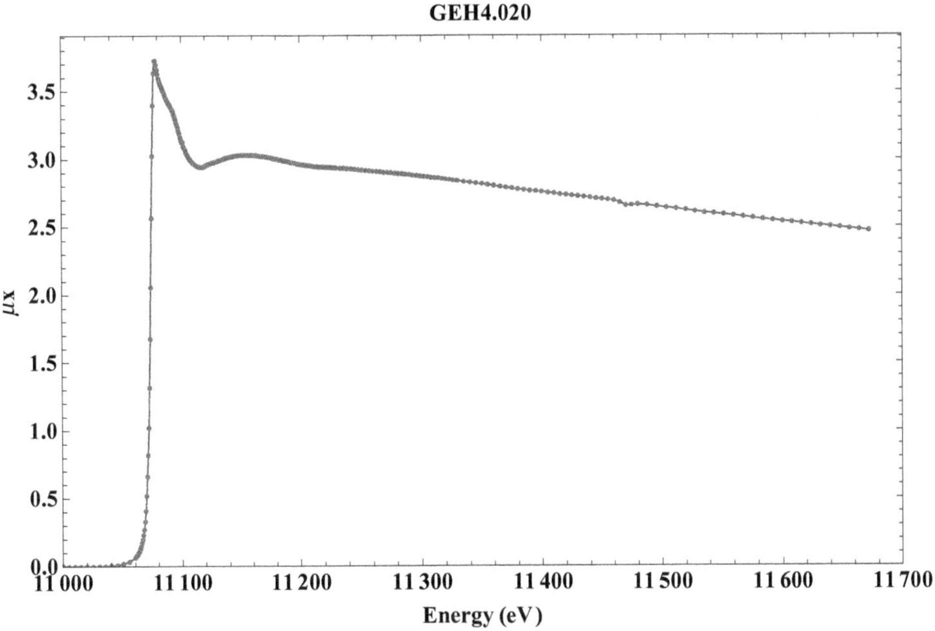

Fig. A4.3 GeH_4 gas at room temperature.

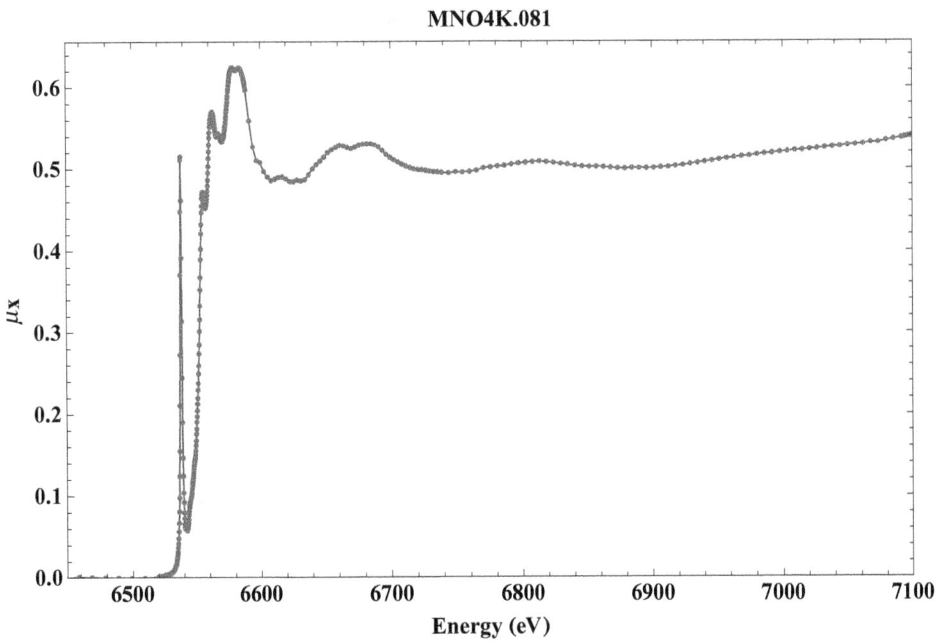

Fig. A4.4 $KMnO_4$ at 80K.

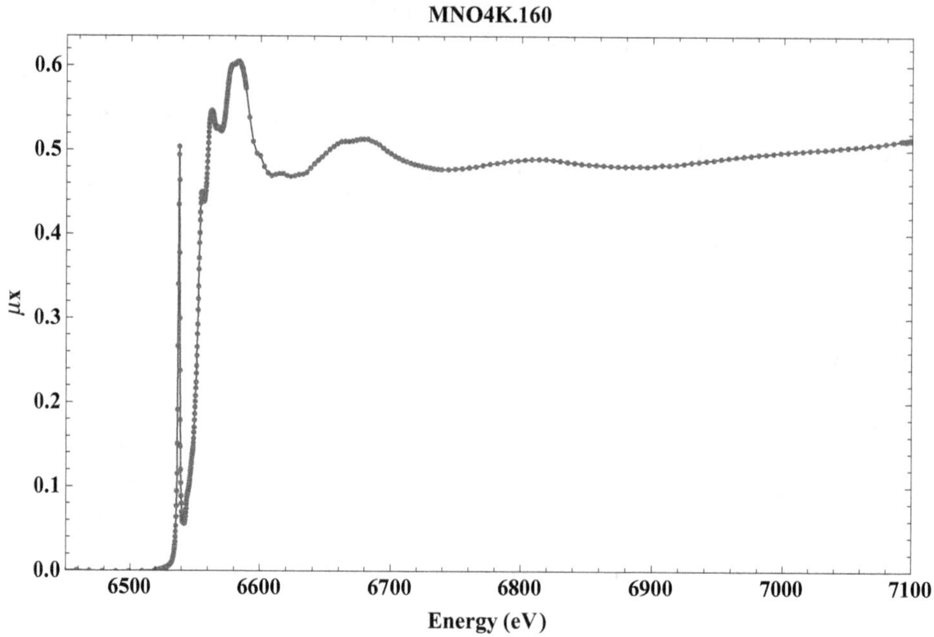

Fig. A4.5 $KMnO_4$ at 160K.

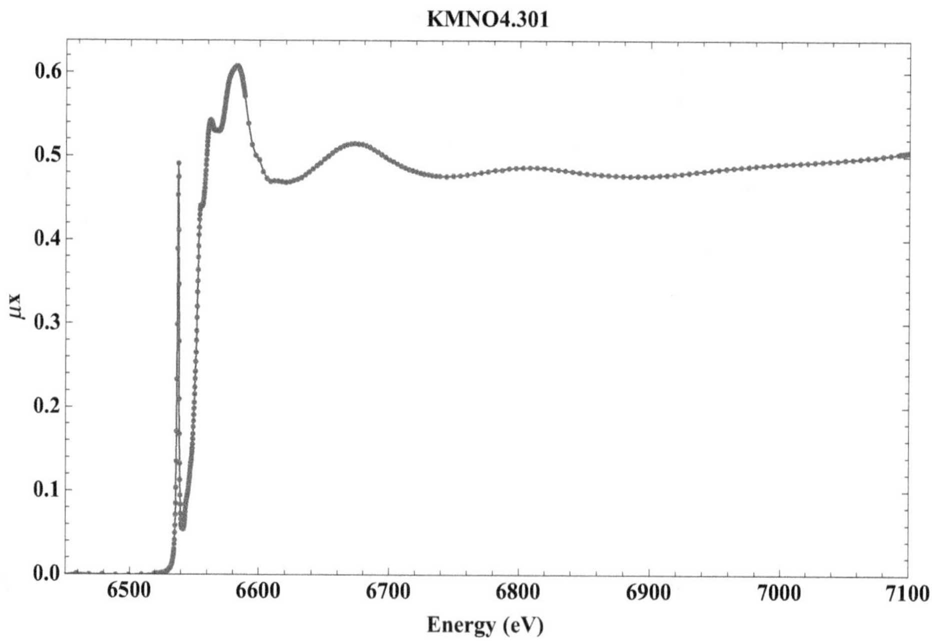

Fig. A4.6 $KMnO_4$ at 300K.

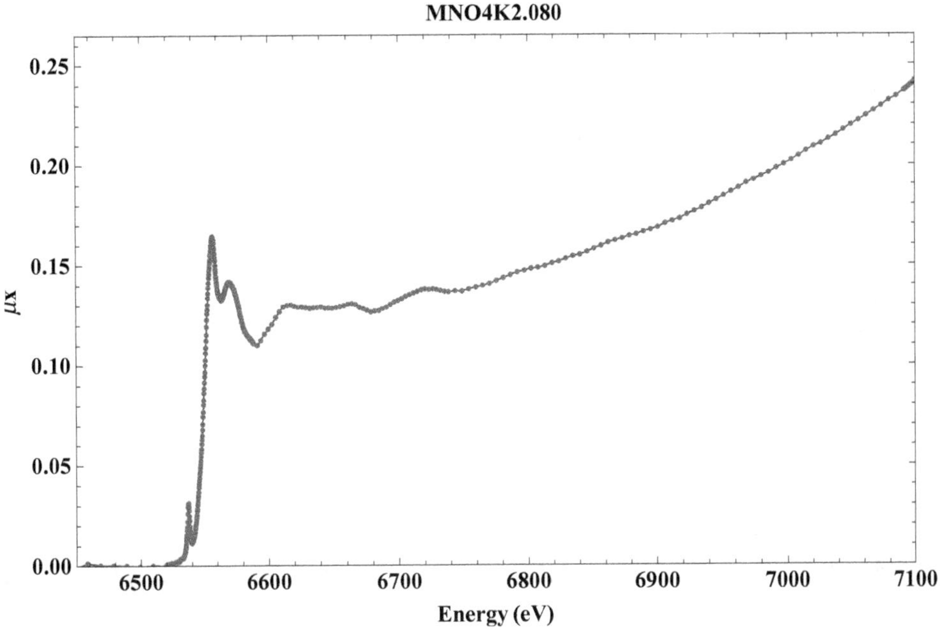

Fig. A4.7 K_2MnO_4 at 80K.

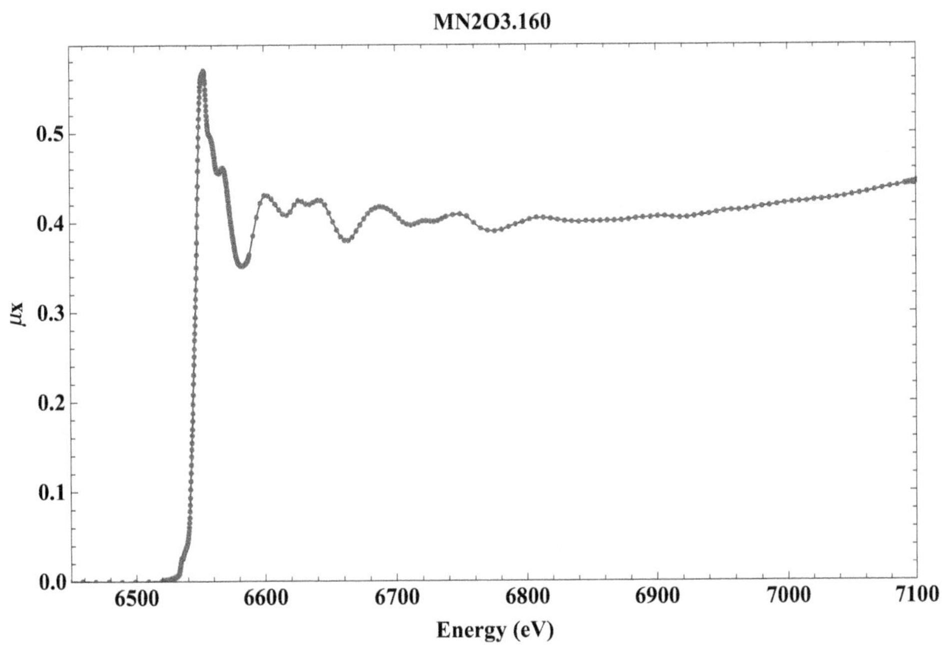

Fig. A4.8 Mn_2O_3 at 160K.

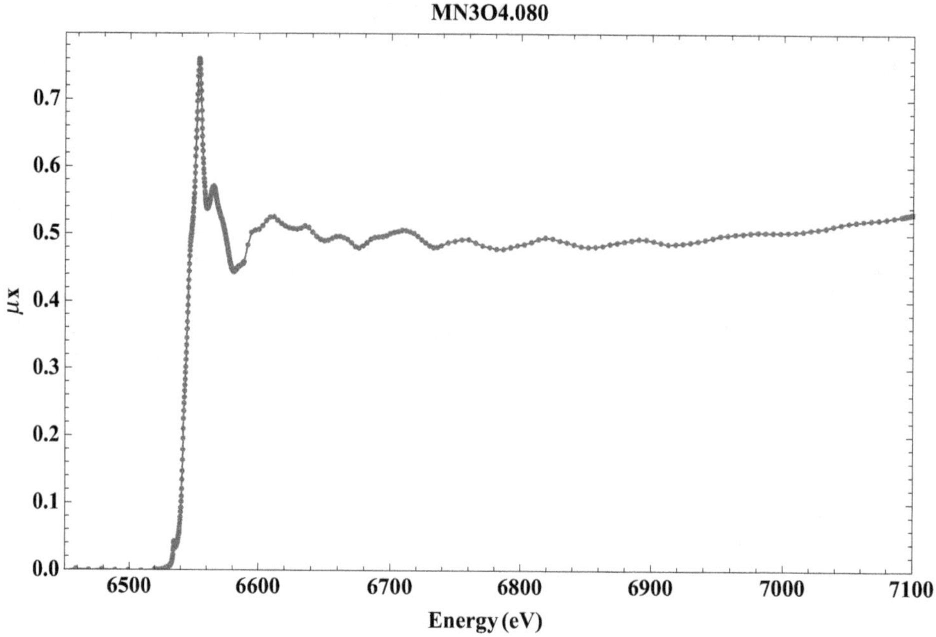

Fig. A4.9 Mn_3O_4 at 80K.

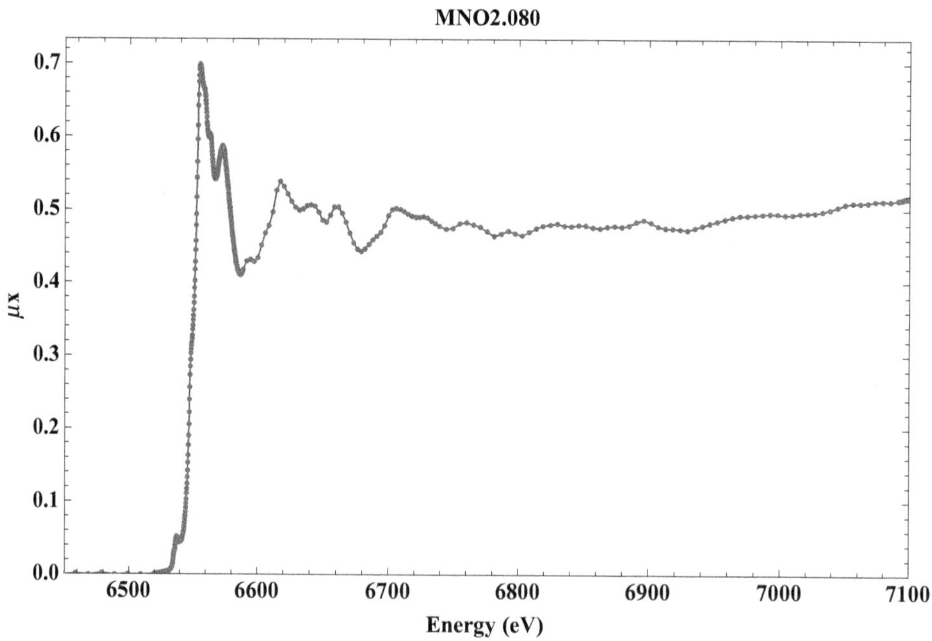

Fig. A4.10 MnO_2 at 80K.

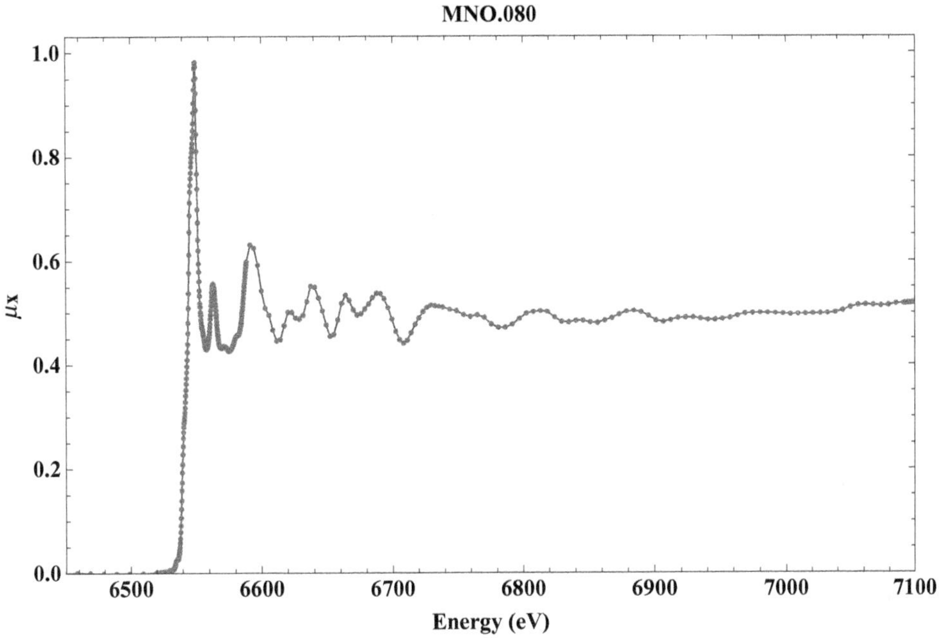

Fig. A4.11 MnO at 80K.

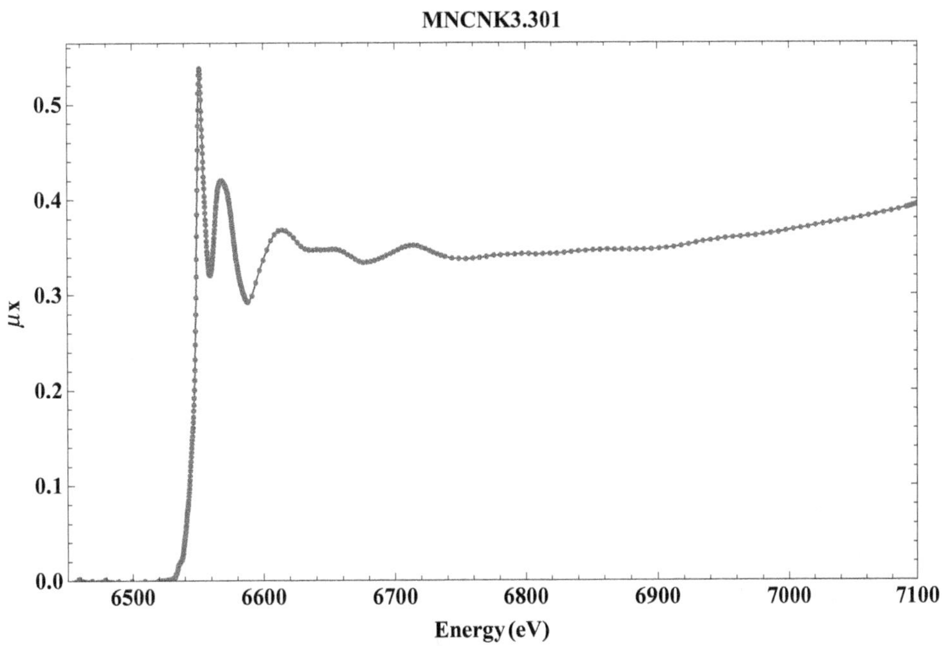

Fig. A4.12 $K_3Mn(CN)_6$ at 300K.

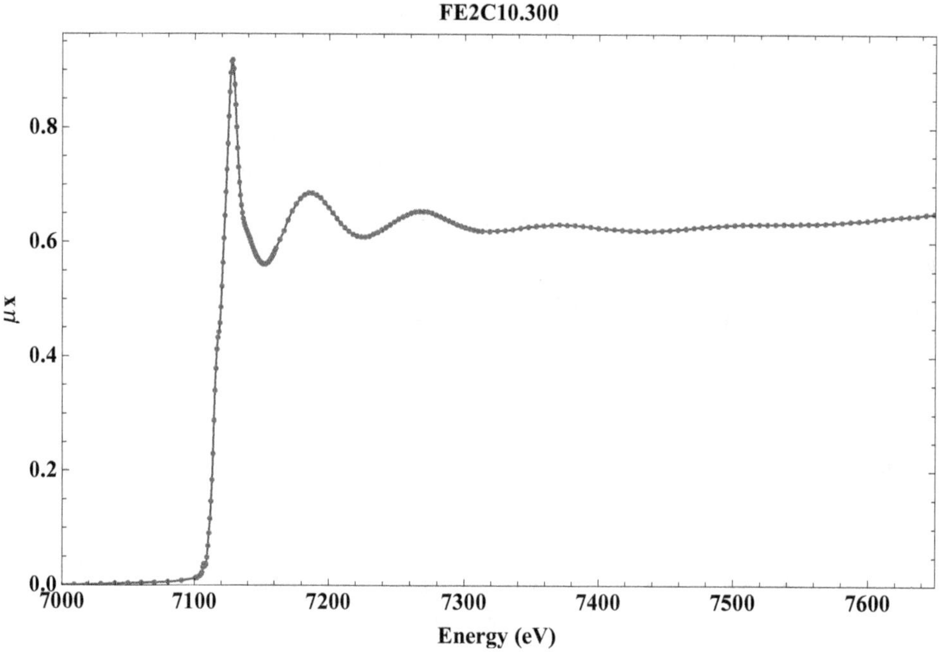

Fig. A4.13 Ferrocene at 300K.

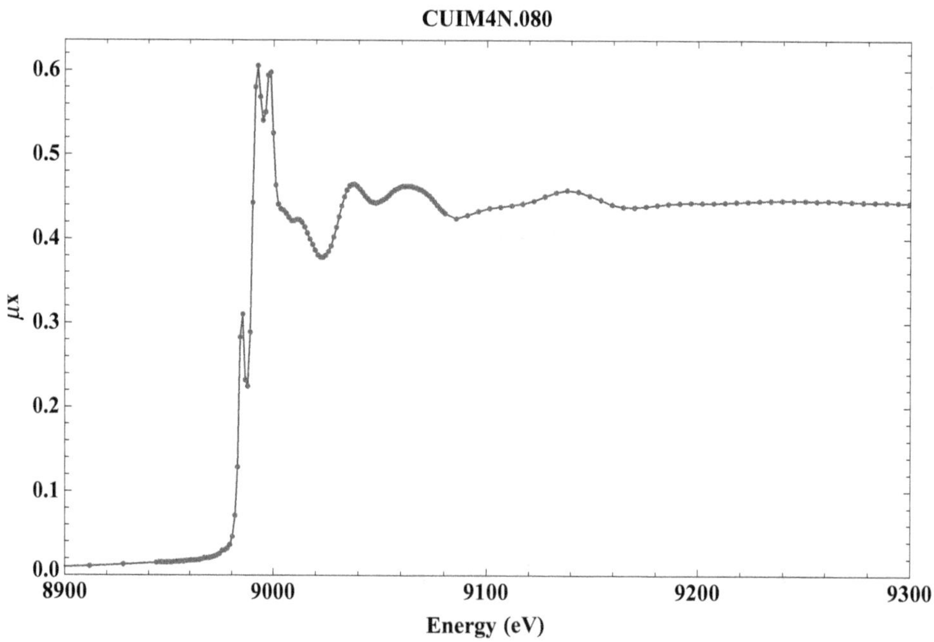

Fig. A4.14 Planar Cu tetraimidazole complex at 80K.

Appendix 5
X-ray tables

The data of Deslattes *et al.* [12] and a comprehensive list of fluorescence energies are available online at `http://physics.nist.gov/XrayTrans`.

The center for X-ray optics at LBL also has posted the X-ray data booklet [9] online at `http://xdb.lbl.gov/`.

Table A5.1 *Correspondence between X-ray diagram levels and electron configuration [133]*

Level	Configuration	Level	Configuration	Level	Configuration
K	$1s^{-1}$	N_1	$4s^{-1}$	O_1	$5s^{-1}$
L_1	$2s^{-1}$	N_2	$4p_{1/2}^{-1}$	O_2	$5p_{1/2}^{-1}$
L_2	$2p_{1/2}^{-1}$	N_3	$4p_{3/2}^{-1}$	O_3	$5p_{3/2}^{-1}$
L_3	$2p_{3/2}^{-1}$	N_4	$4d_{3/2}^{-1}$	O_4	$5d_{3/2}^{-1}$
M_1	$3s^{-1}$	N_5	$4d_{5/2}^{-1}$	$O5$	$5d_{5/2}^{-1}$
M_2	$3p_{1/2}^{-1}$	N_6	$4f_{5/2}^{-1}$	O_6	$5f_{5/2}^{-1}$
M_3	$3p_{3/2}^{-1}$	N_7	$4f_{7/2}^{-1}$	O_7	$5f_{7/2}^{-1}$
				M_4	$3d_{3/2}^{-1}$
				M_5	$3d_{5/2}^{-1}$

Table A5.2 *Siegbahn and IUPAC nomenclature for X-ray transitions [133]*

Siegbahn	IUPAC	Siegbahn	IUPAC	Siegbahn	IUPAC
$K\alpha_1$	$K - L_3$	$L\alpha_1$	$L_3 - M_5$	$L\gamma_1$	$L_2 - N_4$
$K\alpha_2$	$K - L_2$	$L\alpha_2$	$L_3 - M_4$	$L\gamma_2$	$L_1 - N_2$
$K\beta_1$	$K - M_3$	$L\beta_1$	$L_2 - M_4$	$L\gamma_3$	$L_1 - N_3$
$K^I\beta_2$	$K - N_3$	$L\beta_2$	$L_3 - N_5$	$L\gamma_4$	$L_1 - O3$
$K^{II}\beta_2$	$K - N_2$	$L\beta_3$	$L_1 - M_3$	$L\gamma_4'$	$L_1 - O2$
$K\beta_3$	$K - M_2$	$L\beta_4$	$L_1 - M_2$	$L\gamma_5$	$L_2 - N_1$
$K^I\beta_4$	$K - N_5$	$L\beta_5$	$L_3 - O4,5$	$L\gamma_6$	$L_2 - O4$
$K^{II}\beta_4$	$K - N_4$	$L\beta_6$	$L_3 - N_1$	$L\gamma_8$	$L_2 - O1$
$K\beta_{4x}$	$K - N_4$	$L\beta_7$	$L_3 - O1$	$L\gamma_8'$	$L_2 - N_{6(7)}$
$K^I\beta_5$	$K - M_5$	$L\beta_7'$	$L_3 - N_{6,7}$	$L\eta$	$L_2 - M_1$
$K^{II}\beta_5$	$K - M_4$	$L\beta_9$	$L_1 - M_5$	Ll	$L_3 - M_1$
		$L\beta_{10}$	$L_1 - M_4$	Ls	$L_3 - M_3$
		$L\beta_{15}$	$L_3 - N_4$	Lt	$L_3 - M_2$
		$L\beta_{17}$	$L_2 - M_3$	Lu	$L_3 - N_{6,7}$
				Lv	$L_2 - N_{6(7)}$
$M\alpha_1$	$M_5 - N_7$				
$M\alpha_2$	$M_5 - N_6$				
$M\beta$	$M_4 - N_6$				
$M\gamma$	$M_3 - N_5$				
$M\zeta$	$M_{4,5} - N_{2,3}$				

Table A5.3 K-*edge energies from Deslattes [12]*

Element	Z	Theory (eV)	Combined (eV)
Ne	10	870.73 ± 0.16	870.23 ± 0.18
Na	11	1080.15 ± 0.15	1071.52 ± 0.13
Mg	12	1312.3 ± 0.14	1303.33 ± 0.27
Al	13	1569.56 ± 0.13	1559.53 ± 0.27
Si	14	1850.26 ± 0.13	1839.13 ± 0.37
P	15	2154.24 ± 0.12	2144.5 ± 2.5
S	16	2481.71 ± 0.12	2471.63 ± 0.7
Cl	17	2832.76 ± 0.12	2822.64 ± 0.71
Ar	18	3207.44 ± 0.12	3206.14 ± 0.54
K	19	3616.22 ± 0.12	3608.49 ± 0.16
Ca	20	4049.35 ± 0.12	4038.34 ± 0.14
Sc	21	4501.68 ± 0.12	4489.37 ± 0.47
Ti	22	4977.92 ± 0.12	4964.88 ± 0.059
V	23	5478.28 ± 0.12	5464.43 ± 0.26
Cr	24	5995.66 ± 0.12	5989.16 ± 0.48
Mn	25	6552.12 ± 0.12	6537.68 ± 0.14
Fe	26	7125.87 ± 0.13	7110.86 ± 0.4
Co	27	7724.26 ± 0.13	7708.75 ± 0.8
Ni	28	8347.42 ± 0.14	8331. ± 1.4
Cu	29	8987.96 ± 0.15	8980.5 ± 1.
Zn	30	9668.55 ± 0.15	9660.7 ± 1.2
Ga	31	10377.8 ± 0.16	10368.3 ± 0.44
Ge	32	11113.8 ± 0.16	11103.6 ± 0.55
As	33	11876.7 ± 0.18	11867.2 ± 0.85
Se	34	12666.7 ± 0.19	12656.7 ± 0.54
Br	35	13483.9 ± 0.19	13474.1 ± 0.65
Kr	36	14328.1 ± 0.2	14327.2 ± 0.13
Rb	37	15207.7 ± 0.22	15201.5 ± 2.
Sr	38	16115.3 ± 0.23	16105.6 ± 0.81
Y	39	17047.9 ± 0.24	17036.6 ± 0.55
Zr	40	18008.2 ± 0.26	17996.2 ± 0.79
Nb	41	18990.7 ± 0.27	18983.6 ± 0.85
Mo	42	20008.8 ± 0.28	20000.5 ± 2.1
Tc	43	21050.5 ± 0.3	21045.7 ± 0.85
Ru	44	22127.7 ± 0.32	22117.9 ± 0.55
Rh	45	23230.2 ± 0.32	23220.1 ± 0.44
Pd	46	24357.6 ± 0.36	24350.9 ± 0.51
Ag	47	25523.7 ± 0.39	25515.5 ± 0.48
Cd	48	26720.6 ± 0.41	26712.9 ± 0.71
In	49	27949.7 ± 0.44	27940.7 ± 0.69
Sn	50	29209.8 ± 0.47	29200.9 ± 0.92
Sb	51	30501.3 ± 0.49	30492. ± 0.91
Te	52	31824.3 ± 0.52	31815. ± 1.1
I	53	33179.5 ± 0.54	33169.7 ± 0.89
Xe	54	34566.5 ± 2.6	34565.1 ± 0.33

Table A5.4 K-*edge energies from Deslattes (continued)*

Element	Z	Theory (eV)	Combined (eV)
Cs	55	35991.9 ± 0.62	35985.6 ± 6.1
Ba	56	37450.2 ± 0.63	37440. ± 0.34
La	57	38939.5 ± 0.67	38929.3 ± 4.2
Ce	58	40446.6 ± 0.71	40444.7 ± 1.7
Pr	59	41994.1 ± 0.75	41988.8 ± 0.78
Nd	60	43575.3 ± 0.79	43571.9 ± 0.6
Pm	61	45189.8 ± 0.87	
Sm	62	46839. ± 0.91	46837.7 ± 1.5
Eu	63	48523.8 ± 0.96	48517. ± 0.63
Gd	64	50251.7 ± 0.97	50243.4 ± 1.1
Tb	65	51999.5 ± 1.1	51996.4 ± 2.6
Dy	66	53792.3 ± 1.1	53786.2 ± 2.5
Ho	67	55620.8 ± 1.1	55614.6 ± 1.2
Er	68	57487.4 ± 1.2	57486.3 ± 1.3
Tm	69	59391.1 ± 1.3	59389. ± 1.3
Yb	70	61333.3 ± 1.3	61330.8 ± 6.8
Lu	71	63322.7 ± 1.4	63315.5 ± 0.64
Hf	72	65352. ± 1.4	65350.4 ± 0.98
Ta	73	67431.9 ± 1.5	67411.2 ± 0.79
W	74	69533. ± 1.6	69524.8 ± 2.5
Re	75	71687.5 ± 1.7	71677.4 ± 0.74
Os	76	73884. ± 1.7	73876.4 ± 1.2
Ir	77	76117.5 ± 1.8	76112.4 ± 0.6
Pt	78	78404.2 ± 1.9	78398.7 ± 2.3
Au	79	80734.7 ± 2.1	80725.6 ± 1.5
Hg	80	83111.3 ± 2.1	83104.5 ± 3.2
Tl	81	85538.2 ± 2.2	85530.1 ± 1.3
Pb	82	88012.8 ± 2.3	88004.7 ± 0.69
Bi	83	90536.5 ± 2.4	90528.6 ± 0.96
Po	84	93109.9 ± 2.5	93107.2 ± 4.
At	85	95733.5 ± 2.7	95729 ± 15
Rn	86	98408.1 ± 2.8	98404 ± 24

Table A5.5 L_1-*edge energies from Deslattes*

Element	Z	Theory (eV)	Combined (eV)
Ne	10	53.04 ± 0.4	48.445 ± 0.05
Na	11	75.16 ± 0.35	63.57 ± 0.2
Mg	12	100.75 ± 0.29	88.62 ± 0.2
Al	13	130.62 ± 0.25	117.87 ± 0.3
Si	14	163.72 ± 0.2	149.8 ± 0.5
P	15	199.79 ± 0.16	189. ± 1.
S	16	239.15 ± 0.11	
Cl	17	281.8 ± 0.15	270. ± 1.
Ar	18	327.31 ± 0.23	326.3 ± 0.1
K	19	386.25 ± 0.22	378.6 ± 0.3
Ca	20	450.46 ± 0.2	438.5 ± 0.5
Sc	21	510.11 ± 0.29	498. ± 0.3
Ti	22	573.33 ± 0.33	561.05 ± 0.6
V	23	639.78 ± 0.37	626.86 ± 0.6
Cr	24	702.31 ± 0.4	696.37 ± 0.6
Mn	25	782.94 ± 0.44	769.48 ± 0.9
Fe	26	859.8 ± 0.48	848.6 ± 2.
Co	27	940.18 ± 0.51	925.26 ± 0.59
Ni	28	1024.13 ± 0.55	1008.4 ± 1.1
Cu	29	1103.12 ± 0.59	1098. ± 1.5
Zn	30	1203.31 ± 0.58	1196.7 ± 1.4
Ga	31	1309.87 ± 0.65	1302.6 ± 1.5
Ge	32	1421.74 ± 0.73	1412.9 ± 1.9
As	33	1540.04 ± 0.7	1532.2 ± 2.5
Se	34	1663.23 ± 0.73	1653.6 ± 2.9
Br	35	1791.81 ± 0.76	1781.8 ± 3.5
Kr	36	1925.49 ± 0.79	1920.4 ± 1.2
Rb	37	2072.55 ± 0.81	2066.07 ± 0.37
Sr	38	2225.51 ± 0.83	2216.17 ± 0.63
Y	39	2380.76 ± 0.85	2370.78 ± 0.24
Zr	40	2541.1 ± 0.87	2530.9 ± 0.21
Nb	41	2700.53 ± 0.88	2695.46 ± 0.25
Mo	42	2873.84 ± 0.88	2867.2 ± 0.26
Tc	43	3046.63 ± 0.91	
Ru	44	3232.69 ± 0.93	3225.1 ± 1.4
Rh	45	3420.89 ± 0.92	3412.4 ± 1.6
Pd	46	3609.87 ± 0.93	3604.74 ± 0.64
Ag	47	3814.27 ± 0.97	3807.41 ± 0.34
Cd	48	4026.07 ± 0.98	4019.68 ± 0.22
In	49	4246.17 ± 0.99	4238.3 ± 0.27
Sn	50	4473.2 ± 0.99	4465.02 ± 0.11
Sb	51	4707.3 ± 1.	4699.15 ± 0.12
Te	52	4948.5 ± 1.	4939.65 ± 0.99
I	53	5197.2 ± 1.	5188.38 ± 0.81
Xe	54	5453.7 ± 1.3	5452.57 ± 0.17

Table A5.6 L_1-*edge energies from Deslattes (continued)*

Element	Z	Theory (eV)	Combined (eV)
Cs	55	5721.4 ± 1.	5719.8 ± 7.6
Ba	56	5997.7 ± 1.	5990.4 ± 4.5
La	57	6279. ± 1.	6271.17 ± 0.9
Ce	58	6550.4 ± 1.1	6548.9 ± 2.5
Pr	59	6836.8 ± 1.1	6832. ± 1.2
Nd	60	7130.2 ± 1.1	7129.47 ± 0.72
Pm	61	7430.2 ± 1.1	
Sm	62	7737.9 ± 1.2	7739.29 ± 0.58
Eu	63	8053.7 ± 1.2	8047.53 ± 0.88
Gd	64	8385.9 ± 1.2	8381.7 ± 1.4
Tb	65	8708.1 ± 1.2	8713.8 ± 6.
Dy	66	9048. ± 1.2	9046.1 ± 2.7
Ho	67	9395.2 ± 1.2	9395.8 ± 2.
Er	68	9750.6 ± 1.3	9751.4 ± 1.1
Tm	69	10114.4 ± 1.3	10111.9 ± 2.3
Yb	70	10486.8 ± 1.4	10483.5 ± 5.4
Lu	71	10877.2 ± 1.4	10871.7 ± 2.1
Hf	72	11277.2 ± 1.4	11270.3 ± 2.5
Ta	73	11696.9 ± 1.4	11679.8 ± 3.3
W	74	12106.9 ± 1.4	12102.4 ± 4.1
Re	75	12537.5 ± 1.4	12527.8 ± 2.7
Os	76	12978.4 ± 1.5	12971.3 ± 5.
Ir	77	13423.3 ± 1.5	13419 ± 18
Pt	78	13888.7 ± 1.4	13879 ± 21
Au	79	14362.2 ± 1.6	14352.9 ± 7.5
Hg	80	14850.8 ± 1.5	14840.2 ± 7.6
Tl	81	15353.1 ± 1.5	15345.6 ± 2.9
Pb	82	15867.7 ± 1.5	15860.5 ± 4.8
Bi	83	16395.4 ± 1.6	16389 ± 12
Po	84	16936. ± 1.6	16911 ± 31
At	85	17489.7 ± 1.6	
Rn	86	18056.8 ± 1.6	
Fr	87	18642.4 ± 1.7	
Ra	88	19242.3 ± 1.7	19237.5 ± 4.6
Ac	89	19855.2 ± 1.7	
Th	90	20481.6 ± 1.7	20470. ± 3.6
Pa	91	21113.7 ± 1.7	21100.3 ± 4.2
U233	92	21766.1 ± 1.8	
U238	92	21766.1 ± 1.8	21756. ± 1.5
Np	93	22433.6 ± 1.8	
Pu244	94	23111.8 ± 1.8	
Pu239	94	23111.9 ± 1.8	
Am243	95	23812.4 ± 1.9	
Am241	95	23812.5 ± 1.9	
Cm248	96	24535.2 ± 1.9	
Cm245	96	24535.4 ± 1.9	
Bk250	97	25270.9 ± 2.	
Bk249	97	25271. ± 2.	
Cf251	98	26032.9 ± 2.	
Cf250	98	26033.3 ± 2.	
Cf249	98	26033.4 ± 2.	
Es	99	26792.1 ± 2.1	
Fm	100	27584.4 ± 2.2	

Table A5.7 L_2-*edge energies from Deslattes*

Element	Z	Theory (eV)	Combined (eV)
Ne	10	21.63 ± 0.39	21.661 ± 0.01
Na	11	38.38 ± 0.37	30.6 ± 0.1
Mg	12	58.16 ± 0.35	49.5 ± 0.1
Al	13	82.44 ± 0.34	72.75 ± 0.2
Si	14	110.59 ± 0.34	99.2 ± 0.1
P	15	141.1 ± 0.33	136. ± 1.
S	16	174.7 ± 0.33	163.6 ± 0.3
Cl	17	211.49 ± 0.33	202. ± 1.
Ar	18	251.55 ± 0.32	250.57 ± 0.27
K	19	304.25 ± 0.33	297.28 ± 0.23
Ca	20	361.79 ± 0.32	350.4 ± 0.39
Sc	21	416.25 ± 0.33	403.62 ± 0.22
Ti	22	473.85 ± 0.33	460. ± 0.23
V	23	534.69 ± 0.33	519.72 ± 0.19
Cr	24	591.6 ± 0.33	583.57 ± 0.48
Mn	25	665.92 ± 0.33	649.88 ± 0.19
Fe	26	736.35 ± 0.34	719.8 ± 0.39
Co	27	810.18 ± 0.34	793.38 ± 0.71
Ni	28	887.46 ± 0.34	870. ± 1.4
Cu	29	959.58 ± 0.34	952.5 ± 1.
Zn	30	1052.33 ± 0.36	1044.94 ± 0.58
Ga	31	1152.66 ± 0.35	1143.62 ± 0.31
Ge	32	1258.15 ± 0.35	1248.08 ± 0.34
As	33	1368.9 ± 0.35	1359.74 ± 0.26
Se	34	1484.9 ± 0.35	1474.2 ± 1.
Br	35	1606.17 ± 0.35	1596.31 ± 0.71
Kr	36	1732.49 ± 0.36	1730.9 ± 0.5
Rb	37	1871.98 ± 0.36	1865.9 ± 1.7
Sr	38	2017.25 ± 0.36	2007.44 ± 0.23
Y	39	2164.89 ± 0.36	2153.47 ± 0.31
Zr	40	2317.53 ± 0.36	2305.68 ± 0.64
Nb	41	2469.32 ± 0.35	2462.54 ± 0.31
Mo	42	2634.63 ± 0.36	2625.98 ± 0.33
Tc	43	2799.19 ± 0.36	2794.43 ± 0.79
Ru	44	2977.03 ± 0.37	2967.45 ± 0.55
Rh	45	3156.74 ± 0.37	3146.39 ± 0.44
Pd	46	3337.01 ± 0.37	3330.66 ± 0.46
Ag	47	3533.04 ± 0.38	3525.24 ± 0.26
Cd	48	3736.1 ± 0.39	3728.54 ± 0.33
In	49	3947.38 ± 0.39	3938.71 ± 0.59
Sn	50	4165.49 ± 0.39	4156.23 ± 0.78
Sb	51	4390.54 ± 0.4	4381.22 ± 0.79
Te	52	4622.56 ± 0.4	4613.05 ± 0.15
I	53	4861.84 ± 0.38	4852.01 ± 0.59
Xe	54	5108.1 ± 0.37	5106.72 ± 0.2

Table A5.8 L_2-*edge energies from Deslattes (continued)*

Element	Z	Theory (eV)	Combined (eV)
Cs	55	5367.05 ± 0.39	5359.2 ± 4.
Ba	56	5633.67 ± 0.39	5623.32 ± 0.32
La	57	5905.22 ± 0.39	5895.1 ± 0.29
Ce	58	6167.01 ± 0.39	6165.8 ± 0.44
Pr	59	6443.6 ± 0.39	6437.2 ± 1.2
Nd	60	6727.09 ± 0.4	6724.63 ± 0.59
Pm	61	7017.09 ± 0.44	
Sm	62	7314.76 ± 0.44	7314.92 ± 0.57
Eu	63	7620.28 ± 0.45	7614.32 ± 0.98
Gd	64	7942.02 ± 0.41	7934.3 ± 1.1
Tb	65	8253.93 ± 0.46	8254.9 ± 4.4
Dy	66	8583.26 ± 0.42	8580.2 ± 1.2
Ho	67	8919.77 ± 0.42	8913.9 ± 1.7
Er	68	9264.4 ± 0.42	9266.9 ± 1.8
Tm	69	9617.34 ± 0.43	9615.5 ± 2.
Yb	70	9978.7 ± 0.44	9971.46 ± 0.56
Lu	71	10357.4 ± 0.44	10349.7 ± 0.52
Hf	72	10745.6 ± 0.45	10734.9 ± 9.8
Ta	73	11153.2 ± 0.46	11132.9 ± 1.4
W	74	11551.2 ± 0.47	11540.8 ± 4.1
Re	75	11969.2 ± 0.47	11959.6 ± 2.3
Os	76	12397.4 ± 0.48	12388.7 ± 0.71
Ir	77	12829.4 ± 0.5	12824.8 ± 2.4
Pt	78	13281.7 ± 0.51	13275.2 ± 3.8
Au	79	13741.7 ± 0.52	13735.8 ± 1.8
Hg	80	14216.9 ± 0.53	14212.2 ± 4.3
Tl	81	14705.5 ± 0.56	14698.6 ± 1.3
Pb	82	15206.1 ± 0.56	15198.7 ± 3.4
Bi	83	15719.7 ± 0.57	15712.1 ± 1.8
Po	84	16245.8 ± 0.59	16244.2 ± 2.8
At	85	16785. ± 0.6	16784.7 ± 3.
Rn	86	17337.4 ± 0.62	17337.7 ± 3.2
Fr	87	17907.9 ± 0.64	17906.4 ± 3.4
Ra	88	18492.6 ± 0.67	18483.8 ± 3.3
Ac	89	19090. ± 0.69	19083.4 ± 3.8
Th	90	19701.6 ± 0.72	19690.5 ± 3.4
Pa	91	20318.1 ± 0.73	20313.5 ± 3.
U233	92	20954.7 ± 0.77	
U238	92	20955. ± 0.77	20945.7 ± 2.3
Np	93	21606.6 ± 0.79	
Pu239	94	22269.4 ± 0.79	
Pu244	94	22269.5 ± 0.79	
Am243	95	22954. ± 0.8	
Am241	95	22954. ± 0.8	
Cm248	96	23661. ± 0.8	
Cm245	96	23661. ± 0.8	
Bk250	97	24380.7 ± 0.9	
Bk249	97	24380.7 ± 0.9	
Cf251	98	25126.6 ± 0.9	
Cf250	98	25126.9 ± 0.9	
Cf249	98	25127. ± 0.9	
Es	99	25869.9 ± 0.9	
Fm	100	26646.5 ± 1.	

Table A5.9 L_3-*edge energies from Deslattes*

Element	Z	Theory (eV)	Combined (eV)
Ne	10	21.55 ± 0.38	21.564 ± 0.01
Na	11	38.21 ± 0.37	30.4 ± 0.2
Mg	12	57.9 ± 0.35	49.79 ± 0.061
Al	13	82.03 ± 0.34	72.87 ± 0.061
Si	14	109.1 ± 0.34	99.34 ± 0.061
P	15	139.51 ± 0.32	130.01 ± 0.061
S	16	172.91 ± 0.32	163.82 ± 0.061
Cl	17	209.09 ± 0.31	200. ± 1.
Ar	18	249.54 ± 0.31	248.46 ± 0.53
K	19	301.62 ± 0.31	294.55 ± 0.22
Ca	20	358.37 ± 0.3	346.611 ± 0.066
Sc	21	411.53 ± 0.31	398.55 ± 0.47
Ti	22	467.55 ± 0.31	453.979 ± 0.061
V	23	526.5 ± 0.31	512.21 ± 0.22
Cr	24	581.78 ± 0.3	574.36 ± 0.13
Mn	25	654.02 ± 0.31	638.89 ± 0.14
Fe	26	722.74 ± 0.31	706.86 ± 0.14
Co	27	794.62 ± 0.31	778.36 ± 0.21
Ni	28	869.7 ± 0.31	852.74 ± 0.34
Cu	29	939.85 ± 0.31	932.68 ± 0.45
Zn	30	1029.45 ± 0.31	1021.8 ± 1.2
Ga	31	1125.86 ± 0.31	1116.57 ± 0.3
Ge	32	1227.15 ± 0.31	1217.33 ± 0.55
As	33	1333.26 ± 0.31	1323.92 ± 0.41
Se	34	1444.18 ± 0.31	1434.24 ± 0.13
Br	35	1559.84 ± 0.31	1549.98 ± 0.58
Kr	36	1680.06 ± 0.31	1679.07 ± 0.39
Rb	37	1812.69 ± 0.31	1806.2 ± 1.8
Sr	38	1950.46 ± 0.31	1940.48 ± 0.26
Y	39	2089.76 ± 0.31	2078.26 ± 0.54
Zr	40	2233.28 ± 0.32	2221.29 ± 0.59
Nb	41	2375.01 ± 0.3	2368.24 ± 0.44
Mo	42	2529.71 ± 0.31	2521.1 ± 1.6
Tc	43	2682.91 ± 0.31	2677.9 ± 0.54
Ru	44	2848.19 ± 0.32	2838.62 ± 0.21
Rh	45	3014.48 ± 0.31	3004. ± 0.37
Pd	46	3180.38 ± 0.32	3173.75 ± 0.51
Ag	47	3360.71 ± 0.32	3352.58 ± 0.48
Cd	48	3546.84 ± 0.32	3538.88 ± 0.45
In	49	3739.91 ± 0.33	3730.84 ± 0.48
Sn	50	3938.45 ± 0.33	3929.51 ± 0.87
Sb	51	4142.58 ± 0.33	4132.99 ± 0.91
Te	52	4352.2 ± 0.34	4342.35 ± 0.53
I	53	4567.52 ± 0.32	4557.12 ± 0.54
Xe	54	4788.22 ± 0.32	4786.47 ± 0.17

Table A5.10 L_3-*edge energies from Deslattes (continued)*

Element	Z	Theory (eV)	Combined (eV)
Cs	55	5019.87 ± 0.32	5012.98 ± 0.33
Ba	56	5257.36 ± 0.32	5246.71 ± 0.22
La	57	5497.83 ± 0.32	5487.05 ± 0.49
Ce	58	5726.47 ± 0.32	5724.5 ± 1.
Pr	59	5967.84 ± 0.33	5962.35 ± 0.62
Nd	60	6213.9 ± 0.33	6211.15 ± 0.6
Pm	61	6464.15 ± 0.36	
Sm	62	6719.67 ± 0.37	6718.8 ± 1.5
Eu	63	6980.47 ± 0.37	6974.53 ± 0.63
Gd	64	7254.88 ± 0.33	7246.66 ± 0.9
Tb	65	7516.62 ± 0.39	7513.2 ± 2.
Dy	66	7792.9 ± 0.34	7786. ± 1.6
Ho	67	8073.31 ± 0.34	8071.1 ± 1.8
Er	68	8358.66 ± 0.34	8359.4 ± 1.5
Tm	69	8648.97 ± 0.34	8648.1 ± 1.6
Yb	70	8944.26 ± 0.35	8946.6 ± 6.5
Lu	71	9253.24 ± 0.35	9245.28 ± 0.64
Hf	72	9567.88 ± 0.36	9558 ± 11
Ta	73	9898.04 ± 0.36	9878.7 ± 2.7
W	74	10214.3 ± 0.36	10200.9 ± 8.3
Re	75	10546.4 ± 0.36	10536.3 ± 2.5
Os	76	10884.1 ± 0.37	10875.6 ± 2.2
Ir	77	11221. ± 0.38	11216.7 ± 2.2
Pt	78	11573.1 ± 0.38	11565.7 ± 3.8
Au	79	11927.8 ± 0.39	11920.6 ± 3.
Hg	80	12292.3 ± 0.39	12285.5 ± 3.7
Tl	81	12664.4 ± 0.4	12657.5 ± 3.3
Pb	82	13042.6 ± 0.4	13035.4 ± 3.
Bi	83	13427.6 ± 0.41	13420.1 ± 1.6
Po	84	13818.7 ± 0.42	13813.6 ± 1.3
At	85	14216. ± 0.45	
Rn	86	14619.5 ± 0.43	
Fr	87	15033.8 ± 0.44	15027 ± 50
Ra	88	15454.3 ± 0.45	15444.6 ± 0.84
Ac	89	15879.5 ± 0.46	
Th	90	16310.3 ± 0.47	16300. ± 2.1
Pa	91	16736.9 ± 0.48	16732.9 ± 2.2
U233	92	17174.2 ± 0.49	
U238	92	17174.5 ± 0.49	17164.7 ± 1.2
Np	93	17616.9 ± 0.5	
Pu239	94	18060. ± 0.51	
Pu244	94	18060.1 ± 0.51	
Am243	95	18514.4 ± 0.5	
Am241	95	18514.4 ± 0.5	
Cm245	96	18979.7 ± 0.5	
Cm248	96	18979.7 ± 0.5	
Bk249	97	19445.8 ± 0.5	
Bk250	97	19445.8 ± 0.5	
Cf251	98	19925.6 ± 0.5	
Cf250	98	19925.8 ± 0.5	
Cf249	98	19926. ± 0.5	
Es	99	20389.4 ± 0.6	
Fm	100	20872.8 ± 0.6	

References

[1] Sayers, D. E., Stern, E. A., and Lytle, F. W. (1971) New technique for investigating noncrystalline structures: Fourier analysis of the extended x-ray-absorption fine structure, *Phys. Rev. Lett.*, **27**, 8, 1204–1207.

[2] Stern, E. A. (1974) Theory of the extended x-ray-absorption fine structure, *Phys. Rev.*, **B10**, 8, 3027–3037.

[3] Lytle, F. W., Sayers, D. E., and Stern, E. A. (1975) Extended x-ray-absorption fine-structure technique. II. Experimental practice and selected results, *Phys. Rev.*, **B11**, 12, 4825–4835.

[4] Stern, E. A., Sayers, D. E., and Lytle, F. W. (1975) Extended x-ray-absorption fine-structure technique. III. Determination of physical parameters, *Phys. Rev.*, **B11**, 12, 4836–4846.

[5] Stumm von Bordwehr, R. (1989) *Ann. Phys. Fr.*, **14**, 377–466.

[6] Koch, E. (ed.) (1983) *Handbook of Synchrotron Radiation*, North-Holland Publishing Company, 955–1014.

[7] Als-Nielsen, J. and McMorrow, D. (2001) *Elements of Modern X-ray Physics*, Wiley.

[8] Duke, P. J. (2009) *Synchrotron Radiation*, Oxford University Press.

[9] Lawrence Berkeley Labs, Center for X-ray Optics, X-ray data booklet, [online] http://xdb.lbl.gov/.

[10] Koningsberger, D. C. and Prins, R. (1988) *X-ray Absorption: Principles, Applications, Techniques of EXAFS, SEXAFS, and XANES*, Wiley.

[11] Krause, M. O. and Oliver, J. H. (1979) Natural widths of atomic K and L levels, K_α x-ray lines and several KLL Auger lines, *J. Phys. Chem. Ref. Data*, **8**, 2, 329–338.

[12] Deslattes, R. D., Kessler, E. G., Indelicato P., *et al.* (2003) X-ray transition energies: new approach to a comprehensive evaluation, *Rev. Mod. Phys*, **75**, 35–99.

[13] Knoll, G. F. (2002) *Radiation Detection and Measurement*, Wiley.

[14] Bateman, J. E. (2000) The effect of beam time structure on counting detectors in SRS experiments, *J. Synchrotron Rad.*, **7**, 307–312.

[15] Rossi, B. B. and Staub H. H. (1949) *Ionization Chambers and Counters*, McGraw-Hill.

[16] Stern, E. A. and Heald, S. M. (1979) X-ray filter assembly for fluorescence measurements of x-ray absorption fine structure, *Rev. Sci. Instrum.*, **50**, 1579–1582.

[17] Landau, L. D. and Lifshitz, E. M. (1958) *Quantum Mechanics: Non-Relativistic Theory*, Elsevier Science Ltd.

[18] Lee, P. A. and Pendry, J. B. (1975) Theory of the extended x-ray absorption fine structure, *Phys. Rev.*, **B11**, 8, 2795–2811.

[19] Griffiths, D. J. (1999) *Introduction to Electrodynamics*, third edition, Prentice Hall.

[20] Jackson, J. D. (1999) *Classical Electrodynamics*, third edition, John Wiley and Sons.

[21] Zachariasen, W. H. (1967) *Theory of X-ray Diffraction in Crystals*, Dover Publications.

[22] James, R. W. (1962) *The Optical Principles of the Diffraction of X-rays*, Oxbow Press.

[23] Authier, A. (2001) *Dynamical Theory of X-Ray Diffraction*, Vol. 11 of IUCr Monographs on Crystallography, Oxford University Press.

[24] Batterman, B. W. and Cole, H. (1964) Dynamical diffraction of x rays by perfect crystals, *Rev. Mod. Phys.* **36**, 681–717.

[25] McMaster, W. H., Del Grande, N. K., Mallett, J. H. and Hubbell, J. H. (1969) *Compilation of X-ray Cross Sections*, Lawrence Livermore Lab., Report UCRL-50174.

[26] Chantler, C. T., Olsen, K., Dragoset, R. A., *et al.* (2005) X-Ray Form Factor, Attenuation and Scattering Tables (version 2.1) [online] http://physics.nist.gov/ffast.

[27] Stern, E. A. (1979) personal communication.

[28] Elam, W. T., Ravel, B. D., and Sieber, J. R. (2002) A new atomic database for X-ray spectroscopic calculations, *Radiat. Phys. Chem.*, **63**, 121–128.

[29] Matsushita, T. and Hashizume, H. (1983) in *Handbook on Synchrotron Radiation*, volume 1A, Koch, E. E. (ed.), North-Holland Publishing Company.

[30] Suzuki, M., Kawamura, N., Lytle, F., and Ishikawa, T., (2002) Fast multigrid fluorescent ion chamber with 0.1 ms time response, *J. Synchrotron Rad.*, **9**, 99–102.

[31] Booth, C. H. and Bridges, F. (2005) Improved self-absorption correction for fluorescence measurements of extended x-ray absorption fine-structure, *Phys. Scripta*, **T115**, 202204.

[32] Bedzyk, M. J. and Cheng, L. (2002) X-ray standing wave studies of minerals and mineral surfaces: principles and applications, *Rev. Mineral. Geochem.*, **49**, 221–266.

[33] Brown, G. E. Jr. and Sturchio, N. C. (2002) An overview of synchrotron radiation applications to low temperature geochemistry and environmental science, *Reviews in Mineralogy and Geochemistry*, **49**, 1–115.

[34] Waychunas, G. (2002) Grazing-incidence x-ray absorption and emission spectroscopy, *Rev. Mineral. Geochem.*, **49**, 267–315.

[35] Heald, S. M., Tranquada, J. M., and Chen, H. (1988) Glancing-angle extended x-ray-absorption fine structure and reflectivity studies of interfacial regions, *Phys. Rev. B*, **38**, 1016–1026.

[36] Borthen, P. and Strehblow, H.-H. (1995) X-ray-reflectivity fine structure and EXAFS, *Phys. Rev. B*, **52**, 3017–3019.

[37] D'Acapito, F., Davoli, I., Ghigna, P., and Mobilio, S. (2003) The ReflEXAFS station at the GILDA beamline (BM08) of ESRF, *J. Synchrotron Rad.*, **10**, 260–264.

[38] Zhang K, Stern E. A., Rehr, J. J., and Ellis, F. (1991) Double electron

excitation in atomic Xe, *Phys. Rev. B*, **44**, 2030–2039.

[39] Kodre, A., Arčon, I, Padežnik, J., *et al.* (2002) Multielectron excitations in x-ray absorption spectra of Rb and Kr, *J. Phys. B*, **35**, 34973513.

[40] Zhang, K., Rosenbaum, G., and Bunker, G. (1993) The correction of the dead time loss of the germanium detector in x-ray absorption spectroscopy, *Jpn. J. Appl. Phys.*, **32**, Suppl. 32-2, 147–149.

[41] Bateman, J. E. (2000) The effect of beam time structure on counting detectors in SRS experiments, *J. Synchrotron. Rad.*, **7**, 307–312.

[42] Ciatto, G., d'Acapito, F., Boscherini, F., and Mobilio, S. (2004) Treatment of EXAFS data taken in the fluorescence mode in non-linear conditions, *J. Synchrotron Rad.*, **11**, 278–283.

[43] Pease, D. M., Daniel, M., Budnick, J. I., *et al.* (2000) Log spiral of revolution highly oriented pyrolytic graphite monochromator for fluorescence x-ray absorption edge fine structure, *Rev. Sci. Instrum.*, **71**, 3267.

[44] Adams, B. W. and Attenkofer, K (2008) An active-optic x-ray fluorescence analyzer with high energy resolution, large solid angle coverage, and a large tuning range, *Rev. Sci. Instrum.*, **79**, 023102-1–12.

[45] Zhang, K., Rosenbaum, G., Liu, R., *et al.* (2004) Development of multilayer analyzer array detectors for x-ray fluorescence at the third generation synchrotron source, Eighth International Conference on Synchrotron Radiation Instrumentation. *AIP Conference Proceedings*, **705**, 957–960.

[46] Zhong, Z., Chapman, L., Bunker, B., *et al.* (1999) A bent laue analyzer for fluorescence EXAFS detection, *J. Synchrotron Rad.*, **6**, 212.

[47] Karanfil, C. (2003) Bent crystal X-ray optics for synchrotron radiation research, Illinois Institute of Technology, Ph.D. thesis.

[48] Parratt, L. G., Hempstead, C. F., and Jossem, E. L. (1957) "Thickness effect" in absorption spectra near absorption edges, *Phys. Rev.*, **105**, 4, 1228–1232.

[49] Heald, S. M. and Stern, E. A. (1977) Anisotropic x-ray absorption in layered compounds, *Phys. Rev. B*, **16**, 12, 5549–5559.

[50] Manceau, A., Chateigner, D., and Gates, W. P. (1998) Polarized EXAFS, distance-valence least-squares modeling (DVLS), and quantitative texture analysis approaches to the structural refinement of Garfield nontronite, *Phys. Chem. Miner.*, **25**, 5, 347–365.

[51] Stern, E. A. and Kim, K. (1981) Thickness effect on the extended x-ray absorption fine structure amplitude, *Phys. Rev. B*, **23**, 8, 3781–3787.

[52] Bunker, G. B. (1984) An x-ray absorption study of transition metal oxides, University of Washington, Ph.D. Thesis.

[53] Bunker, G. (1983) Application of the ratio method of EXAFS analysis to disordered systems, *Nucl. Instr. and Meth.*, **207**, 437–444.

[54] Kubo, R. (1962) Generalized cumulant expansion method, *J. Phys. Soc. Jpn.*, **17**, 7, 1100–1120.

[55] Bouldin, C. E. and Stern, E. A. (1982) Extended x-ray absorption fine-structure study of Kr-Grafoil submonolayers, *Phys. Rev. B*, **25**, 3462–3473.

[56] Frenkel, A. I., Pease, D. M., Budnick, J. I., Shanthakumar, P., and Huang, T. (2007) Application of glancing-emergent-angle fluorescence for polarized XAFS studies of single crystals, *J. Synchrotron. Rad.*, **14**, 272275.

[57] Pettifer, R. F., Brouder, C., Benfatto, M., *et al.* (1990) Magic-angle theorem in powder x-ray-absorption spectroscopy, *Phys. Rev. B*, **42**, 1, 37–42.

[58] Pascarelli, S., Mathon, O., Munoz, M., Mairs, T., and Susini J. (2006) Energy dispersive absorption spectroscopy for hard x-ray micro-XAS applications, *J.*

Synchrotron Rad., **13**, 351.

[59] Dirac, P. A. M. (1981) *The Principles of Quantum Mechanics*, fourth edition, Oxford at the Clarendon Press.

[60] Fermi, E. (1974) Nuclear Physics: A Course Given by Enrico Fermi at the University of Chicago, University of Chicago Press.

[61] Stern, E. A. and Heald, S. M. (1983) Basic principles and applications of EXAFS, in *Handbook of Synchrotron Radiation*, Koch, E. (ed.), North-Holland, 955–1014.

[62] Rehr, J. J. and Ankudinov, A. L. (2005) Progress in the theory and interpretation of XANES, *Coordin. Chem. Rev.*, **249**, 131140.

[63] de Groot, F. (2005) Multiplet effects in X-ray spectroscopy, *Coordin. Chem. Rev.*, **249**, 3163.

[64] Rehr, J. J. and Albers, R. C. Theoretical approaches to x-ray absorption fine structure, *Rev. Mod. Phys.*, **72**, 3, 621–654.

[65] Zabinsky, S. I., Rehr, J. J., Ankudinov, J. J., Albers, R. C., and Eller, M. J. (1995) Multiple scattering calculations of x-ray absorption spectra, *Phys. Rev. B*, **52**, 2995–3009.

[66] Rehr, J. J. and Albers, R. C. (1990) Scattering matrix formulation of curved wave multiple scattering theory: application to x-ray-absorption fine structure, *Phys. Rev. B*, **41**, 8139–8149.

[67] Fujikawa, T. (1993) Basic features of the short-range-order multiple scattering XANES theory, *J. Phys. Soc. Jpn.*, **62**, 6, 2155–2165.

[68] de Groot, F. and Kotani, A. (2008) *Core Level Spectroscopy of Solids*, CRC Group, Taylor and Francis.

[69] Martin, M. (2004) *Electronic Structure: Basic Theory and Practical Methods*, Cambridge University Press.

[70] Newton, R. G. (1990) Korringa-Kohn-Rostoker spectral-band theory for general potentials, *Phys. Rev. Lett.*, **65**, 16, 2031–2034.

[71] Taillefumier, M., Cabaret, D., Flank, A.-M., and Mauri, F. (2002) X-ray absorption near-edge structure calculations with the pseudopotentials: Application to the K edge in diamond and a-quartz, *Phys. Rev. B*, **66**, 195107-1-8.

[72] Schiff, L. I. (1968) *Quantum Mechanics*, third edition, McGraw-Hill.

[73] Gottfried, K. (1966) *Quantum Mechanics*, W.A. Benjamin.

[74] Quantum Mechanics (1999) Dover Publications.

[75] Joly, Y., Cabaret, D., Renevier, H., and Natoli, C. R. (1999) Electron population analysis by full-potential x-ray absorption simulations, *Phys. Rev. Lett.*, **82,** 11, 2398–2401.

[76] Hahn, J. E., Scott, R. A., Hodgson, K. O., *et al.* (1982) Observation of an electric quadrupole transition in the x-ray absorption-spectrum of a Cu(Ii) complex, *Chem. Phys. Lett.*, **88**, 595–598.

[77] Solomon, E. I. (ed.) (2005) Synchrotron radiation in inorganic and bioinorganic chemistry, *Coordin. Chem. Rev.*, **249**.

[78] Brouder, C. (1990) Angular dependence of x-ray absorption spectra, *J. Phys. Condensed Matter*, **2**, 701–738.

[79] Harris, D. C. and Bertolucci, M. D. (1989) *Symmetry and Spectroscopy: An Introduction to Vibrational and Electronic Spectroscopy*, Dover Publications.

[80] Bishop, D. M. (1973) *Group Theory and Chemistry*, Dover Publications.

[81] Sevillano, E., Meuth, H., and Rehr, J. J. (1979) Extended X-ray absorption fine structure Debye-Waller factors. I: monatomic crystals, *Phys. Rev. B*, **20**, 12, 4908–4911.

[82] Poiarkova, A. V. and Rehr, J. J. (1999) Multiple-scattering x-ray-absorption fine-structure Debye-Waller factor calculations, *Phys. Rev. B*, **59**, 2, 948–957.
[83] Loeffen, P. W. and Pettifer, R. F. (1996) An EXAFS calculation using known four-body correlations, *Phys. Rev. Lett.*, **76**, 4, 636–639.
[84] Dimakis, N. and Bunker, G. (1998) Ab initio single- and multiple-scattering EXAFS Debye-Waller factors: Raman and infrared data, *Phys. Rev. B*, **58**, 5, 2467–2475.
[85] Dimakis, N. and Bunker, G. (2002) Group-fitted ab initio single- and multiple-scattering EXAFS Debye-Waller factors, *Phys. Rev. B*, **65**, 201103-1-4.
[86] Dimakis, N. and Bunker, G. (2004) XAFS Debye-Waller factors for Zn metalloproteins, *Phys. Rev. B*, **70**, 195114-1-12.
[87] Dimakis, N. and Bunker, G. (2006) *Biophys. J.*, **91**, L87.
[88] Dimakis, N., Junaid, M., Garza, E., and Bunker, G. Zinc cysteine active sites of metalloproteins: A density functional theory and x-ray absorption fine structure study, *J. Chem. Phys.*, **128**, 115104.
[89] Dalba, G., Fornasini, P., Grisenti, R., and Purans, J. (1999) Sensitivity of extended x-ray-absorption fine structure to thermal expansion, *Phys. Rev. Lett.*, **82**, 21, 4240–4243.
[90] Frenkel, A. I. and Rehr, J. J. (1993) Thermal expansion and x-ray-absorption fine-structure cumulants, *Phys. Rev. B*, **48**, 1.
[91] Holland, B. W., Pendry, J. B., Pettifer, R. F., and Bordas, J. (1978) Atomic origin of structure in EXAFS experiments, *J. Phys. C.*, **11**.
[92] van der Eerden, Ad M. J., Visser, T., Nijhuis, T.A., *et al.* (2005) Atomic XAFS as a tool to probe the electronic properties of supported noble metal nanoclusters, *J. Am. Chem. Soc.*, **127**, 10, 3272–3273.
[93] Ankudinov, A. L., Rehr, J. J., Low, J. J., and Bare, S. R. (2001) L-edge XANES as a probe of Pt clusters, *J. Synchrotron Rad.*, **8**, 578–580.
[94] [online]`http://gnxas.unicam.it/XASLABwww/pag_gnxas.html`
[95] [online] `http://leonardo.phys.washington.edu/feff/`
[96] Newville, M. (2001) IFEFFIT: interactive EXAFS analysis and FEFF fitting, *J. Synchrotron Rad.*, **8**, 322–324.
[97] [online] `http://www.cse.scitech.ac.uk/cmg/EXCURV`
[98] Filliponi, A., Di Cicco, A., and Natoli, C. R. (1995) X-ray-absorption spectroscopy and n-body distribution functions in condensed matter. I. Theory, *Phys. Rev. B*, **52**, 21, 15122–15134.
[99] Filliponi, A. and Di Cicco, A. (1995) X-ray-absorption spectroscopy and n-body distribution functions in condensed matter. II. Data analysis and applications, *Phys. Rev. B*, **52**, 21, 15135–15149.
[100] Pettifer, R. F. and Cox, A. D. (1983) The reliability of ab initio calculations in extracting structural information from EXAFS, in *EXAFS and Near Edge Structure*, Bianconi, A. Incoccia, L. and Stipcich, S. (eds.), Springer-Verlag.
[101] Smolentsev, G., Soldatov, A. V., and Feiters, M. C. (2007) Three-dimensional local structure refinement using a full-potential XANES analysis Three-dimensional local structure refinement using a full-potential XANES analysis, *Phys. Rev. B*, **75**, 144106(1-5).
[102] Bunker, G. and Stern, E. A. (1984) Experimental study of multiple scattering in x-ray-absorption near-edge structure, *Phys. Rev. Lett.*, **52**, 1990–1993.
[103] Bunker, G., unpublished.
[104] Durham, P. J., Pendry, J. B., and Hodges, C. H. (1982) Calculation of x-ray absorption near-edge structure, XANES, *Comput. Phys. Commun.*, **25**,

193–206.

[105] Muoz, M., Argoul, P., and Farges, F. (2003) Continuous Cauchy wavelet transform analyses of EXAFS spectra: A qualitative approach, *Am. Mineralogist*, **88**, 4, 694–700.

[106] Tannazi, F. (2004) Illinois Institute of Technology, Ph.D. thesis.

[107] Tannazi, F. and Bunker, G. (2005) Determination of chemical speciation by XAFS, *Phys. Scripta*, **T115**, 953956.

[108] Wasserman, S. R., Allen, P. G., Shuh, D. K., Bucher, J. J., and Edelstein, N. M. (1999) EXAFS and principal component analysis: a new shell game, *J. Synchrotron Rad.*, **6**, 284–286.

[109] Bouldin, C. E., Bunker G., McKeown, D. A., Forman, R. A., and Ritter, J. J. (1988) Multiple scattering in the x-ray-absorption near-edge structure of tetrahedral Ge gases, *Phys. Rev. B*, **38**, 10816–10819.

[110] Weng, T.-Cl., Waldo, G. S., Penner-Hahn, J. E. (2005) A method for normalization of X-ray absorption spectra, *J. Synchrotron Rad.*, **12**, 506510.

[111] Press, W. H., Teukolsky, S. A., Vetterling, W. T., and Flannery, B. P. (2007) *Numerical Recipes 3rd Edition: The Art of Scientific Computing*, Cambridge University Press.

[112] James, F. (2006) *Statistical Methods in Experimental Physics*, 2nd edition, World Scientific.

[113] Gregory, P. (2005) *Bayesian Logical Data Analysis for the Physical Sciences*, Cambridge University Press.

[114] Krappe, H. J. and Rossner, H. H. (2000) Error analysis of XAFS measurements, *Phys. Rev. B*, **61**, 10, 6596–6610.

[115] Krappe, H. J. and Rossner, H. H. (2002) Bayes-Turchin approach to x-ray absorption fine structure data analysis, *Phys. Rev. B*, **66**, 184303-1-20.

[116] Krappe, H. J. and Rossner, H. H. (2004) Bayesian approach to background subtraction for data from the extended x-ray-absorption fine structure, *Phys. Rev. B*, **70**, 104102-1-7.

[117] Rossner, H. H., Schmitz, D., Imperia, P., Krappe, H. J., and Rehr, J. J. (2006) Bayes-Turchin analysis of x-ray absorption data above the Fe L2,3-edges, *Phys. Rev. B*, **74**, 134107-1-12.

[118] Klementev, K. V. (2001) Statistical evaluations in the fitting problems, *J. Synchrotron Rad.*, **8**, 270–272.

[119] Klementev, K. V. (2001) Deconvolution problems in x-ray absorption fine structure spectroscopy, *J. Phys. D*, **34**, 22412247.

[120] Klementev, K. V. (2001) Extraction of the fine structure from x-ray absorption spectra, *J. Phys. D*, **34**, 209217.

[121] Kirkpatrick, S., Gelatt, C. D., and Vecchi, M. P. (1983) Optimization by simulated annealing, *Science*, **220**, 4598, 671–680.

[122] Nelder, J. A. and Mead, R. (1965) A simplex method for function minimization, *Computer Journal*, **7**, 308–313.

[123] Storn, R. and Price, K. (1997) Differential evolution – a simple and efficient heuristic for global optimization over continuous spaces, *J. Glob. Optim.*, **11**, 341359.

[124] McGreevy, R. L. and Pusztai, L. (1988) Reverse Monte Carlo simulation: A new technique for the determination of disordered structures, *Mol. Simul.*, **1**, 359367.

[125] Stern, E. A. (1993) Number of relevant independent points in x-ray-absorption fine-structure spectra, *Phys. Rev. B*, **48**, 13, 9825–9827.

[126] Babanov, Yu. A., Vasin, V. V., Ageev, A. L., and Ershov, N. V. (1981) A new interpretation of EXAFS spectra in real space: I. General formalism, *Phys. Stat. Sol.*, **105**, 747–754.

[127] Babanov, Yu. A., Vasin, V. V., Ageev, A. L., and Ershov, N. V. (1981) A new interpretation of EXAFS spectra in real space: II. A comparison of the regularization technique with the Fourier transformation method, *Phys. Stat. Sol.*, **108**, 103–111.

[128] Yang, D. S. and Bunker, G. (1996) Improved R-space resolution of EXAFS spectra using combined regularization methods and nonlinear least-squares fitting, *Phys. Rev. B*, **54**, 3169–3172.

[129] Khelashvili, G. and Bunker, G. (1999) Practical regularization methods for analysis of EXAFS spectra, *J. Synchrotron Rad.*, **6**, 271–273.

[130] Khelashvili, G. (2000) Regularization and determination of structure in disordered materials, Illinois Institute of Technology, Ph.D. thesis.

[131] Vasin, V. V. and Ageev A. L.(1995) *Ill-posed Problems with A Priori Information*, VSP Brill Press.

[132] Ageev, A. L., Korshunov, M. E., Reich, T. Ye., Reich, T. and Moll, H. Regularization methods for the analysis of EXAFS spectra of chemical complexes, *J. Inv. Ill-posed Problems.*

[133] Jenkins, R., Manne, R., Robin, R., and Senemaud, C. (1991) Nomenclature system for X-ray spectroscopy, *Pure Appl. Chem.*, **63**, 5, 735–746.

Index

www.ingramcontent.com/pod-product-compliance
Lightning Source LLC
LaVergne TN
LVHW060055220926
848484LV00006BA/1353
* 9 7 8 0 5 2 1 7 6 7 7 5 0 *